AF388921

# Progress in Volatile Organic Compounds Research

# Progress in Volatile Organic Compounds Research

Special Issue Editor

**Igor Jerković**

MDPI • Basel • Beijing • Wuhan • Barcelona • Belgrade • Manchester • Tokyo • Cluj • Tianjin

*Special Issue Editor*
Igor Jerković
University of Split
Croatia

*Editorial Office*
MDPI
St. Alban-Anlage 66
4052 Basel, Switzerland

This is a reprint of articles from the Special Issue published online in the open access journal *Molecules* (ISSN 1420-3049) (available at: https://www.mdpi.com/journal/molecules/special_issues/volatile).

For citation purposes, cite each article independently as indicated on the article page online and as indicated below:

LastName, A.A.; LastName, B.B.; LastName, C.C. Article Title. *Journal Name* **Year**, *Article Number*, Page Range.

ISBN 978-3-03928-871-7 (Hbk)
ISBN 978-3-03928-872-4 (PDF)

# Contents

# About the Special Issue Editor

**Igor Jerković** is employed at the University of Split, Faculty of Chemistry and Technology, Department of Organic Chemistry as full professor (https://www.ktf.unist.hr/index.php/obavijesti-2/obavijesti-poslijediplomski-studij/172-djelatnici/cv/187-cv59). He was leader of 4 research projects at the national level and has published more than 115 papers in Web of Science Core Collection Journals that were cited more than 2000 times with an h-index of 24 (https://orcid.org/0000-0002-0727-6662). He has reviewed more than 200 research papers and 5 research projects. Awards: Medal of the president of Republic of Croatia for science and promotion of science within the Republic of Croatia and world (2019), Award of the University of Split for natural sciences (2018), Award Rudjer Boskovic of the University of Split for science (2013), and Award of Faculty of Chemistry and Technology for achievements in science and education (2011).

# Preface to "Progress in Volatile Organic Compounds Research"

Volatile organic compounds (VOCs) have been intensively investigated in the last few decades. Their origins differ: plant secondary metabolites, food/beverages aromas, fungal/bacterial volatiles, and others. VOCs typically occur as a complex mixture of compounds (e.g., monoterpenes, sesquiterpenes, norisoprenoids, aliphatic/aromatic compounds, sulfur containing compounds, and others). They form through different biochemical pathways and can be modified or created during drying or maturation, thermal treatment, and others. Different conventional or modern methods of VOCs isolation, followed by analysis with chromatographic and spectroscopic techniques, usually provide different chemical profiles and have been under constant modification and upgrade. The ecological interactions are mediated by VOCs (inter- and intra-organismic communication) and they can act as pheromones, attractants, or alleochemicals. Among them, chemical biomarkers of botanical origin or chemotaxonomic markers can be found. Many VOCs possess different biological activities, such as antioxidant, antimicrobial, antiviral, anticancer, and other activities.

VOCs from different sources is still needed to report their distribution and chemical profiles, and to discover new compounds. This Special Issue aims to attract up-to-date contributions on all aspects of VOCs chemistry, from challenges in their isolation to the analysis, and on unlocking their biological activities or other useful properties.

**Igor Jerković**
*Special Issue Editor*

*Communication*

# Composition of Essential Oils from Roots and Aerial Parts of *Carpesium divaricatum*, a Traditional Herbal Medicine and Wild Edible Plant from South-East Asia, Grown in Poland

**Anna Wajs-Bonikowska [1], Janusz Malarz [2] and Anna Stojakowska [2,*]**

[1]  Institute of General Food Chemistry, Faculty of Biotechnology and Food Sciences, Łódź University of Technology, Stefanowskiego street 4/10, 90-924 Łódź, Poland; anna.wajs@p.lodz.pl
[2]  Maj Institute of Pharmacology, Polish Academy of Sciences, Department of Phytochemistry, Smętna street 12, 31-343 Kraków, Poland; malarzj@if-pan.krakow.pl
*  Correspondence: stoja@if-pan.krakow.pl; Tel.: +481-26-623-254

Received: 31 October 2019; Accepted: 2 December 2019; Published: 3 December 2019

**Abstract:** *Carpesium divaricatum* Sieb. and Zucc. has long been used both as traditional medicine and seasonal food. The most extensively studied specialized metabolites synthesized by the plant are sesquiterpene lactones of germacrane-type. Low-molecular and volatile terpenoids produced by *C. divaricatum*, however, have never been explored. In this work, compositions of essential oils distilled from roots and shoots of *C. divaricatum* plants, cultivated either in the open field or in the glasshouse have been studied by GC-MS-FID supported by NMR spectroscopy. The analyses led to the identification of 145 compounds in all, 112 of which were localized in aerial parts and 80 in roots of the plants grown in the open field. Moreover, remarkable differences in composition of oils produced by aerial and underground parts of *C. divaricatum* have been observed. The major volatiles found in the shoots were: α-pinene (40%), nerol (4%) and neryl-isobutyrate (3%), whereas predominant components of the root oil were 10-isobutyryloxy-8,9-epoxythymyl-isobutyrate (29%), thymyl-isobutyrate (6%) and 9-isobutyryloxythymyl-isobutyrate (6%). In the analyzed oils, seventeen thymol derivatives were identified. Among them eight compounds were specific for roots. Roots of the plants cultivated in the glasshouse were, in general, a poor source of essential oil in comparison with those of the plants grown in the open field. Chemophenetic relationships with other taxa of the Inuleae-Inulineae were also briefly discussed.

**Keywords:** alpha-pinene; *Carpesium divaricatum*; Inuleae; monoterpenoids; thymol derivatives

---

## 1. Introduction

Plant genera of the subtribe Inuleae-Inulinae (family Asteraceae), e.g., *Inula, Pulicaria, Telekia, Dittrichia, Blumea* and *Chiliadenus,* are known to produce essential oils containing biologically active mono- and sesquiterpenoids [1–3]. Essential oils from *Carpesium* spp. are less studied. To our knowledge, only two communications on essential oil from the herb of *C. abrotanoides* L., the species included in the Chinese pharmacopoeia, have been published to date [4,5]. *C. divaricatum* Sieb. and Zucc. is a medicinal and food plant rich in terpenoid metabolites [6–9]. Recently, hydroxycinnamates and biologically active oxylipin from aerial parts of the plant have been also described [10]. According to the recent taxonomic studies [11,12], *Telekia speciosa* (Schreb.) Baumg. as well as some species of the genus *Inula,* known as essential oil bearing plants, are closely related to *C. divaricatum.* However, the content and composition of essential oils from *C. divaricatum* has remained unknown until now. The aim of the present study was to investigate the volatile compounds from roots and aerial parts of *C. divaricatum* and to compare the newly generated data with those reported previously for the related species.

## 2. Results

Unlike in its natural habitat, in our climate *C. divaricatum* is an annual plant. Moreover, due to late flowering, the plants grown in the open field failed to produce seeds. Fertile seeds were obtained only from the plants cultivated in a glasshouse. Yields of essential oils produced by the aerial parts of the plants were low (<0.02%, see Table 1) and except for the variations in percentages of individual compounds, oils distilled from shoots of the field grown plants and from the aerial parts of plants cultivated in a glasshouse demonstrated some minor qualitative differences in their composition (112 versus 89 identified constituents). The major compounds found in the essential oil from shoots of *C. divaricatum* were: α-pinene (c. 40% of oil), nerol (2.1%–3.7%) and neryl isobutyrate (3.2%–3.9%). Identified thymol derivatives (compounds: **54, 73, 84, 85, 111, 130, 142, 148** and **149**, see Figure 1) constituted c. 6% of the oil. Roots of the plant turned out to be much better source of volatile terpenoids (yield of essential oil—0.15%). In contrast to the aerial parts (Figure 2), they contained only low amount of α-pinene (up to 1.8% of the essential oil). Thymol derivatives (17 identified structures, see Figure 1) accounted for over 60% and 44% of the essential oil from roots of the garden grown plants and plants cultivated in the glasshouse, respectively. 10-Isobutyryloxy-8,9-epoxythymyl isobutyrate was the major constituent of the analyzed root oils (18.1%–29.2%).

**Table 1.** Chemical composition of essential oils from aerial parts and roots of *Carpesium divaricatum*.

| No | Compound | Amount (%) | | | | RI [c] exp. | RI [d] lit. | Identification Method |
|---|---|---|---|---|---|---|---|---|
| | | Aerial Parts | | Roots | | | | |
| | | OF [a] | G [b] | OF [a] | G [b] | | | |
| 1 | hexanal | - | - | 0.1 | - | 771 | 771 | RI [e], MS [f] |
| 2 | (E)-hex-2-enal | 0.2 | - | - | - | 825 | 832 | RI, MS |
| 3 | hexan-1-ol | 0.1 | - | tr. | - | 852 | 837 | RI, MS |
| 4 | tricyclene | tr. | - | - | - | 917 | 927 | RI, MS |
| 5 | α-thujene | 0.1 | - | - | - | 922 | 932 | RI, MS |
| 6 | **α-pinene** | **40.2** | **21.8** | 0.1 | **1.8** | 930 | 936 | RI, MS |
| 7 | camphene | 0.3 | - | - | - | 940 | 950 | RI, MS |
| 8 | sabinene | tr. | - | - | - | 944 | 973 | RI, MS |
| 9 | 6-methylhept-5-en-2-one | 0.2 | 0.2 | - | - | 962 | 978 | RI, MS |
| 10 | **β-pinene** | **0.5** | **0.3** | - | tr. | 966 | 978 | RI, MS |
| 11 | 2-pentylfuran | 0.4 | 0.3 | 0.2 | - | 976 | 981 | RI, MS |
| 12 | *trans*-2-(pent-2-enyl)furan | 0.1 | 0.1 | - | - | 984 | 984 | RI, MS |
| 13 | α-phellandrene | - | - | tr. | - | 991 | 1002 | RI, MS |
| 14 | δ-car-3-ene | tr. | - | - | - | 1005 | 1010 | RI, MS |
| 15 | m-cymene | 0.2 | tr. | tr. | - | 1006 | 1013 | RI, MS |
| 16 | p-cymene | - | - | tr. | - | 1007 | 1015 | RI, MS |
| 17 | β-phellandrene | - | - | tr. | - | 1014 | 1023 | RI, MS |
| 18 | limonene | 0.2 | tr. | - | 0.1 | 1018 | 1025 | RI, MS |
| 19 | γ-terpinene | 0.1 | 0.2 | - | - | 1047 | 1051 | RI, MS |
| 20 | *trans*-linalool oxide (furanoid) | tr. | - | tr. | - | 1055 | 1058 | RI, MS |
| 21 | camphen-6-ol | tr. | - | - | - | 1066 | 1082 | RI, MS |
| 22 | terpinolene | tr. | tr. | - | - | 1077 | 1082 | RI, MS |
| 23 | n-nonanal | - | 0.1 | - | - | 1080 | 1076 | RI, MS |
| 24 | **linalool** | **2.1** | **3.4** | 0.1 | tr. | 1083 | 1086 | RI, MS |
| 25 | 145/89/143/115 M? | 0.1 | - | - | - | 1091 | - | RI, MS |
| 26 | limona ketone | - | - | tr. | - | 1100 | 1105 | RI, MS |
| 27 | **α-campholenal** | **0.6** | 0.1 | - | - | 1101 | 1105 | RI, MS |
| 28 | cis-p-menth-2-en-1-ol | 0.1 | tr. | 0.1 | 0.1 | 1104 | 1108 | RI, MS |
| 29 | trans-p-menth-2-en-1-ol | - | - | 0.1 | 0.1 | 1119 | 1116 | RI, MS |
| 30 | **trans-pinocarveol** | **0.7** | 0.2 | - | - | 1120 | 1126 | RI, MS |
| 31 | cis-verbenol | - | tr. | - | - | 1121 | 1132 | RI, MS |
| 32 | trans-verbenol | 0.3 | 0.2 | - | - | 1124 | 1134 | RI, MS |
| 33 | 2-hydroxy-3-methyl-benzaldehyde | - | - | 0.1 | 0.2 | 1126 | 1135 | RI, MS |
| 34 | (E)-non-2-enal | 0.1 | tr. | 0.1 | 0.2 | 1133 | 1136 | RI, MS |
| 35 | **nerol oxide** | - | - | 0.2 | **1.0** | 1134 | 1137 | RI, MS |

**Table 1.** *Cont.*

| No | Compound | Amount (%) | | | | RI [c] exp. | RI [d] lit. | Identification Method |
|----|----------|------|---|------|---|---|---|---|
| | | Aerial Parts | | Roots | | | | |
| | | OF [a] | G [b] | OF [a] | G [b] | | | |
| 36 | pinocarvone | 0.5 | 0.1 | - | - | 1135 | 1137 | RI, MS |
| 37 | *p*-mentha-1,5-dien-8-ol | 0.1 | 0.2 | - | - | 1143 | 1138 | RI, MS |
| 38 | geijeren | 0.4 | tr. | 4.2 | 3.1 | 1148 | 1139 | RI, MS |
| 39 | terpinen-4-ol | 0.7 | 0.9 | - | - | 1159 | 1164 | RI, MS |
| 40 | myrtenal | - | tr. | - | - | 1165 | 1172 | RI, MS |
| 41 | α-terpineol | 0.8 | 0.7 | 0.1 | 0.1 | 1171 | 1176 | RI, MS |
| 42 | *cis*-piperitol | - | - | tr. | tr. | 1176 | 1181 | RI, MS |
| 43 | myrtenol | 0.1 | - | - | - | 1177 | 1178 | RI, MS |
| 44 | n-decanal | 0.6 | 1.2 | - | - | 1182 | 1180 | RI, MS |
| 45 | *trans*-piperitol | - | - | tr. | tr. | 1186 | 1193 | RI, MS |
| 46 | 2-ethenyl-3-methyloanisol | 0.2 | - | 0.7 | 0.9 | 1190 | 1196 | RI, MS |
| 47 | β-cyclocitral | 0.2 | 0.3 | - | - | 1193 | 1195 | RI, MS |
| 48 | *trans*-carveol | 0.1 | - | - | - | 1195 | 1200 | RI, MS |
| 49 | nerol | 3.7 | 2.1 | 1.4 | 1.2 | 1210 | 1210 | [1]H, RI, MS |
| 50 | thymol methyl ether | - | - | 0.4 | 0.1 | 1211 | 1215 | [1]H, RI, MS |
| 51 | geraniol | 0.2 | 0.4 | - | - | 1233 | 1235 | RI, MS |
| 52 | α-jonene | 0.1 | - | - | - | 1241 | 1258 | RI, MS |
| 53 | cuminol | tr. | - | 0.3 | 0.3 | 1245 | 1266 | RI, MS |
| 54 | thymol | 0.1 | - | 0.1 | 0.1 | 1258 | 1267 | RI, MS |
| 55 | carvacrol | 0.9 | - | 0.3 | 0.3 | 1264 | 1278 | RI, MS |
| 56 | dihydroedulan II | 0.1 | 0.1 | - | - | 1278 | 1290 | RI, MS |
| 57 | (*E,E*)-deca-2,4-dienal | 0.1 | - | - | 0.1 | 1286 | 1291 | RI, MS |
| 58 | 4,6-dimethyl-2,3-2H-benzofuran-2-one | - | - | 0.2 | 0.2 | 1317 | - | RI, MS |
| 59 | 7αH-silphiperfol-5-ene | 0.5 | 0.1 | 0.7 | 2.5 | 1323 | 1329 | RI, MS |
| 60 | presilphiperfol-7-ene | 0.2 | - | 0.2 | 0.3 | 1332 | 1342 | RI, MS |
| 61 | 7βH-silphiperfol-5-ene | 0.9 | 0.1 | 1.1 | 3.4 | 1342 | 1352 | RI, MS |
| 62 | α-cubebene | tr. | 0.1 | - | - | 1344 | 1355 | RI, MS |
| 63 | α-longipinene | 0.2 | tr. | 0.3 | 0.4 | 1348 | 1360 | RI, MS |
| 64 | (*E*)-tridec-6-en-4-yn | 0.2 | 0.1 | - | - | 1363 | - | RI, MS |
| 65 | viburtinal | - | - | 0.5 | 1.2 | 1367 | - | RI, MS |
| 66 | longicyclene | 0.4 | - | 0.3 | 0.4 | 1371 | 1372 | RI, MS |
| 67 | cyclosativene | 0.4 | - | - | - | 1372 | 1378 | RI, MS |
| 68 | α-copaene | - | 0.5 | - | - | 1375 | 1379 | RI, MS |
| 69 | silphiperfol-6-ene | - | - | 0.3 | 0.4 | 1376 | 1379 | RI, MS |
| 70 | modephene | 0.2 | - | 0.3 | 1.0 | 1377 | 1383 | RI, MS |
| 71 | α-isocomene | 0.4 | 0.3 | 0.4 | 1.3 | 1383 | 1389 | RI, MS |
| 72 | 137/121/95/136 M204 | 0.6 | 0.1 | 0.5 | 0.8 | 1391 | - | RI, MS |
| 73 | 6-methoxythymol methyl-ether | 2.1 | 0.5 | 2.7 | 1.4 | 1394 | 1398 | RI, MS |
| 74 | β-isocomene | 0.6 | 0.2 | 0.7 | 2.0 | 1402 | 1411 | RI, MS |
| 75 | α-cedrene | 0.1 | - | tr. | tr. | 1409 | 1418 | RI, MS |
| 76 | α-gurjunene | - | tr. | - | - | 1410 | 1418 | RI, MS |
| 77 | α-santalene | 1.0 | 0.5 | 0.6 | 0.8 | 1413 | 1422 | RI, MS |
| 78 | *trans*-geranylacetone | 0.3 | 0.3 | - | - | 1423 | 1430 | RI, MS |
| 79 | *trans*-α-bergamotene | 0.6 | 0.1 | 0.3 | 0.2 | 1428 | 1434 | RI, MS |
| 80 | *epi*-β-santalene | 1.5 | 0.4 | 0.9 | 0.8 | 1438 | 1446 | RI, MS |
| 81 | α-himachalene | 0.5 | 0.1 | - | - | 1441 | 1450 | RI, MS |
| 82 | aromadendrene | 0.2 | - | 0.3 | 0.4 | 1442 | 1449 | RI, MS |
| 83 | α-humulene | 0.1 | - | tr. | tr. | 1446 | 1455 | RI, MS |
| 84 | 8,9-didehydrothymyl-isobutyrate | 0.9 | 0.2 | 1.6 | 0.8 | 1461 | 1458 | RI, MS |
| 85 | thymyl-isobutyrate | 2.0 | 1.0 | 6.3 | 3.5 | 1467 | 1462 | [1]H, RI, MS |
| 86 | β-jonone | 0.9 | 1.5 | - | - | 1468 | 1468 | RI, MS |
| 87 | neryl isobutyrate | 3.2 | 3.9 | 4.1 | 3.9 | 1475 | 1468 | [1]H, RI, MS |
| 88 | γ-himachalene | 0.3 | 0.1 | - | - | 1480 | 1479 | RI, MS |
| 89 | 123/93/94/121 M204 | 0.8 | 0.1 | 0.5 | 0.7 | 1484 | - | RI, MS |
| 90 | (3*E*,6*Z*)-α-farnesene | 1.8 | 5.5 | - | - | 1487 | 1475 | RI, MS |
| 91 | α-terpinyl isovalerate | - | - | 0.2 | 0.7 | 1489 | 1488 | RI, MS |
| 92 | γ-muurolene | tr. | 0.2 | - | - | 1493 | 1494 | RI, MS |
| 93 | elixene (4-isopropylidene-1-vinyl-*o*-menth-8-ene) | - | - | 0.2 | 0.2 | 1498 | 1493 | RI, MS |
| 94 | ledene | 1.5 | 2.2 | tr. | 0.2 | 1499 | 1491 | RI, MS |
| 95 | α-muurolene | - | 1.1 | - | - | 1500 | 1496 | RI, MS |
| 96 | (*E,E*)-α-farnesene | 0.7 | 1.1 | - | - | 1502 | 1498 | RI, MS |

**Table 1.** *Cont.*

| No | Compound | Amount (%) | | | | RI [c] exp. | RI [d] lit. | Identification Method |
|----|----------|---|---|---|---|---|---|---|
| | | Aerial Parts | | Roots | | | | |
| | | OF [a] | G [b] | OF [a] | G [b] | | | |
| 97 | β-bisabolene | 1.6 | 0.6 | 0.5 | 0.4 | 1507 | 1503 | RI, MS |
| 98 | γ-cadinene | 1.1 | 3.2 | - | - | 1511 | 1507 | RI, MS |
| 99 | cameronan-7α-ol | - | - | tr. | 0.1 | 1513 | 1513 | RI, MS |
| 100 | α-photosantalol | - | - | 0.1 | 0.2 | 1514 | 1514 | RI, MS |
| 101 | isolongifolan-8-ol | - | - | 0.1 | 0.5 | 1517 | 1515 | RI, MS |
| 102 | *cis/trans*-calamenene | 0.2 | 0.4 | 0.2 | 0.3 | 1526 | 1517 | RI, MS |
| 103 | δ-cadinene | 1.4 | 5.3 | - | - | 1520 | 1520 | RI, MS |
| 104 | β-cadinene | 0.1 | 0.3 | - | - | 1523 | 1526 | RI, MS |
| 105 | 9-methoxycalamenene | 0.1 | - | - | - | 1524 | - | RI, MS |
| 106 | 147/162/121/177 M206 | - | - | tr. | 0.1 | 1531 | - | RI, MS |
| 107 | 121/163/93/134 M218 | - | - | 0.2 | 0.1 | 1534 | - | RI, MS |
| 108 | α-cadinene | 0.3 | 0.5 | - | - | 1535 | 1534 | RI, MS |
| 109 | (*E*)-α-bisabolene | 0.1 | 0.2 | 0.1 | 0.7 | 1536 | 1530 | RI, MS |
| 110 | (*E*)-nerolidol | 2.2 | 8.6 | 0.6 | 0.6 | 1543 | 1553 | [1]H, RI, MS |
| 111 | thymyl-2-methylbutyrate | 0.1 | - | 0.2 | 0.2 | 1546 | - | RI, MS |
| 112 | neryl-α-methylbutyrate | 1.7 | 1.6 | 3.6 | 3.4 | 1551 | 1565 | RI, MS |
| 113 | neryl isovalerate | 1.6 | 1.3 | 2.3 | 1.8 | 1557 | 1579 | RI, MS |
| 114 | caryophyllene oxide | 0.4 | 0.4 | 1.1 | 2.1 | 1565 | 1578 | [1]H, RI, MS |
| 115 | viridiflorol | 0.5 | 1.4 | - | - | 1577 | 1592 | RI, MS |
| 116 | isoaromadendreneepoxide | 0.3 | 0.1 | - | - | 1584 | 1590 | RI, MS |
| 117 | ledol | 0.3 | 0.5 | - | - | 1588 | 1600 | RI, MS |
| 118 | humulene II epoxide | - | - | 0.4 | 1.1 | 1595 | 1602 | RI, MS |
| 119 | 1,10-di-*epi*-cubenol | 0.1 | 0.2 | - | - | 1597 | 1615 | RI, MS |
| 120 | 135/146/159/71 M218 | - | - | 0.2 | 0.3 | 1602 | - | RI, MS |
| 121 | muurola-4,10(14)-dien-1β-ol | 0.2 | 0.2 | - | - | 1605 | 1620 | RI, MS |
| 122 | gossonorol | - | - | 0.1 | 0.1 | 1613 | 1626 | RI, MS |
| 123 | 1-*epi*-cubenol | 0.2 | 0.5 | - | - | 1614 | 1623 | RI, MS |
| 124 | α-acorenol | tr. | - | 0.4 | 0.5 | 1620 | 1623 | RI, MS |
| 125 | τ-cadinol | 1.4 | 4.1 | - | - | 1625 | 1633 | [1]H, RI, MS |
| 126 | τ-muurolol | 0.3 | 0.4 | - | - | 1628 | 1633 | RI, MS |
| 127 | β-eudesmol | 0.6 | 0.2 | 3.4 | 3.8 | 1631 | 1641 | RI, MS |
| 128 | α-cadinol | 1.3 | 3.8 | 0.4 | 0.3 | 1638 | 1643 | [1]H, RI, MS |
| 129 | 5β,7βH,10α-eudesm-11-en-1α-ol | - | - | 0.3 | - | 1653 | - | RI, MS |
| 130 | 6-methoxythymyl isobutyrate | 0.9 | 0.7 | 3.8 | 4.9 | 1657 | 1658 | [1]H,[13]C, RI, MS |
| 131 | 6-methoxy-8,9-didehydrothymyl isobutyrate | tr. | - | 0.4 | 0.2 | 1665 | 1676 | RI, MS |
| 132 | 10-isobutyryloxy-8,9-didehydrothymol-methyl-ether | - | - | 0.4 | 0.3 | 1666 | 1684 | [1]H,[13]C, RI, MS |
| 133 | α-bisabolol | 0.1 | 0.4 | 0.3 | 1.3 | 1668 | 1683 | RI, MS |
| 134 | 145/162/71/115 M232 | - | - | 0.3 | 0.5 | 1681 | - | RI, MS |
| 135 | aromadendrene oxide | 0.1 | 0.1 | - | - | 1702 | 1672 | RI, MS |
| 136 | 135/148/133/91 M236 | - | - | 0.1 | 0.1 | 1725 | - | RI, MS |
| 137 | 135/164/71/91 M234 | - | - | 0.2 | 0.1 | 1733 | - | RI, MS |
| 138 | fenantrene (artifact) | 0.1 | 0.3 | - | - | 1741 | 1744 | RI, MS |
| 139 | diisobutylphtalate (artifact) | 0.9 | 2.5 | 0.3 | 0.6 | 1817 | 1819 | RI, MS |
| 140 | hexahydrofarnesylacetone | 0.2 | 0.8 | - | - | 1820 | 1830 | RI, MS |
| 141 | alantolactone | 0.1 | - | 0.2 | tr. | 1854 | 1878 | RI, MS |
| 142 | 9-isobutyryloxythymyl-isobutyrate | 0.2 | 0.8 | 5.6 | 5.7 | 1879 | 1891 | [1]H,[13]C, RI, MS |
| 143 | 10-isobutyryloxy-8,9-didehydrothymyl-isobutyrate | - | - | 3.0 | 3.1 | 1882 | 1891 | RI, MS |
| 144 | (5*E*,9*E*)-farnesylacetone | 0.1 | 0.2 | - | - | 1889 | 1895 | RI, MS |
| 145 | dibutylphtalate (artifact) | 0.2 | 1.8 | - | - | 1906 | 1909 | RI, MS |
| 146 | 7-isobutyryloxythymyl-isobutyrate | - | - | 0.6 | 0.7 | 1914 | 1924 | RI, MS |
| 147 | 9-(2-methylbutyryloxy)thymyl-isobutyrate | - | 1.0 | 1.4 | | 1964 | 1970 | RI, MS |

**Table 1.** *Cont.*

| No | Compound | Amount (%) | | | | RI [c] exp. | RI [d] lit. | Identification Method |
|---|---|---|---|---|---|---|---|---|
| | | Aerial Parts | | Roots | | | | |
| | | OF [a] | G [b] | OF [a] | G [b] | | | |
| 148 | 10-(2-methylbutyryloxy)-8,9-didehydrothymyl-isobutyrate | 0.1 | 0.4 | 0.3 | 0.4 | 1967 | 1970 | RI, MS |
| 149 | **10-isobutyryloxy-8,9-epoxythymyl-isobutyrate** | 0.2 | **0.6** | **29.2** | **18.1** | 2002 | 2036 | [1]H,[13]C,RI,MS |
| 150 | **71/177/150/135 M290** | - | - | **0.9** | **0.5** | 2048 | - | RI, MS |
| 151 | **10-(2-methylbutyryloxy)-8,9-epoxythymyl-isobutyrate** | - | - | **4.4** | **3.6** | 2077 | 2056 | RI, MS |
| 152 | 10-isovaleroxy-8,9-epoxythymyl-isobutyrate | - | - | 0.3 | 0.1 | 2097 | 2122 | RI, MS |
| 153 | **fitol** | 0.3 | **1.7** | - | - | 2098 | - | RI, MS |
| 154 | 57/177/71/85 M304 | - | - | 0.1 | 0.1 | 2149 | - | RI, MS |
| 155 | tricosane | 0.1 | 0.2 | - | - | 2286 | 2300 | RI, MS |
| 156 | tetracosane | 0.1 | - | - | - | 2386 | 2400 | RI, MS |
| 157 | **pentacosane** | **0.6** | 0.3 | - | - | 2489 | 2500 | RI, MS |
| 158 | hexacosane | tr. | - | - | - | 2589 | 2600 | RI, MS |
| 159 | heptacosane | 0.2 | 0.1 | - | - | 2685 | 2700 | RI, MS |
| | **Sum of Identified** | **96.7** | **97.5** | **94.3** | **91.7** | | | |
| | **Yield of Essential Oil (%)** | **0.016** | **0.014** | **0.150** | **0.059** | | | |

[a] Essential oils isolated from aerial parts and roots of *Carpesium divaricatum* cultivated in the open field. [b] Essential oils isolated from aerial parts and roots of *Carpesium divaricatum* cultivated in a greenhouse. [c] Experimental retention index measured on non-polar column. [d] Literature retention index from non-polar column. [e] Identification based on retention index. [f] Identification based on mass spectrum. Tr.—<0.05%.

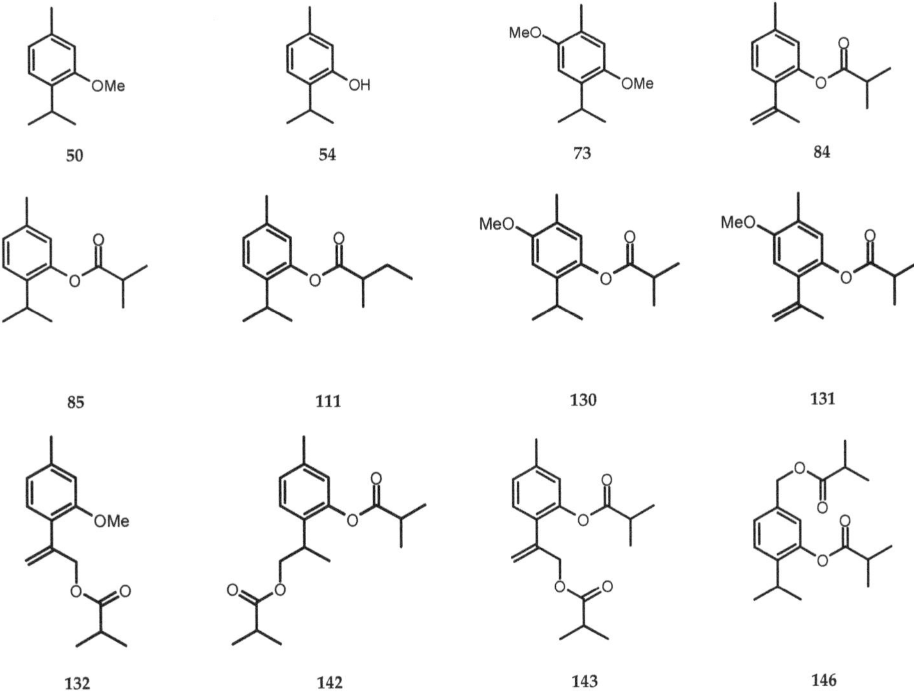

**Figure 1.** *Cont.*

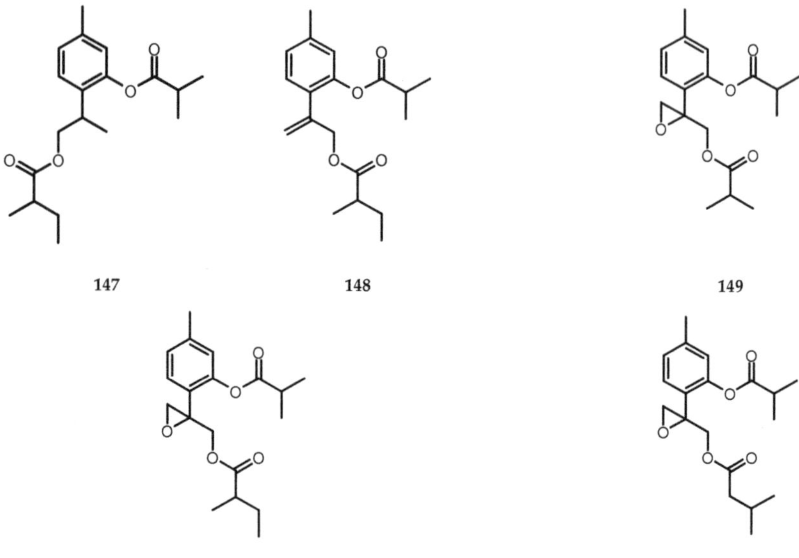

147      148      149

151           152

**Figure 1.** Structures of thymol derivatives identified in essential oils from roots of *Carpesium divaricatum*. **50**: thymol-methyl-ether; **54**: thymol; **73**: 6-methoxythymol-methyl-ether; **84**: 8,9-didehydrothymyl-isobutyrate; **85**: thymyl-isobutyrate; **111**: thymyl-2-methylbutyrate; **130**: 6-methoxythymyl-isobutyrate; **131**: 6-methoxy-8,9-didehydrothymyl-isobutyrate; **132**: 10-isobutyryloxy-8,9-didehydrothymol-methyl-ether; **142**: 9-isobutyryloxythymyl-isobutyrate; **143**: 10-isobutyryloxy-8,9-didehydrothymyl-isobutyrate; **146**: 7-Isobutyryloxythymyl-isobutyrate; **147**: 9-(2-methylbutyryloxy)-thymyl-isobutyrate; **148**: 10-(2-methylbutyryloxy)-8,9-didehydrothymyl-isobutyrate; **149**: 10-isobutyryloxy-8,9-epoxythymyl-isobutyrate; **151**: 10-(2-methylbutyryloxy)-8,9-epoxythymyl-isobutyrate; **152**: 10-isovaleryloxy-8,9-epoxythymyl-isobutyrate.

The essential oils from *C. divaricatum* contained some volatiles, which were difficult to identify based on GC-MS only. Flash chromatography (FC), monitored by thin-layer chromatography (TLC), was used to obtain fractions of oils rich in components of interest (purity 19%–63%, by GC-FID). The fractions were subsequently subjected to NMR analysis and the experimental chemical shifts of the chosen volatiles were compared to the literature data (see Supplementary Material).

Structures of 11 components remained unresolved, due to the small available amounts of the compounds, insufficient to perform full spectral analysis. MS spectra and retention indices of the compounds are shown in Supplementary Material.

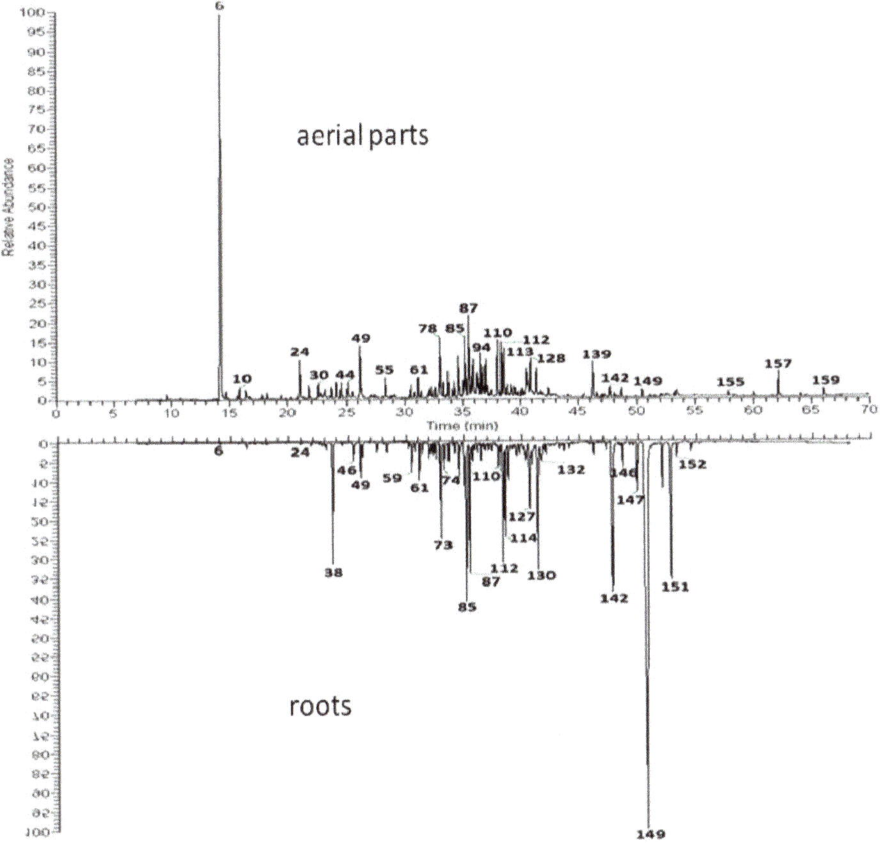


**Figure 2.** Gas chromatograms of essential oils from aerial parts and roots of *Carpesium divaricatum*. The numbering of the compounds corresponds to that in Table 1.

## 3. Discussion

Though the essential oil content in aerial parts of *C. divaricatum* was very low, the occurrence of α-pinene (40% of the oil) is worth to note. The compound demonstrated anxiolytic and moderate anti-inflammatory effect in mice [13,14]. Essential oils obtained from plants of different provenience can markedly vary in their composition. Aerial parts of *Pulicaria gnaphalodes* (Vent.) Boiss., collected in four different locations, contained extremely different quantities of α-pinene (0.0–34.1% of the essential oil) [15]. Thus, some data on the composition of essential oils from *C. divaricatum* plants grown in their natural habitat would be of interest, to establish whether or not the high α-pinene content is typical of *C. divaricatum* aerial parts. Not much is known from the literature on essential oils from plants of the genus *Carpesium*. To date, only two studies on volatiles from the whole herb of *C. abrotanoides* have been published [4,5]. However, the authors managed to identify 14–44 components of the oils and neither α-pinene nor thymol derivatives have been detected. The major constituents were β-bisabolene (7.3–24.7%), caryophyllene-oxide (c. 13%) and eudesma-5,11(13)-dien-8,12-olide (c. 22%). Volatile constituents from other species of the Inuleae-Inulinae subtribe are better investigated. Thymol and its derivatives seem to be widespread within the plants of the subtribe, except for *Blumea* spp. [16–18]. The genus *Pulicaria* comprises species with essential oils rich in thymol and its methyl ether, like *Pulicaria vulgaris* Gaertn. [19] and *Pulicaria sicula* (L.) Moris [15] together with some species devoid of thymol derivatives [20]. The content of thymol derivatives in essential oil from aerial parts of *C. divaricatum* (6.4%) is similar to those detected in oils from aerial parts of other species of the subtribe, e.g., *Schizogyne sericea* (L.F.) DC. [21,22], *Telekia speciosa* (Schreb.) Baumg. [23,24] and *Limbarda crithmoides* (L.) Dumort. (formerly *Inula crithmoides* L.) [25]. Structural diversity of the compounds was also similar, with numerous thymol esters.

Essential oils from roots of the Inuleae-Inulinae plants have rarely been studied. Literature data on a few species are available, including *Dittrichia viscosa* (L.) Greuter (formerly *Inula viscosa* (L.) Aiton) [26], *Inula racemosa* Hook. f. [27,28], *Inula helenium* L. [1], *Pulicaria mauritanica* Coss. [29] and *T. speciosa* [24,30]. The common feature of essential oils from *I. helenium*, *I. racemosa* and *T. speciosa* is a very high content of eudesmane-type sesquiterpene lactones (up to 82%). Such composition of essential oils seems to be correlated with a presence of resin canals in roots of the plants. Thymol derivatives were not described as constituents of essential oil from roots of *I. racemosa*. The compounds, however, were found in the oils from the remaining species. Juvenile roots of *I. helenium* and *I. viscosa* contained higher amounts of the monoterpenoids than the old ones [26,31]. Two derivatives of thymol methyl ether constituted nearly 80% of the volatile fraction from *I. viscosa* roots [26]. Thymol, thymol methyl ether and eight thymyl ester derivatives accounted for c. 5.5% of the essential oil from roots of *T. speciosa*. *P. mauritanica* root oil contained c. 16% of the structurally related compounds. Though there are no any data on essential oils from roots of *Carpesium* spp., some thymol derivatives were described as constituents of methanol extract from aerial parts of *C. divaricatum* [32]. All of the compounds were found in essential oils from the plants analyzed in this study. Volatile fraction from roots of *C. divaricatum* is exceptional, in respect of both thymol derivatives content (over 60%) and their structural diversity (17 compounds; for MS spectra see Supplementary Material). 10-Isobutyryloxy-8,9-epoxy-thymyl isobutyrate, major constituent of the analyzed essential oil, demonstrated moderate activity against *Staphylococcus aureus* and *Candida albicans* [33].

## 4. Materials and Methods

### 4.1. General Experimental Procedures

GC-MS-FID analyses of essential oils and their fractions were performed on a Trace GC Ultra Gas Chromatograph coupled with DSQII mass spectrometer (Thermo Electron, Waltham, MA, USA). Simultaneous GC-FID and GC-MS analysis were performed using a MS-FID splitter (SGE Analytical Science, Ringwood, VIC, Australia). Mass range was 33–550 amu, ion source-heating: 200 °C; ionization energy: 70 eV. One microliter of essential oil solution (80% *v/v*) diluted in pentane:diethyl ether was injected in split mode at split ratios (50:1). Operating conditions: capillary column Rtx-1 MS (60 m × 0.25 mm i.d., film thickness 0.25 μm), and temperature program: 50 °C (3 min)—300 °C (30 min) at 4 °C/min. Injector and detector temperatures were 280 °C and 300 °C, respectively. Carrier gas was helium (constant pressure: 300 kPa). The relative composition of each essential oil sample was calculated from GC peak areas according to total peak normalization—the most popular method used in the essential oil analysis. $^1$H-NMR (250 MHz) and $^{13}$C-NMR (62.90 MHz) spectra for components of essential oils were recorded with a Bruker DPX 250 Avance spectrometer in CDCl$_3$, with TMS as an internal standard.

### 4.2. Plant Material

Seeds of *Carpesium divaricatum* Sieb. and Zucc, provided by the Research Center for Medicinal Plant Resources, National Institute of Biomedical Innovation, Tsukuba (Japan), were sown in the end of March 2015, into multipots with garden soil. In the stage of 4–5 mature leaves, the plants were transferred to plastic pots with a substrate composed of garden soil, peat and sand (2:1:1, *v/v*). Plants were grown in a glasshouse of the Garden of Medicinal Plants, Maj Institute of Pharmacology PAS in Krakow, under controlled conditions (temperatures by day 18–38 °C; by night 12–18 °C), without any chemical treatment. In the third week of May, the plants were divided into two groups. First group was left in the glasshouse for further growth and the second one was transplanted into the open field. Data on cultivation conditions (type of soil, average annual temperature, annual rainfall and agrotechnical procedures applied) are available elsewhere [34]. Aerial parts and roots of the plants were collected in the beginning of flowering period (August/September) and dried under shade at room temperature. Voucher specimen (3/15) was deposited in the collection kept at the Garden of Medicinal Plants, Institute of Pharmacology, Kraków, Poland. The dry plant material was stored no longer than five months.

### 4.3. Isolation of Essential Oil

Essential oils from aerial (dried leaf, branches, flowers) and underground parts (dried roots) of *C. divaricatum* were obtained by hydrodistillation using a Clevenger-type apparatus. Each hydrodistillation was conducted for 4 h using 100–300 g of plant material. The yellowish essential oils were dried over anhydrous magnesium sulphate, and stored at 4 °C in the dark, until tested and analyzed.

### 4.4. Isolation and NMR Analysis of Volatile Components

To isolate the volatiles of interest, the essential oils from aerial parts (i.e., dried leaves with petioles, stems and flowers, 504 mg) and from roots (dried plant material, 973 mg) of the plants grown in the open field were separately flash-chromatographed (FC) on a glass column (500 × 30 mm) filled with silica gel 60 (0.040–0.063 mm, Merck, EM Science, NJ USA), starting the elution with *n*-hexane and gradually increasing the polarity by addition of diethyl ether. The elution was accelerated by means of pressurized nitrogen (flow rate 100 mL/min). The separation was monitored by TLC and GC-MS. Twenty fractions (1a–20a) of essential oil distilled from the aerial parts of the plant and twenty fractions (1b–20b) of root essential oil were obtained and analyzed by GC-MS-FID. Structures of 11 volatiles from the following fractions were confirmed using NMR spectroscopy ($^1$H and/or $^{13}$C; see Supplementary Material): 1a: (42 mg) neryl-isobutyrate (26%); 13a: (22mg) (*E*)-nerolidol (25%); 15a: (22 mg) τ-cadinol (21%); 17a: (58 mg) nerol (25%); 18a: (48 mg) α-cadinol (23%); 3b: (17 mg) thymol-methyl-ether (33%); 7b: (32 mg) thymyl isobutyrate (57%); 8b: (51 mg) 6-methoxythymyl-isobutyrate (62%); 13b: (13 mg) caryophyllene-oxide (52%); 14b: (36%) 10-isobutyryloxy-8,9-didehydrothymyl-isobutyrate (46%); 15b: (65 mg) 9-isobutyryloxythymyl-isobutyrate (55%); 17b: (53 mg) 10-isobutyryloxy-8,9-epoxythymyl-isobutyrate (63%); 18b: (25 mg) nerol (43%).

### 4.5. Identification of Essential Oil Constituents

Constituents of the essential oils were identified based on their MS spectra and their comparison with those from mass spectra libraries: NIST 2012, Wiley Registry of Mass Spectral Data 8th edition and MassFinder 4.1, along with the relative retention indices (RI) on DB-1 column (available from MassFinder 4.1) and on Rtx-1MS column found in the literature [35]. Isolated compounds were also identified by the comparison of their $^1$H-NMR and $^{13}$C-NMR spectral data with those of the compounds isolated previously in our laboratory or those from the literature.

## 5. Conclusions

This was the first study on composition of essential oils from *C. divaricatum*. Aerial parts of *C. divaricatum* occurred to be a poor source of volatiles. Essential oil from roots of the plant was rich in thymyl ester derivatives of various structures. As some of the compounds, according to the literature [36], demonstrated moderate antibacterial, antifungal and anti-inflammatory activities, the essential oil from roots of *C. divaricatum* as well as its components are worth further studies.

**Supplementary Materials:** The following are available online: Figure S1: Mass spectra and retention indices (RI) together with chemical structures of thymol derivatives detected in *C. divaricatum* essential oils, Figure S2: Mass spectra and experimental retention indices (RI) of unidentified compounds from *C. divaricatum* essential oils, Figure S3: Results of NMR analyses of crude fractions (obtained by flash chromatography) from *C. divaricatum* essential oils.

**Author Contributions:** Conceptualization, A.S.; methodology, A.W.-B. and A.S.; investigation, A.W.-B., J.M., A.S.; resources, A.W.-B., J.M., A.S.; data curation, A.W.-B. and A.S.; writing—original draft preparation, A.W.-B., J.M., A.S.; writing—review and editing, A.S.; project administration, A.S.

**Funding:** This research received no external funding.

**Acknowledgments:** We are greatly indebted to the workers of the Garden of Medicinal Plants, Maj Institute of Pharmacology PAS in Kraków, for cultivation of plants.

**Conflicts of Interest:** The authors declare no conflict of interest.

## References

1. Stojanović-Radić, Z.; Čomić, L.j.; Radulović, N.; Blagojević, P.; Denić, M.; Miltojević, A.; Rajkowić, J.; Mihajilov-Krstev, T. Antistaphylococcal activity of *Inula helenium* L. root essential oil: Eudesmane sesquiterpene lactones induce cell membrane damage. *Eur. J. Clin. Microbiol. Infect. Dis.* **2012**, *31*, 1015–1025. [CrossRef] [PubMed]

2. Awadh Ali, N.A.; Crouch, R.A.; Al-Fatimi, M.A.; Arnold, N.; Teichert, A.; Setzer, W.N.; Wessjohann, L. Chemical composition, antimicrobial, antiradical and anticholinesterase activity of the essential oil of *Pulicaria stephanocarpa* from Soqotra. *Nat. Prod. Commun.* **2012**, *7*, 113–116.

3. Pang, Y.; Wang, D.; Hu, X.; Wang, H.; Fu, W.; Fan, Z.; Chen, X.; Yu, F. Effect of volatile oil from *Blumea balsamifera* (L.) DC. leaves on wound healing in mice. *J. Tradit. Chin. Med.* **2014**, *34*, 716–724. [CrossRef]

4. Kameoka, H.; Sagara, K.; Miyazawa, M. Components of essential oils of Kakushitsu (*Daucus carota* L. and *Carpesium abrotanoides* L.). *Nippon Nōgeikagaku Kaishi* **1989**, *63*, 185–188. [CrossRef]

5. Wang, Q.; Pan, L.; Lin, L.; Zhang, R.; Du, Y.; Chen, H.; Huang, M.; Guo, K.; Yang, X. Essential oil from *Carpesium abrotanoides* L. induces apoptosis via activating mitochondrial pathway in hepatocellular carcinoma cells. *Curr. Med. Sci.* **2018**, *38*, 1045–1055. [CrossRef]

6. Zhang, J.-P.; Wang, G.-W.; Tian, X.-H.; Yang, Y.-X.; Liu, Q.-X.; Chen, L.-P.; Li, H.-L.; Zhang, W.-D. The genus *Carpesium*: A review of its ethnopharmacology, phytochemistry and pharmacology. *J. Ethnopharmacol.* **2015**, *163*, 173–191. [CrossRef]

7. Kim, H.; Song, M.-J. Ethnobotanical analysis for traditional knowledge of wild edible plants in North Jeolla Province (Korea). *Genet. Resour. Crop Evol.* **2013**, *60*, 1571–1585. [CrossRef]

8. Geng, Y.; Zhang, Y.; Ranjitkar, S.; Huai, H.; Wang, Y. Traditional knowledge and its transmission of wild edibles used by the Naxi in Baidi Village, northwest Yunnan province. *J. Ethnobiol. Ethnomed.* **2016**, *12*, 1–21. [CrossRef]

9. Zhang, T.; Chen, J.-H.; Si, J.-G.; Ding, G.; Zhang, Q.-B.; Zhang, H.-W.; Jia, H.-M.; Zou, Z.-M. Isolation, structure elucidation, and absolute configuration of germacrane isomers from *Carpesium divaricatum*. *Sci. Rep.* **2018**, *8*, 12418. [CrossRef] [PubMed]

10. Kłeczek, N.; Michalak, B.; Malarz, J.; Kiss, A.K.; Stojakowska, A. *Carpesium divaricatum* Sieb. & Zucc. revisited: New constituents from aerial parts of the plant and their possible contribution to the biological activity of the plant. *Molecules* **2019**, *24*, 1614. [CrossRef]

11. Anderberg, A.A.; Eldenäs, P.; Bayer, R.J.; Englund, M. Evolutionary relationships in the Asteraceae tribe Inuleae (incl. Plucheeae) evidenced by DNA sequences of *ndh*F; with notes on the systematic positions of some aberrant genera. *Org. Divers. Evol.* **2005**, *5*, 135–146. [CrossRef]

12. Gutiérrez-Larruscain, D.; Santos-Vicente, M.; Anderberg, A.A.; Rico, E.; Martínez-Ortega, M.M. Phylogeny of the *Inula* group (Asteraceae: Inuleae): Evidence from nuclear and plastid genomes and a recircumscription of *Pentanema*. *Taxon* **2018**, *67*, 149–164. [CrossRef]

13. Orhan, I.; Küpeli, E.; Aslan, M.; Kartal, M.; Yesilada, E. Bioassay-guided evaluation of anti-inflammatory and antinociceptive activities of pistachio, *Pistacia vera* L. *J. Ethnopharm.* **2006**, *105*, 235–240. [CrossRef]

14. Satou, T.; Kasuya, H.; Maeda, K.; Koike, K. Daily inhalation of α-pinene in mice: Effects on behavior and organ accumulation. *Phytother. Res.* **2014**, *28*, 1284–1287. [CrossRef] [PubMed]

15. Maggio, A.; Riccobono, L.; Spadaro, V.; Campisi, P.; Bruno, M.; Senatore, F. Volatile constituents of the aerial parts of *Pulicaria sicula* (L.) Moris growing wild in Sicily: Chemotaxonomic volatile markers of the genus *Pulicaria* Gaertn. *Chem. Biodiversity* **2015**, *12*, 781–799. [CrossRef] [PubMed]

16. Owolabi, M.S.; Lajide, L.; Villanueva, H.E.; Setzer, W.N. Essential oil composition and insecticidal activity of *Blumea perrottetiana* growing in Southwestern Nigeria. *Nat. Prod. Commun.* **2010**, *5*, 1135–1138. [PubMed]

17. Satyal, P.; Chhetri, B.K.; Dosoky, N.S.; Shrestha, S.; Poudel, A.; Setzer, W.N. Chemical composition of *Blumea lacera* essential oil from Nepal. Biological activities of the essential oil and (Z)-lachnophyllum ester. *Nat. Prod. Commun.* **2015**, *10*, 1749–1750. [CrossRef] [PubMed]

18. Yuan, Y.; Huang, M.; Pang, Y.-X.; Yu, F.-L.; Chen, C.; Liu, L.-W.; Chen, Z.-X.; Zhang, Y.-B.; Chen, X.-L.; Hu, X. Variations in essential oil yield, composition, and antioxidant activity of different plant organs from *Blumea balsamifera* (L.) DC. at different growth times. *Molecules* **2016**, *21*, 1024. [CrossRef]

19. Sharifi-Rad, J.; Miri, A.; Hoseini-Alfatemi, S.M.; Sharifi-Rad, M.; Setzer, W.N.; Hadjiakhoondi, A. Chemical composition and biological activity of *Pulicaria vulgaris* essential oil from Iran. *Nat. Prod. Commun.* **2014**, *9*, 1633–1636. [CrossRef] [PubMed]

20. Chaib, F.; Allali, H.; Bennaceur, M.; Flamini, G. Chemical composition and antimicrobial activity of essential oils from the aerial parts of *Asteriscus graveolens* (Forssk.) Less. and *Pulicaria incisa* (Lam.) DC.: Two Asteraceae herbs growing wild in the Hoggar. *Chem. Biodiversity* **2017**, *14*, e1700092. [CrossRef]

21. Venditti, A.; Bianco, A.; Muscolo, C.; Zorzetto, C.; Sánchez-Mateo, C.; Rabanal, R.M.; Quassinti, L.; Bramucci, M.; Damiano, S.; Iannarelli, R.; et al. Bioactive secondary metabolites from *Schizogyne sericea* (Asteraceae) endemic to Canary Islands. *Chem. Biodiversity* **2016**, *13*, 826–836. [CrossRef] [PubMed]

22. Zorzetto, C.; Sánchez-Mateo, C.; Rabanal, R.M.; Iannarelli, R.; Maggi, F. Chemical analysis of the essential oils from *Schizogyne sericea* growing in different areas of Tenerife (Spain). *Biochem. Syst. Ecol.* **2016**, *65*, 192–197. [CrossRef]

23. Radulović, N.; Blagojević, P.; Palić, R.; Zlatković, B. Volatiles of *Telekia speciosa* (Schreb.) Baumg. (Asteraceae) from Serbia. *J. Essent. Oil Res.* **2010**, *22*, 250–254. [CrossRef]

24. Wajs-Bonikowska, A.; Stojakowska, A.; Kalemba, D. Chemical composition of essential oils from a multiple shoot culture of *Telekia speciosa* and different plant organs. *Nat. Prod. Commun.* **2012**, *7*, 625–628. [CrossRef]

25. Andreani, S.; De Cian, M.-C.; Paolini, J.; Desjobert, J.-M.; Costa, J.; Muselli, A. Chemical variability and antioxidant activity of *Limbarda crithmoides* L. essential oil from Corsica. *Chem. Biodiversity* **2013**, *10*, 2061–2077. [CrossRef] [PubMed]

26. Shtacher, G.; Kashman, Y. Chemical investigation of volatile constituents of *Inula viscosa* Ait. *Tetrahedron* **1971**, *27*, 1343–1349. [CrossRef]

27. Bokadia, M.M.; MacLeod, A.J.; Mehta, S.C.; Mehta, B.K.; Patel, H. The essential oil of *Inula racemosa*. *Phytochemistry* **1986**, *25*, 2887–2888. [CrossRef]

28. Choudhary, A.; Sharma, R.J.; Singh, I.P. Determination of major sesquiterpene lactones in essential oil of *Inula racemosa* and *Saussurea lappa* using a qNMR. *J. Essent. Oil Bear. Pl.* **2016**, *19*, 20–31. [CrossRef]

29. Xu, T.; Gherib, M.; Bekhechi, C.; Atik-Bekkara, F.; Casabianca, H.; Tomi, F.; Casanova, J.; Bighelli, A. Thymyl esters derivatives and a new natural product modhephanone from *Pulicaria mauritanica* Coss. (Asteraceae) root oil. *Flavour Fragr. J.* **2015**, *30*, 83–90. [CrossRef]

30. Cilović, E.; Brantner, A.; Tran, H.T.; Arsenijević, J.; Maksimović, Z. Methanol extracts and volatiles of *Telekia speciosa* (Schreb.) Baumg. from Bosnia and Herzegovina. *Technol. Acta* **2019**, *12*, 9–13.

31. Stojakowska, A.; Michalska, K.; Malarz, J. Simultaneous quantification of eudesmanolides and thymol derivatives from tissues of *Inula helenium* and *I. royleana* by reversed-phase high-performance liquid chromatography. *Phytochem. Anal.* **2006**, *17*, 157–161. [CrossRef] [PubMed]

32. Zee, O.P.; Kim, D.K.; Lee, K.R. Thymol derivatives from *Carpesium divaricatum*. *Arch. Pharm. Res.* **1998**, *21*, 618–620. [CrossRef] [PubMed]

33. Stojakowska, A.; Kędzia, B.; Kisiel, W. Antimicrobial activity of 10-isobutyryloxy-8,9-epoxythymol isobutyrate. *Fitoterapia* **2005**, *76*, 687–690. [CrossRef] [PubMed]

34. Piszczek, P.; Kuszewska, K.; Błaszkowski, J.; Sochacka-Obruśnik, A.; Stojakowska, A.; Zubek, S. Associations between root-inhabiting fungi and 40 species of medicinal plants with potential applications in the pharmaceutical and biotechnological industries. *Appl. Soil Ecol.* **2019**, *137*, 69–77. [CrossRef]

35. Bonikowski, R.; Paoli, M.; Szymczak, K.; Krajewska, A.; Wajs-Bonikowska, A.; Tomi, F.; Kalemba, D. Chromatographic and spectral characteristic of some esters of a common monoterpene alcohols. *Flav. Fragr. J.* **2016**, *31*, 290–292. [CrossRef]
36. Talavera-Alemán, A.; Rodríguez-García, G.; López, Y.; García-Gutiérrez, H.A.; Torres-Valencia, J.M.; del Río, R.E.; Cerda-García-Rojas, C.M.; Joseph-Nathan, P.; Gómez-Hurtado, M.A. Systematic evaluation of thymol derivatives possessing stereogenic or prostereogenic centers. *Phytochem. Rev.* **2016**, *15*, 251–277. [CrossRef]

**Sample Availability:** Samples of the compounds are not available from the authors.

 *molecules*

Communication

# Characterization of Volatile Compounds in Four Different *Rhododendron* Flowers by GC×GC-QTOFMS

Chen-Yu Qian [1,2], Wen-Xuan Quan [2], Zhang-Min Xiang [1,*] and Chao-Chan Li [2,*]

[1] Guangdong Provincial Key Laboratory of Emergency Test for Dangerous Chemicals/Guangdong Engineering and Technology Research Center for Ambient Mass Spectrometry, Guangdong Institute of Analysis, Guangzhou 510070, China; qianchenyu94@163.com

[2] Guizhou Provincial Key Laboratory of Mountainous Environmental Protection, Guizhou Normal University, Guiyang 550001, China; wenxuanq@gznu.edu.cn

* Correspondence: xiangzm@live.com (Z.-M.X.); chaochanl@gznu.edu.cn (C.-C.L.)

Academic Editor: Igor Jerković
Received: 14 August 2019; Accepted: 11 September 2019; Published: 12 September 2019

**Abstract:** Volatile compounds in flowers of *Rhododendron delavayi, R. agastum, R. annae,* and *R. irroratum* were analyzed using comprehensive two-dimensional gas chromatography-mass spectrometry (GC×GC) coupled with high-resolution quadrupole time-of-flight mass spectrometry (QTOFMS). A significantly increased number of compounds was separated by GC×GC compared to conventional one-dimensional GC (1DGC), allowing more comprehensive understanding of the volatile composition of *Rhododendron* flowers. In total, 129 volatile compounds were detected and quantified. Among them, hexanal, limonene, benzeneacetaldehyde, 2-nonen-1-ol, phenylethyl alcohol, citronellal, isopulegol, 3,5-dimethoxytoluene, and pyridine are the main compounds with different content levels in all flower samples. 1,2,3-trimethoxy-5-methyl-benzene exhibits significantly higher content in *R. irroratum* compared to in the other three species, while isopulegol is only found in *R. irroratum* and *R. agastum*.

**Keywords:** *Rhododendron* flowers; volatile compounds; comprehensive two-dimensional gas chromatography-mass spectrometry; quadrupole time-of-flight mass spectrometry; odor description

---

## 1. Introduction

The evergreen woody shrub genus *Rhododendron* is one of the largest genera in the family Ericaceae, and more than 1000 species are currently recognized; of these, 567 species representing 6 subgenera are known from China [1,2]. *Rhododendrons* are not only of high ornamental value but also good medicinal plants. Flowers of *Rhododendron* provide a large number of bioactive natural chemical products, including diterpenoids [3], flavonoids [4], and phenols [5], which are known to be effective for the treatment of rheumatism [6] and to have anti-inflammatory [7], anti-cancer [8], and antioxidant [9] properties. Volatile compounds from flowers also provide some ecological functions [10], including in the role of pollinators [11] and as defenders against nectar-thieving ants [12]. Aside from their ecological functions, flower volatiles have some aesthetic and emotional benefits for humans [13]. On the other hand, different volatile compounds may influence the odor, both in an individual and in a synergistic or antagonistic way, which in turn could be related to one or more chemical compounds or compound classes [14]. In order to investigate the aroma characteristics of *Rhododendron* flowers, it is important to research specific volatile constituents as thoroughly as possible. *Rhododendron irroratum, R. delavayi, R. annae,* and *R. agastum* are ecologically and horticulturally important alpine flowers and are also the pioneer and constructive species in Baili *Rhododendron* National Forest Reserve in the Guizhou province of China [15]. *R. delavayi* belongs to the subsection Arborea, while *R. irroratum,*

*R. annae*, and *R. agastum* belong to the subsection Irrorata. *R. irroratum* is one of the large-flowered *Rhododendron* species [16]. These species were chosen for the investigation of volatile odor constituents in different *Rhododendron* flowers.

Gas chromatography–mass spectrometry (GC-MS) has long been the primary technique used to detect the aroma components of various plants [17,18]. However, GC-MS can only identify a limited number of separable compounds due to its insufficient peak capacity, limited resolved power, and low sensitivity [19]. The combination of gas chromatography with high-resolution quadrupole time-of-flight mass spectrometry (QTOFMS) has been demonstrated for analysis in different fields, including flavor research [20] and volatile profiling [21], and has proved to be a powerful analytical tool. However, the limited chromatographic separation power inevitably causes co-elution problems for complex samples. Compared with traditional one-dimensional gas chromatography (1DGC), comprehensive two-dimensional gas chromatography (GC×GC), which has appeared as a new analytical technique based on the application of two GC columns with different stationary phases, provides substantially enhanced resolving power and peak capacity. GC×GC leads to linear distributions of homologous series in 2D chromatograms, thus greatly reducing the coelution problem [22]. GC×GC can thus be a more suitable tool for analysis of the complex chemical systems of plant aroma, where the number of volatile aroma compounds is large and some of them are present at trace levels [23]. Recently, GC×GC technology has been successfully applied for the assessment of various plants such as teas [24], berries [25], and tobacco [26]. To date, few reports have studied the volatile chemical components in *Rhododendron* flowers by 1DGC. With the 1DGC technique, 9,12,15-octadecatrienoic acid,[Z,Z,Z]-, phytol, and *n*-hexadecanoic acid were found to be the major compounds in flowers of *R. mucronatum* and *R. simii* [27]; while *R. ponticum* comprises mostly α-pinene, β-pinene, and linalool [28], in flowers of *R. schlippenbachii*, only 39 hydrophilic compounds could be detected by 1DGC [29]. A previous study reported the volatile compounds in the leaves, stems, and roots of six *Rhododendron* species [15]. However, the volatile components in flowers of the four *Rhododendron* species in the present study have never been investigated by the GC×GC approach before. Therefore, it is necessary to study the flowers' volatile compounds in order to explore odor characterizations.

In this study, GC×GC-QTOFMS was used in combination with headspace solid-phase microextraction (HS-SPME) to conduct in-depth analysis of the volatile aroma constituents in different *Rhododendron* flowers. The advantages of GC×GC–QTOFMS were exploited for high-throughput, untargeted chromatographic profiling of complex samples. The volatile compounds and their corresponding contents in various representative *Rhododendron* samples were examined. The obtained results provide useful information for establishing a volatile aroma chemical database from *Rhododendron* flowers.

## 2. Results and Discussion

### 2.1. Comparison of 1DGC and GC×GC

In typical 1DGC analysis, it is often difficult to achieve pure mass spectra for compounds in a co-elution peak, thus leading to unreliable results. With improved separation power and enhanced sensitivity, the GC×GC technique is able to resolve and detect more volatile aroma compounds in a complex sample compared to conventional one-dimensional GC-MS [30]. A clear illustration demonstrating the employment of GC×GC is presented in Figure 1. Both the chromatogram obtained by GC×GC–TOF/MS using a 4 s modulation period and the total ion chromatography by 1DGC are shown. As can be seen in the partial chromatograms obtained by 1DGC and GC×GC-QTOFMS, linalool (Peak 1, $^1t_R$ = 19.783 min, $^2t_R$ = 1.405 s) and 2-nonen-1-ol (Peak 2, $^1t_R$ = 19.849 min, $^2t_R$ = 1.447 s) were responsible for the two peaks detected between retention times of 19.380 min and 19.850 min on the HP-5 MS column. However, three other minor compounds in addition to these two peaks were further separated as they exhibited different polarities on the DB-17 MS column; these were linalool oxide (Peak 3, $^1t_R$ = 19.383 min, $^2t_R$ = 1.467 s), *p*-cymenene (Peak 4, $^1t_R$ = 19.450 min, $^2t_R$ = 1.627 s), and benzoic acid, methyl ester (Peak 5, $^1t_R$ = 19.716 min, $^2t_R$ = 2.017 s). The co-eluted compounds in the peak region were interfered with by dominating compounds and would usually be ignored due to their low concentrations. In summary, GC×GC successfully resolved a total of 129 compounds, while only 45 compounds were separated in 1DGC (Table S1). The results revealed the great advantages of GC×GC analysis, which is suitable for the investigation of volatile compounds in complex samples.

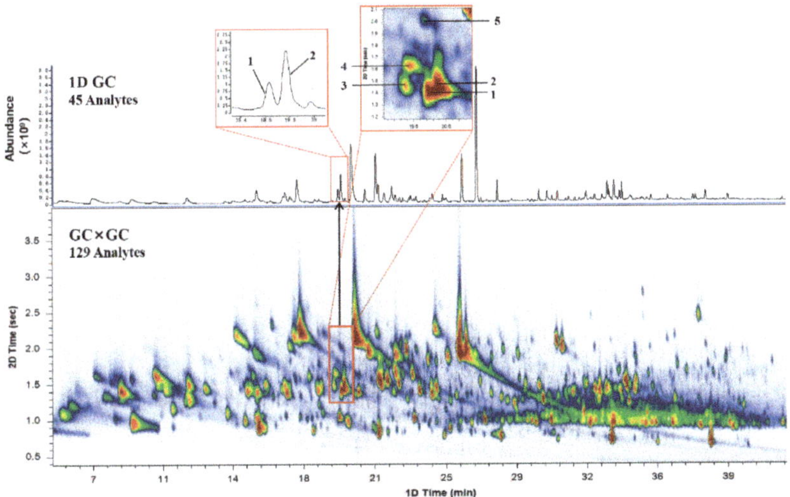


**Figure 1.** Chromatographic analysis of *Rhododendron* by GC-quadrupole time-of-flight mass spectrometry (QTOFMS) and a comprehensive two-dimensional gas chromatography–mass spectrometry (GC×GC)-QTOFMS color diagram (1: linalool; 2: 2-nonen-1-ol; 3: linalool oxide; 4: *p*-cymenene; 5: benzoic acid, methyl ester).

## 2.2. Identification of Common Volatile Components

GC×GC–QTOFMS was used to characterize the detailed chemical composition of all the samples. Several hundred peaks were generated in the GC×GC contour plot with a peak detection threshold of S/N > 3. In total, 129 volatile compounds were tentatively identified in four *Rhododendron* samples based on (1) spectral similarity (both match and reverse match scores of >750), (2) comparison with molecular ions (within 5 ppm), if they existed, and (3) retention index (RI, ±35). Table S1 lists the complete information of the 129 volatile constituents detected by GC×GC-QTOFMS.

Figure 2 introduces the identification process of two examples (1,2-dimethoxybenzene and lilac aldehyde D). First, the National Insititute of Standards and Technology (NIST) library search for Peak 162 and Peak 189 resulted in 7 and 5 possible compounds, respectively, with match factor >750. Then, only exact mass analyses within a mass accuracy of <5 ppm were considered. For Peak 162, the measured accurate mass was 138.0676, which corresponds to a formula of $C_8H_{10}O_2$. The accurate mass reduced the number of possible compounds to two isomers (1,2-dimethoxybenzene and 1,4-dimethoxybenzene). Last, their retention indices were reviewed for further confirmation. The GC×GC analysis provided an experimental RI value of 1151 for this peak, which matched 1,2-dimethoxybenzene (literature RI value of 1151) rather than 1,4-dimethoxybenzene ($RI_{lit}$ = 1168). Therefore 1,2-dimethoxybenzene was the final identified compound for Peak 162.

Taking Peak 189 as another example: The NIST library search provided several possible compound matches. Among them, seven possible compounds were screened out according to their relatively high match scores. Subsequently, the mass spectrum provided a measured mass of 168.1148, corresponding to a chemical formula of $C_{10}H_{16}O_2$. This indicated that 2-methyl-2-(2-oxopropyl) cyclohexanone, lilac aldehyde D, and 2-hydroxy-4,4,6,6-tetramethyl-2- cyclohexen-1-one were three possible compounds with the theoretical molecular ion mass of 168.1145. Lastly, the experimental RI value of the peak ($RI_{exp}$ = 1190) confirmed that lilac aldehyde D with $RI_{lit}$ of 1169 was the final identified compound for Peak 189, while the other two candidates, 2-methyl-2-(2-oxopropyl) cyclohexanone ($RI_{lit}$ = 1360) and 2-hydroxy-4,4,6,6-tetramethyl-2-cyclohexen-1-one ($RI_{lit}$ = 1272), were screened out. The results emphasis the importance of applying further confirmation on the base of the spectral library match, since the compound with the highest match factor might be mistaken for the identity of the component [31]. In conclusion, with complementary identification processes, GC×GC coupled with high-resolution QTOFMS produces more precise compound identification results.

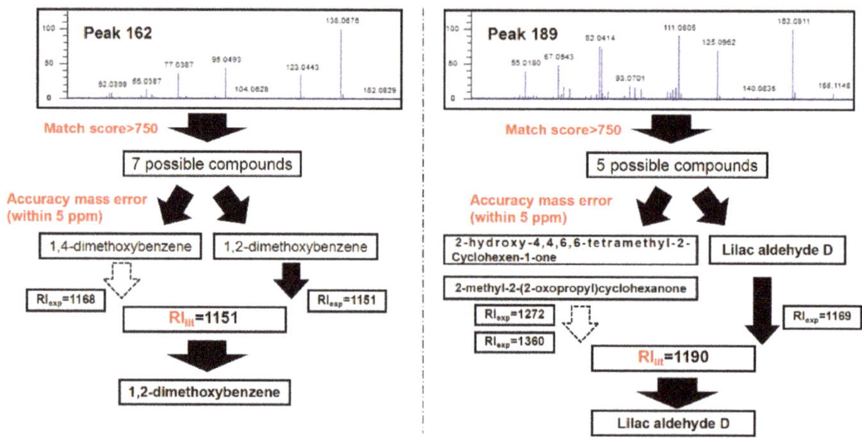


**Figure 2.** Diagram illustrating the process of compound confirmation in GC×GC-QTOFMS.

*2.3. Volatile Component Analysis*

In order to establish the experimental conditions, the mixed sample was analyzed via GC×GC-QTOFMS in triplicate. The intraday precision was evaluated by analyzing three equivalent mixed samples on the same day, and this was then repeated for three consecutive days to determine the interday precision. As shown in Table S2, the intraday and interday precision were expressed as the relative standard deviation (RSD). RSD values of no more than 25% were found in each compound in the mixed sample, demonstrating the good repeatability of the GC×GC-QTOFMS method. Subsequently, the four flower species were analyzed by the established method. The relative contents (%) of compounds in each sample were calculated based on the ratio of the area of the corresponding peak to the total peak area; the averages of the relative contents of each compound in the *Rhododendron* flower samples are tabulated in Table S2. Figure 3 presents the distribution (%) of the major compounds in the four different species of *Rhododendron*. Among them, benzeneacetaldehyde was found in all flower species with high content in *R. irroratum* (6.255% ± 0.951%), *R. delavayi* (7.013% ± 0.059%), and *R. annae* (6.349% ± 0.062%), whereas it presented with relatively low content in *R. agastum* (2.987% ± 0.357%). Citronellal presented the highest content in *R. annae* (7.004% ± 0.028%) and *R. agastum* (7.722% ± 0.303%), and was the second most abundant component in *R. delavayi* (7.944% ± 0.225%), but was slightly low in *R. irroratum* (4.178% ± 0.654%). Both benzeneacetaldehyde and citronellal contribute to the sweet floral profile of these samples. Benzeneacetaldehyde has a grassy odor, while citronellal has a slight hyacinth odor. 1,2,3-trimethoxy-5-methyl-benzene was detected in all species with content ranging from trace (0.243% ± 0.023% in *R. annae*) to abundant (6.046% ± 0.623% in *R. irroratum*). On the other hand, isopulegol was detected only in *R. irroratum* and *R. agastum*, with a highest content of 7.722% ± 0.407% in *R. agastum*. Thus, this compound can be used to discriminate *R. agastum* or *R. irroratum* from other *Rhododendron* species. Phenylethyl alcohol accounted for a significantly high content in *R. delavayi* (up to 8.922% ± 0.061%) compared to in the other three species and was characterized as a dried rose floral aroma. Similarly, 2-nonen-1-ol, with a sweet melon odor, also presented the highest content in *R. delavayi* (5.633% ± 0.813%) but slightly low in *R. irroratum* (4.299% ± 0.288%) and *R. agastum* (4.071% ± 0.378%). Limonene and isopulegol presented in all species with relatively low content compared with other major compounds and with no significant differences between species. Limonene has a sweet citrus or orange odor, while isopulegol has a minty or woody odor. Although the rest of the compounds had relatively low threshold values due to their low contents, they all play a certain role in the odor characterization and finally form the special odor types of different *Rhododendron* varieties.

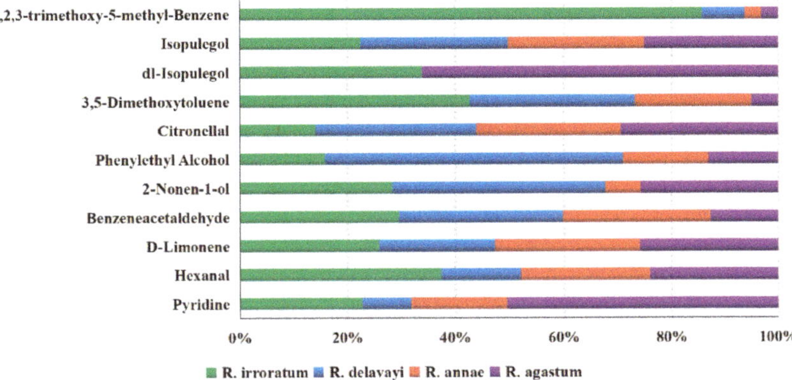

**Figure 3.** Distribution (%) of major compounds presented in four different species of *Rhododendron*.

## 2.4. Odor Analysis

The identified components were classified into various types of compound groups, including alcohols (29), aldehydes (15), alkenes (29), aromatic hydrocarbons (10), esters (19), ketones (10), phenols (4), and others (13)—eight classes in total. Figure 4 shows the relative contents of the chemical classes in the four samples. The predominant groups were aldehydes and alcohols, followed by esters and alkenes. Large amounts of aldehydes were detected in *R. annae* (27.37%) and *R. irroratum* (26.95%). Although alkenes had the same number of compounds compared to alcohols, their contents were far below those of alcohols. Besides this, *R. irroratum* had the lowest content of esters (6.04%), while the other three species had similar proportions.

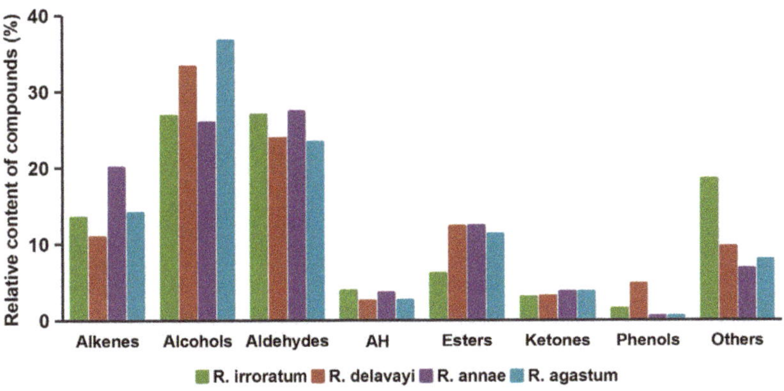

**Figure 4.** Distribution of the chemical classes for *Rhododendron* (AH: aromatic hydrocarbons).

### 2.4.1. Floral and Woody Odor

From an odor perspective, alcohols showed a higher number of compounds with a descriptor of a floral odor. For example, linalool, which is reported to possess a floral and citrus-like aroma [32], was relatively high in the *R. annae* species (4.752% ± 0.114%). Aside from linalool, benzyl alcohol, phenylethyl alcohol, and citronellol are all described as having floral and rose odors. Among them, phenylethyl alcohol is widely used as ingredient for perfumes and produces a rose smell [33]. Citronellol was previously reported as the floral odor compound in lychee juice [34]. Woody odor attributes in *Rhododendron* flowers were mainly associated with alkenes and alcohols. Alkenes showed a higher number of compounds with descriptors of woody and sweet, such as α-pinene (intense woody), β-pinene (dry woody), and α-terpinene (woody, piney), which were previously identified in terebinth fruits [17], but accounted for relatively low contents (0.117% ± 0.056% to 1.456% ± 0.039%) in flowers. Alcohols such as isopulegol and isoborneol also have woody odor characterization and accounted for 1.099% ± 0.091% to 3.328% ± 0.133% in *Rhododendron* flowers, mostly higher than the alkene contents. In addition, β-ionone, well known for its violet odor and described as a complex woody and fruity scent [35], was also found in four flower species.

### 2.4.2. Green and Fresh Odor

Grass odor is sometimes referred to as a fresh note, and the chemicals with this descriptor are predominantly aldehydes with six to nine carbons and C6 alcohols [36,37]. In *Rhododendron* flowers, hexanal was the major such compound in all samples, mainly contributing to the green and grassy odor. Besides this, 2-hexenal, heptanal, octanal, and benzeneacetaldehyde was also found to contribute to the green and fresh odor [34]. Among them, 2-hexenal and heptanal accounted for relatively high proportions in *R. annae* and *R. agastum*. On the other hand, C6 alcohols such as (*E*)-3-hexen-1-ol and 1-hexanol also yielded a green, fresh, and herbal odor [32]. In addition, β-cadinene, 2-pentyl-furan and formic acid, 2-phenylethyl ester are also related to a green odor.

### 2.4.3. Sweet and Fruity Odor

In the *Rhododendron* flowers, the compounds contributing to the sweet and fruity odor mainly included aldehydes and alkenes. Among the aldehydes, citronellal (sweet, citrus), decanal (sweet, orange), and undecanal (floral, citrus) all provide a sweet and fruity odor, especially citronellal with its high contents in the four flower species (4.178% ± 0.654% to 7.944% ± 0.225%). Among the alkenes, limonene is a typical sweet and citrus-like odor compound which was previously identified in lychee [32]. α-Ocimene with a fruity aroma was also reported in a previous study [17]. Some alcohols like major compound 2-nonen-1-ol also have a sweet and melon odor. Besides this, it has been previously reported that α-terpineol is one of the major components providing fruity and floral notes in Pu-erh tea [38]. Other compounds, for example, 2-pentyl-furan, reported to have a fruity, green, and earthy odor [39], accounted for a relatively high proportion in *R. irroratum* (up to 1.643% ± 0.290%) among the four flower species studied.

### 2.4.4. Total Odor Description

As illustrated by the four pie charts shown in Figure 5, the proportion distributions of volatile compounds based on the specific odor characteristics of the *Rhododendron* flowers were surveyed to represent the odor types of compounds in the samples. There was a higher number of chemical compounds with descriptors of floral, woody, sweet, and fresh odor, mainly derived from alkenes, alcohols, esters, and aldehydes, thus comprising the major odor characteristics of *Rhododendron* flowers. Sweet odor represented the highest proportion in *R. annae* (35.96%), *R. irroratum* (27.01%), and *R. agastum* (31.46%), while floral odor was the most abundant in *R. delavayi* (up to 34.29%). Other odors such as herbaceous, piney, and mushroom had relatively low proportions but also contributed to the overall odor characteristics. The different compounds and contents make up the specific *Rhododendron* odors. Volatile aroma components from various species and their content differences determine the flower-specific scent properties. Their odor values and contributions to flower odorant will be further investigated in the future.

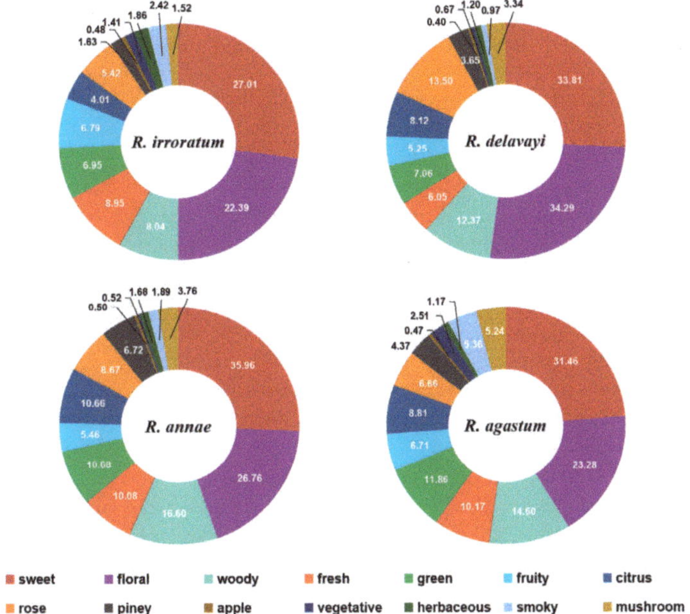



**Figure 5.** Proportions of odor compounds in *Rhododendron*.

## 3. Materials and Methods

### 3.1. Sample Pretreatment

The flowers from four *Rhododendron* species (*R. delavayi*, *R. agastum*, *R. annae*, and *R. irroratum*) were collected in the spring of 2019 (between March and April) in Baili *Rhododendron* National Forest Reserve (E 105°45′~106°04′ 45"; N 27°08′ 30"~27°20′ 00"), located in northwestern Guizhou, China. Flowers were collected and placed in sealed plastic bags, then immediately transported in a cooler with ice to the laboratory. Subsequently, the obtained samples were smashed after being frozen in a vacuum freeze-dryer for a week at −70 °C (FD-1C-80; Boyikang, Beijing, China), then transferred into 50 mL vials [15]. All samples were stored in a freezer at a temperature below −20 °C until analysis. A mixed sample was prepared using all the four flower species in equal quantities and was used for analytical method establishment and repeatability examination.

### 3.2. SPME Methodology

The extraction and concentration of the volatile compounds were carried out using the headspace solid phase microextraction (HS-SPME) technique. As the object of this study was to characterize all volatile compounds, divinylbenzene/carboxen/polydimethylsiloxane (DVB/CAR/PDMS) fiber (50/30 μm) (Supelco, Bellefonte, PA, USA) combining the characteristics of both carboxen and divinylbenzene adsorbents in the coating and thus allowing a wide range of molecules of different sizes to be adsorbed into the coating for natural products [40], was chosen for volatile compound analysis. Quantities of 50 mg of samples were accurately weighed into a 20 mL vial, and then the SPME fiber was exposed to the headspace of the bottle for 20 min at 70 °C. The SPME fiber was then introduced into the GC injector for 3.0 min to allow thermal desorption of the analytes. The established approach for quantitative analysis was validated by studying the repeatability using the mixed sample. All measurements were conducted in triplicate.

### 3.3. Analytical Instrumentation

The system was equipped with simultaneous 1DGC and GC×GC in one instrument which can conduct both techniques at the same time without any change of columns. The system consisted of a gas chromatograph (7890B Agilent Technology, Santa Clara, CA, USA) coupled with a high-resolution quadrupole time-of-flight mass spectrometer (QTOFMS) (mass resolution 20,000 and a mass accuracy specification of 3 ppm) (7250, Agilent Technology). In the presented research, an HP-5 MS (5% phenyl–95% dimethylpolysiloxane, 30 m × 250 μm, 0.25 μm film) was used as the $^1$D column, and a DB-17 MS column (50% phenyl–50% dimethylpolysiloxane, 1.2 m × 180 μm, 0.18 μm film) was used as the $^2$D column. The samples were introduced by a split/splitless injector (SSL) system with an autosampler (PAL RSI 120, CTC Technologies). This study employed a technique to combine GC×GC and 1DGC components into a single system with the column outlet of each component connected at the same three-port splitter prior to the QTOFMS detection. This allowed direct comparison of the GC×GC and 1DGC results and avoided use of a second detector, which is simple and effective.

The 1DGC and GC×GC conditions were the same and were as listed below: the GC injector was kept at 250 °C in splitless mode; helium (99.999%) was used as the carrier gas at a constant flow of 1.2 mL/min; oven temperature was initially set at 50 °C (held for 3 min) , then increased to 250 °C at 4 °C/min (held for 7 min), for a total run time of 60 min. The GC×GC system was coupled with an SSM1800 solid state modulator (J&X Technologies, China). The GC×GC conditions were as follows: The cold zone temperature of the SSM was set at −50 °C. The temperatures of the entry hot zone and exit hot zone were +30 and +120 °C offset relative to oven temperatures, respectively, with a cap temperature of 320 °C for both hot zones. The modulation period was 4 s.

The MS conditions were as follows: The electron ionization and the ion source and transfer line temperatures were set at 70 eV, 250 °C, and 280 °C, respectively. The MS scan rate was 50 Hz. The mass range was set to 50–500 *m/z* in full-scan acquisition mode.

### 3.4. Data Method

The volatile composition was quantified in duplicate by HS-SPME coupled to GC×GC with QTOFMS according to the method of previous reports [41]. The 1DGC data were processed using Agilent Mass Hunter Qualitative Analysis navigator B.08.00; the GC×GC data were analyzed using dedicated Canvas GC×GC data processing software (J&X Technologies, version v1.4.0, Shanghai, China). Tentative compound identification was accomplished by mass spectral match based on the NIST 17 Mass Spectral Library (NIST/EPA/NIH 2017) and then verified using the retention index (RI) and accurate mass. The RI was calculated using a series of *n*-alkanes (C8–C25) analyzed on an HP-5 MS column under the same chromatographic conditions. The odor identification method was performed based on previous studies [42], relying on the Good Scents Company Information System, available online: http://www.thegoodscentscompany.com.

## 4. Conclusions

GC×GC-QTOFMS was applied to identify the volatile aroma compounds in four *Rhododendron* flower species. In total, 129 volatile compounds were separated and confirmed by spectral similarity, exact mass, and retention index. The relative contents of the volatile compounds were profiled for the four species of *Rhododendron* flowers. With the great advantages of the GC×GC technique over traditional 1DGC, this preliminary study improved scientific understanding regarding the volatile components in *Rhododendron* flowers, and the detected compounds could be used to establish the fingerprint signatures of *Rhododendron*.

**Supplementary Materials:** The following are available online at http://www.mdpi.com/1420-3049/24/18/3327/s1, Table S1. Complete compound information for identification; Table S2. Compound contents in *Rhododendron* flowers of different species and their odor description.

**Author Contributions:** Conceptualization, Z.M.X.; Methodology, C.-Y.Q.; Data curation: C.-Y.Q.; Supervision, W.-X.Q.; Writing—original draft preparation, C.-Y.Q.; Writing—review and editing, C.-C.L.

**Funding:** This study was supported by the GDAS' Project of Science and Technology Development (No. 2019GDASYL-0302004, 2018GDASCX-0808 and 2017GDASCX-0104), the National Natural Science Foundation of China (NSFC, No. 31960312), and Joint Fund of the National Natural Science Foundation of China and the Karst Science Research Center of Guizhou province (Grant No. U1812401), the Provincial Natural Science Foundation of Guizhou (Grant No. QKZYD [2017] 4006), special funds of industrial analysis and testing scientific research foundation and capacity in Guangdong province.

**Conflicts of Interest:** The authors declare no conflicts of interest.

# References

1. Zhou, W.; Oh, J.; Li, W.; Kim, D.W.; Yang, M.H.; Jang, J.H.; Ahn, J.S.; Lee, S.H.; Na, M.K. Chemical constituents of the Korean endangered species *Rhododendron Brachycarpum*. *Biochem. Syst. Ecol.* **2014**, *56*, 231–236. [CrossRef]

2. Zhang, L.; Xu, P.W.; Cai, Y.F.; Ma, L.L.; Li, S.F.; Li, S.F.; Xie, W.J.; Song, J.; Peng, L.C.; Yan, H.J.; et al. The draft genome assembly of *Rhododendron delavayi* Franch. var. *delavayi*. *GigaScience* **2017**, *6*, 1–11. [CrossRef] [PubMed]

3. Zhou, S.Z.; Yao, S.; Tang, C.P.; Ke, C.Q.; Li, L.; Lin, G.; Ye, Y. Diterpenoids from the Flowers of *Rhododendron Molle*. *J. Nat. Prod.* **2014**, *775*, 1185–1192. [CrossRef]

4. Jung, S.J.; Kim, D.H.; Hong, Y.H.; Lee, J.H.; Song, H.N.; Rho, Y.D.; Baek, N.I. Flavonoids from the flower of *Rhododendron yedoense* var. *Poukhanense* and their antioxidant activities. *Arch. Pharm. Res.* **2007**, *30*, 146–150. [CrossRef] [PubMed]

5. Kiruba, S.; Mahesh, M.; Nisha, S.R.; Miller Paul, Z.; Jeeva, S. Phytochemical analysis of the flower extracts of *Rhododendron arboreum* Sm. ssp. *nilagiricum* (Zenker) Tagg. *Asian Pac. J. Trop. Biomed.* **2011**, *1*, S284–S286. [CrossRef]

6. Zhu, Y.X.; Zhang, Z.X.; Yan, H.M.; Lu, D.; Zhang, H.P.; Li, L.; Liu, Y.B.; Li, Y. Antinociceptive diterpenoids from the leaves and twigs of *Rhododendron Decorum*. *J. Nat. Prod.* **2018**, *81*, 1183–1192. [CrossRef] [PubMed]

7. Zhou, J.F.; Liu, T.T.; Zhang, H.Q.; Zheng, G.J.; Qiu, Y.; Deng, M.Y.; Zhang, C.; Yao, G.M. Anti-inflammatory grayanane diterpenoids from the leaves of *Rhododendron molle*. *J. Nat. Prod.* **2018**, *81*, 151–161. [CrossRef] [PubMed]

8. Wang, S.J.; Lin, S.; Zhu, C.G.; Yang, Y.C.; Li, S.; Zhang, J.J.; Chen, X.G.; Shi, J.G. Highly acylated diterpenoids with a New 3,4-Secograyanane Skeleton from the Flower Buds of\r *Rhododendron Molle*. *Org. Lett.* **2010**, *12*, 1560–1563. [CrossRef]

9. Li, Y.; Liu, Y.B.; Yu, S.S. Grayanoids from the Ericaceae family: Structures, biological activities and mechanism of action. *Phytochem. Rev.* **2013**, *12*, 305–325. [CrossRef]

10. Zeng, L.T.; Wang, X.Q.; Dong, F.; Watanabe, N.; Yang, Z.Y. Increasing postharvest high-temperatures lead to increased volatile phenylpropanoids/benzenoids accumulation in cut rose (*Rosa hybrida*) flowers. *Postharvest Biol. Tec.* **2019**, *148*, 68–75. [CrossRef]

11. Pichersky, E.; Gershenzon, J. The formation and function of plant volatiles: Perfumes for pollinator attraction and defense. *Curr. Opin. Plant. Biol.* **2002**, *5*, 237–243. [CrossRef]

12. Junker, R.R.; Blüthgen, N. Floral scents repel potentially nectar-thieving ants. *Evol. Ecol. Res.* **2008**, *10*, 295–308.

13. Pichersky, E.; Dudareva, N. Scent engineering: Toward the goal of controlling how flowers smell. *Trends Biotechnol.* **2007**, *25*, 105–110. [CrossRef] [PubMed]

14. Kang, S.Y.; Yan, H.; Zhu, Y.; Liu, X.; Lv, H.P.; Zhang, Y.; Dai, W.D.; Guo, L.; Tan, J.F.; Peng, Q.H.; et al. Identification and quantification of key odorants in the world's four most famous black teas. *Food Res. Int.* **2019**, *121*, 73–83. [CrossRef]

15. Qian, C.Y.; Quan, W.X.; Li, C.C.; Xiang, Z.M. Analysis of volatile terpenoid compounds in *Rhododendron* species by multidimensional gas chromatography with quadrupole time-of-flight mass spectrometry. *Microchem. J.* **2019**, *149*, 104064. [CrossRef]

16. Zha, H.G.; Milne, R.I.; Sun, H. Asymmetric hybridization in *Rhododendron agastum*: A hybrid taxon comprising mainly F1 s in Yunnan, China. *Ann. Bot.* **2010**, *105*, 89–100. [CrossRef] [PubMed]

17. Amanpour, A.; Guclu, G.; Kelebek, H.; Selli, S. Characterization of key aroma compounds in fresh and roasted terebinth fruits using aroma extract dilution analysis and GC-MS-Olfactometry. *Microchem. J.* **2019**, *145*, 96–104. [CrossRef]

18. Qia, D.D.; Miaoa, A.Q.; Cao, J.X.; Wang, W.W.; Chen, W.; Pang, S.; He, X.G.; Ma, C.Y. Study on the effects of rapid aging technology on the aroma quality of whitetea using GC-MS combined with chemometrics: In comparison with naturalaged and fresh white tea. *Food Chem.* **2018**, *265*, 189–199. [CrossRef] [PubMed]

19. Zhu, Y.; Lv, H.P.; Dai, W.D.; Guo, L.; Tan, J.F.; Zhang, Y.; Yu, F.L.; Shao, C.Y.; Peng, Q.H.; Lin, Z. Separation of aroma components in Xihu Longjing tea using simultaneous distillation extraction with comprehensive two-dimensional gas chromatography-time-of-flight mass spectrometry. *Sep. Purif. Technol.* **2016**, *164*, 146–154. [CrossRef]

20. Huang, X.H.; Zheng, X.; Chen, Z.H.; Zhang, Y.Y.; Du, M.; Dong, X.P.; Qin, L.; Zhu, B.W. Fresh and grilled eel volatile fingerprinting by e-Nose, GC-O, GC-MS and GC×GC-QTOF combined with purge and trap and solvent-assisted flavor evaporation. *Food Res. Int.* **2019**, *115*, 32–43. [CrossRef] [PubMed]

21. Liu, F.T.; Li, S.J.; Gao, J.H.; Cheng, K.; Yuan, F. Changes of terpenoids and other volatiles during alcoholic fermentation of blueberry wines made from two southern highbush cultivars. *LWT* **2019**, *109*, 233–240. [CrossRef]

22. Egert, B.; Weinert, C.H.; Kulling, S.E. A peaklet-based generic strategy for the untargeted analysis of comprehensive two-dimensional gas chromatography mass spectrometry data sets. *J. Chromatogr. A.* **2015**, *1405*, 168–177. [CrossRef] [PubMed]

23. Lukić, I.; Carlin, S.; Horvat, I.; Vrhovsek, U. Combined targeted and untargeted profiling of volatile aroma compounds with comprehensive two-dimensional gas chromatography for differentiation of virgin olive oils according to variety and geographical origin. *Food Chem.* **2019**, *270*, 403–414. [CrossRef] [PubMed]

24. Zhang, L.; Zeng, Z.D.; Zhao, C.X.; Kong, H.W.; Lu, X.; Xu, G.W. A comparative study of volatile components in green, oolong and black teas by using comprehensive two-dimensional gas chromatography-time-of-flight mass spectrometry and multivariate data analysis. *J. Chromatogr. A.* **2013**, *1313*, 245–252. [CrossRef] [PubMed]

25. Kupska, M.; Chmiel, T.; Jedrkiewicz, R.; Wardencki, W.; Namies ́nik, J. Comprehensive two-dimensional gas chromatography for determination of the terpenes profile of blue honeysuckle berries. *Food Chem.* **2014**, *152*, 88–93. [CrossRef] [PubMed]

26. Schwanz, T.G.; Bokowski, L.V.V.; Marcelo, M.C.A.; Jandrey, A.C.; Dias, J.C.; Maximiano, D.H.; Canova, L.S.; Pontes, O.F.S.; Sabin, G.P.; Kaiser, S. Analysis of chemosensory markers in cigarette smoke from different tobacco varieties by GC×GC-TOFMS and chemometrics. *Talanta* **2019**, *202*, 74–89. [CrossRef] [PubMed]

27. Zhao, C.X.; Li, X.N.; Liang, Y.Z.; Fang, H.Z.; Huang, L.F.; Guo, F.Q. Comparative analysis of chemical components of essential oils from different samples of *Rhododendron* with the help of chemometrics methods. *Chemom. Intell. Lab. Syst.* **2006**, *82*, 218–228. [CrossRef]

28. Küçük, S.; Kürkçüoğlu, M.; Başer, K.H.C. Morphological, Indumentum and Chemical Characteristics and Analysis of the Volatile Components of the Flowers of *Rhododendron ponticum* L. subsp. *ponticum* (Ericaceae) of Turkish Origin. *Rec. Nat. Prod.* **2018**, *12*, 498–507.

29. Park, C.H.; Yeo, H.J.; Kim, N.S.; Park, Y.E.; Park, S.Y.; Kim, J.K.; Park, S.U. Metabolomic profiling of the white, violet, and red flowers of *Rhododendron schlippenbachii* Maxim. *Molecules* **2018**, *23*, 827. [CrossRef]

30. Purcaro, G.; Cordero, C.; Liberto, E.; Bicchi, C.; Conte, L.S. Toward a definition of blueprint of virgin olive oil by comprehensive two-dimensional gas chromatography. *J. Chromatogr. A.* **2014**, *1334*, 101–111. [CrossRef]

31. Jiang, M.; Kulsing, C.; Marriott, P.J. Comprehensive 2D gas chromatography-time-of-flight mass spectrometry with 2D retention indices for analysis of volatile compounds in frankincense (*Boswellia papyrifera*). *Anal. Bioanal. Chem.* **2018**, *410*, 3185–3196. [CrossRef]

32. Feng, S.; Huang, M.Y.; Crane, J.H.; Wang, Y. Characterization of key aroma-active compounds in lychee (*Litchi chinensis* Sonn.). *J. Food Drug Anal.* **2018**, *26*, 497–503. [CrossRef]

33. Mo, E.K.; Sung, C.K. Phenylethyl alcohol (PEA) application slows fungal growth and maintains aroma in strawberry. *Postharvest Biol. Technol.* **2007**, *45*, 234–239. [CrossRef]

34. An, K.J.; Liu, H.C.; Fu, M.Q.; Qian, M.C.; Yu, Y.S.; Wu, J.J.; Xiao, G.S.; Xu., Y.J. Identification of the cooked off-flavor in heat-sterilized lychee (*Litchi chinensis* Sonn.) juice by means of molecular sensory science. *Food Chem.* **2019**, *301*, 125282. [CrossRef]

35. Gulati, A.; Ravindranath, S.D. Seasonal variations in quality of Kangra tea (*Camellia sinensis* (L) O Kuntze) in Himachal Pradesh. *J. Sci. Food Agr.* **1996**, *71*, 231–236. [CrossRef]
36. Zhu, J.C.; Chen, F.; Wang, L.Y.; Niu, Y.W.; Chen, H.X.; Wang, H.L.; Xiao, Z.B. Characterization of the key aroma volatile compounds in Cranberry (*Vaccinium macrocarpon* Ait.) using gas chromatography-olfactometry (GC-O) and odor activity value (OAV). *J. Agr. Food Chem.* **2016**, *64*, 4990–4999. [CrossRef]
37. Zhu, J.C.; Wang, L.Y.; Xiao, Z.B.; Niu, Y.W. Characterization of the key aroma compounds in mulberry fruits by application of gas chromatography-olfactometry (GC-O), odor activity value (OAV), gas chromatography-mass spectrometry (GC-MS) and flame photometric detection (FPD). *Food Chem.* **2018**, *245*, 775–785. [CrossRef]
38. Lv, H.P.; Zhong, Q.S.; Lin, Z.; Wang, L.; Tan, J.F.; Guo, L. Aroma characterisation of Pu-erh tea using headspace-solid phase microextraction combined with GC/MS and GC–olfactometry. *Food Chem.* **2012**, *130*, 1074–1081. [CrossRef]
39. Feng, Z.H.; Li, Y.F.; Li, M.; Wang, Y.J.; Zhang, L.; Wan, X.C.; Yang, X.G. Tea aroma formation from six model manufacturing processes. *Food Chem.* **2019**, *285*, 347–354. [CrossRef]
40. Silva, É.A.S.; Saboia, G.; Jorge, N.C.; Hoffmann, C.; dos Santos Isaias, R.M.; Soares, G.L.G.; Zini, C.A. Development of a HS-SPME-GC/MS protocol assisted by chemometric tools to study herbivore-induced volatiles in *Myrcia splendens*. *Talanta* **2017**, *175*, 9–20. [CrossRef]
41. Xiang, Z.M.; Chen, X.T.; Zhao, Z.J.; Xiao, X.; Guo, P.R.; Song, H.C.; Yang, X.; Huang, M.H. Analysis of volatile components in *Dalbergia cochinchinensis* Pierre by a comprehensive two-dimensional gas chromatography with mass spectrometry method using a solid-state modulator. *J. Sep. Sci.* **2018**, *41*, 1–8. [CrossRef]
42. Wang, C.C.; Zhang, W.J.; Li, H.D.; Mao, J.S.; Guo, C.Y.; Ding, R.Y.; Wang, Y.; Fang, L.P.; Chen, Z.L.; Yang, G.S. Analysis of Volatile Compounds in Pears by HS-SPME-GC×GC-TOFMS. *Molecules* **2019**, *24*, 1795. [CrossRef]

**Sample Availability:** Samples of the compounds are available from the authors.

 *molecules*

Article

# A Mass Spectrometry-Based Study Shows that Volatiles Emitted by *Arthrobacter agilis* UMCV2 Increase the Content of Brassinosteroids in *Medicago truncatula* in Response to Iron Deficiency Stress

Idolina Flores-Cortez [1], Robert Winkler [2], Arturo Ramírez-Ordorica [1], Ma. Isabel Cristina Elizarraraz-Anaya [2], María Teresa Carrillo-Rayas [2], Eduardo Valencia-Cantero [1] and Lourdes Macías-Rodríguez [1,*]

[1] Instituto de Investigaciones Químico Biológicas, Universidad Michoacana de San Nicolás de Hidalgo. Edifico B3, Ciudad Universitaria, Morelia 58030, Michoacán, Mexico
[2] Department of Biotechnology and Biochemistry, Cinvestav Unidad Irapuato, Irapuato, Km 9.6 Libramiento Norte Carr. Irapuato-León, Guanajuato 36824, Mexico
* Correspondence: lmacias@umich.mx; Tel.: +52-443-3223500 (ext. 4237); Fax: +52-443-3265788 (ext. 103)

Received: 6 July 2019; Accepted: 10 August 2019; Published: 20 August 2019

**Abstract:** Iron is an essential plant micronutrient. It is a component of numerous proteins and participates in cell redox reactions; iron deficiency results in a reduction in nutritional quality and crop yields. Volatiles from the rhizobacterium *Arthrobacter agilis* UMCV2 induce iron acquisition mechanisms in plants. However, it is not known whether microbial volatiles modulate other metabolic plant stress responses to reduce the negative effect of iron deficiency. Mass spectrometry has great potential to analyze metabolite alterations in plants exposed to biotic and abiotic factors. Direct liquid introduction-electrospray-mass spectrometry was used to study the metabolite profile in *Medicago truncatula* due to iron deficiency, and in response to microbial volatiles. The putatively identified compounds belonged to different classes, including pigments, terpenes, flavonoids, and brassinosteroids, which have been associated with defense responses against abiotic stress. Notably, the levels of these compounds increased in the presence of the rhizobacterium. In particular, the analysis of brassinolide by gas chromatography in tandem with mass spectrometry showed that the phytohormone increased ten times in plants grown under iron-deficient growth conditions and exposed to microbial volatiles. In this mass spectrometry-based study, we provide new evidence on the role of *A. agilis* UMCV2 in the modulation of certain compounds involved in stress tolerance in *M. truncatula*.

**Keywords:** legumes; microbial volatiles; Fe deficiency; DLI-ESI-MS

---

## 1. Introduction

Mass spectrometry (MS) is gaining considerable popularity for profiling metabolites in complex biological samples. The increased applications have led to the improvement of MS technology in sample introduction, ionization source, mass analyzer, ion detection, and data acquisition and processing. Direct liquid introduction-electrospray ionization-mass spectrometry (DLI-ESI-MS, the acronym recommended by the Analytical Methods Committee [1]) is a rapid and high-throughput analytical tool that has been successfully applied in medicine and food and biological sciences [2–5]. DLI-ESI-MS does not require preliminary sample separation, and it can be applied to multiple biological matrices. The straightforward sample introduction allows for simultaneous fingerprinting of a vast number of metabolites from different samples within a single period. In addition, different studies support the repeatability of DLI-ESI-MS data, and the quantification of the intensity of the ion signals ($m/z$) with a larger variance in plants because of environment, physiological state, or the genotype can also be performed [5–7]. High analytical performance (sensitivity, selectivity) allows it to be used for untargeted metabolomics screening approaches for different plant extracts; thus it offers an excellent cost-benefit ratio compared to other analytical platforms such as near-infrared reflectance spectroscopy (NIRS), ultra-performance liquid chromatography-mass spectrometry (UPLC-MS), gas chromatography with flame ionization detection (GC-FID), and GC-MS, which are slow and expensive to use for routine plant biochemistry studies [2]. Due to the various benefits reported for DLI-ESI-MS, we decided to conduct a study to determine its usefulness in microbial ecology research, as DLI-ESI-MS provides robust chemical information, is bioinformatically easy to handle, and could help us understand the chemical response of plants to abiotic or biotic factors.

Iron (Fe) is an essential micronutrient for plant growth and crop productivity. Plants acquire Fe mainly from the rhizosphere; therefore, the mechanisms that regulate Fe acquisition and homeostasis in the plant are of interest. The role of specific metabolites such as nitric oxide, ferritin, phenolic compounds, and brassinosteroids (BRs) have been highlighted in Fe-deficient growth conditions, indicating that plants undergo significant metabolic changes during Fe-adaptive processes [8–17].

In an attempt to make agriculture a viable component of a healthy and pleasant ecosystem, the application of plant growth-promoting rhizobacteria (PGPR) to enhance Fe uptake and transport in plants is an excellent biotechnology strategy. PGPR are soil bacteria that colonize the rhizosphere of plants, stimulating plant growth and health through different mechanisms, such as phosphorus solubilization or nitrogen-fixation, and the production of phytohormones or siderophores to capture Fe from the environment in biologically useful forms [18]. In 2003, Ryu et al. reported that some PGPR can modulate the growth and development of plants without physical contact with them. This mechanism involves the production of volatile compounds such as, acetoin and 2,3-butanediol, which modulate the mechanisms of phytohormone signaling and therefore stimulate morphogenesis programs in plants [19]. Six-years later, Zhang et al. (2009) noted, that the same microbial volatiles can modulate Fe uptake in *Arabidopsis* via deficiency-inducible mechanisms [20].

The rhizobacterium *Arthrobacter agilis* UMCV2 used in this study, was isolated from the rhizosphere of maize (*Zea mays*) [21]. It emits a pool of volatiles that promotes the growth of leguminous and monocotyledonous plants with different levels of available Fe [22–24]. Notably, in *Medicago truncatula*, the UMCV2 strain increased the expression of genes involved in Fe uptake (*MtFRO1*, *MtFRO2*, *MtFRO3*, *MtFRO4*, and *MtFRO5*) under Fe-sufficient and -deficient conditions [25]. Nevertheless, those studies focused on elucidating the molecular mechanisms involved in the modulation of Fe acquisition responses. Thus, one question remaining is whether microbial volatiles modulate the production of other primary or secondary metabolites in plants to ameliorate Fe-deficiency stress.

Here, we used a DLI-ESI-MS method as an untargeted mass spectrometry strategy to study the metabolic profiles of *M. truncatula* seedlings grown under Fe-sufficient and -deficient conditions. We focused on the detection of significant differences among MS profiles to determine whether volatiles emitted by *A. agilis* UMCV2 alleviate plant stress and stimulate the accumulation of metabolites involved in abiotic stress tolerance; in addition, we used a complementary GC-MS method to confirm the identification of brassinolide, which is involved in Fe-adaptive processes in plants.

## 2. Results

### 2.1. DLI-ESI-MS in the Analysis of Fe Deficiency in Medicago truncatula Seedlings and Response to Bacterial Volatiles

Under conditions of Fe deficiency, plants adjust their metabolism to maintain cellular Fe homeostasis. Some visual symptoms of Fe deficiency, such as leaf yellowing (Figure 1d–f) and decreased plant size (Figure 1g) were observed in our experiments in comparison to plants grown under Fe sufficiency (Figure 1a–c). Additionally, we studied plants exposed to volatiles from *A. agilis*

UMCV2, a rhizobacterium that induces Fe acquisition in plants (Figure 1c,f), and plants exposed to volatiles from *Bacillus* sp. L264, a commensal rhizobacterium (Figure 1b,e). As we expected, volatiles from the UMCV2 strain, had a stimulatory effect on plant growth under Fe-sufficient and -deficient growth conditions (Figure 1g).

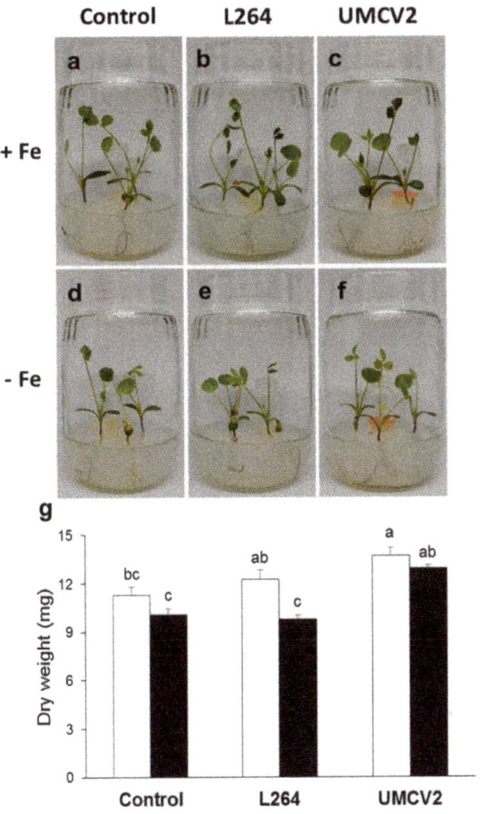


**Figure 1.** Interaction between *Medicago truncatula* and rhizobacteria through the emission of volatile compounds. A 4 mL glass vial with 2 mL nutritive agar medium was inserted in each system; in the control system, 20 µL water was added into the vial instead of the bacterial inoculum. The interaction lasted for 10 days. Uninoculated 12-day-old plants grown under conditions of iron (Fe) sufficiency (a) and deficiency (d). (b) Plants were inoculated with the commensal strain *Bacillus* sp. L264 grown under Fe-sufficient and -deficient conditions (e). Inoculated plants exposed to volatiles from *A. agilis* UMCV2 and under Fe-sufficient (c) and -deficient conditions (f). (g) Dry weights of control plants and plants during interactions with bacterial volatile compounds. Data shown are means ± standard error (n = 15). White and black bars indicate Fe-sufficient and -deficient growth conditions, respectively. Different letters indicate significant differences ($p \leq 0.05$) among treatments determined with two-way ANOVA and Tukey's test.

These plants were further analyzed by DLI-ESI-MS. The quadrupole analyzer allowed the collection of MS data with satisfactory spectral quality. Typical mass spectra of the broad range of molecular weights of compounds that are produced when plants undergo Fe stress, and microbial volatiles exposure are shown in Figure 2a,b, respectively. In total, 737 ions were obtained, mainly within the range 55.90–1592.52 *m/z*. All metabolite signals were extracted from a database with the MALDIquant package in the RStudio interface. Following purification, alignment, and normalization, a principal component analysis (PCA) was performed (Figure 3). The PCA (highly significant results $p < 0.001$, by permutational multivariate analysis of variance, PERMANOVA) showed that control plants grown under conditions of Fe sufficiency and those exposed to volatiles from L264 had similar mass spectra, since treatments were grouped together. Similarly, plants grown under Fe-deficiency stress and those exposed to volatiles from L264 had the same metabolic fingerprinting, and both treatments presented an overlap, indicating that only the absence of Fe affected the metabolic profile of the plants. The ion profiles of plants grown under conditions of Fe sufficiency and deficiency, and following exposure to UMCV2 volatiles were similar, indicating that UMCV2 promotes metabolic changes in plants under both growing conditions.

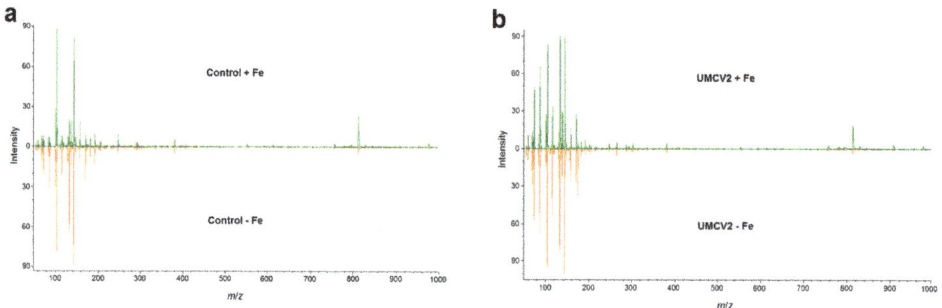


**Figure 2.** Non-targeted metabolomic profiling normalized from leaves of *Medicago truncatula* obtained by DLI-ESI-MS. (**a**) Control plants grown under Fe-sufficient (green) and -deficient conditions (orange). (**b**) Plants exposed to volatile compounds from *A. agilis* UMCV2 for 10 days and grown under iron-sufficient (green) and -deficient conditions (orange).

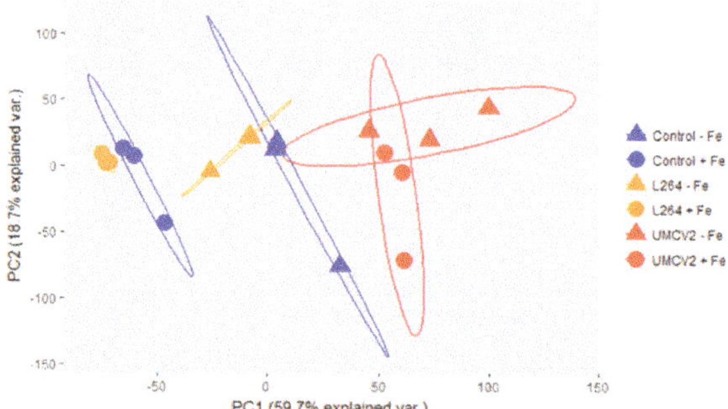


**Figure 3.** Principal component analysis (PCA) obtained from DLI-ESI mass spectra of *Medicago truncatula* leaves under Fe-sufficient and -deficient conditions and following exposure to microbial volatiles. Blue indicates control plants, orange represents plants exposed to volatiles emitted by L264 strain, and red shows the plants exposed to volatiles from *A. agilis* UMCV2. Circles (O) and triangles (Δ) indicate Fe sufficiency and deficiency, respectively. The ellipses represent 95% confidence intervals. Differences between groups were compared with a PERMANOVA test ($p < 0.001$).

### 2.2. Classification of Random Forest Model for Differentiating Plants Grown under Two Different Fe Conditions and Rhizobacterial Inoculation

In order to reduce data complexity, the Random Forest (RF) algorithm was used to generate decision trees and extract the 30 most important ions, defining whether seedling samples were Fe-sufficient or -deficient (Figure 4a). Ions with the greatest mean decrease in Gini index were putatively identified using the PlantCyc database and SpiderMass software. The *m/z* from each ion was compared to the monoisotopic mass (Da) of metabolites previously reported for *M. truncatula*, which provided knowledge about participating metabolites in response to Fe-deficiency stress. Among the most important ions, we identified compounds involved in riboflavin metabolism at 185.22 *m/z* (1-deoxy-L-glycero-tetrulose 4-phosphate), 299.18 *m/z* [5-amino-6-(D-ribitylamino) uracil], and 808.38 *m/z* (flavin adenine dinucleotide, FAD); lipid metabolism at 797.49 *m/z* (1-18:3-2-18:3-monogalactosyldiacylglycerol), 859.98 *m/z* (butanoyl-CoA); and chlorophyll metabolism at 222.07 *m/z* (phosphonothreonine), 613.36 *m/z* (protochlorophyllide a) [11,26,27]. In addition, compounds that alleviate abiotic stresses in plants were also identified at 189.26 *m/z* (norspermine), 269.06 *m/z* ((+)-marmesin), 300.18 *m/z* ((*S*)-*N*-methylcoclaurine), 314.26 *m/z* (9,10-epoxy-18-hydroxystearate), 351.23 *m/z* (crocetin), and 371.07 *m/z* (chelerythrine) (Table 1). Of these ions, 185.22, 189.26, 222.07, 269.06, 299.18, and 351.23 *m/z* (Figure 5a) were detected at a higher intensity under Fe deficiency conditions. The remaining 17 ions selected by the RF algorithm for the Fe condition could not be identified.

The RF model also helped to identify the 30 most important ions including those that differentiated uninoculated seedlings, and those inoculated with L264 or UMCV2 strains (Figure 4b). Twenty ions were identified (Table 2). Six of these, 87.41 *m/z* (3-pentanone), 88.08 *m/z* (pyruvate), 88.33 *m/z* (4-aminobutanal), 97.83 *m/z* (glycolate), 98.04 *m/z* (*N*-monomethylethanolamine), and 287.19 *m/z* (kaempferol) showed stronger detectable signal intensities under UMCV2 treatment (Figure 5b), suggesting that these ions are responsible for the discrimination between the sample groups, and revealing the associated chemical modulations made by the UMCV2 strain. Thus, DLI-ESI-MS displays great potential for determining whether volatiles emitted from other rhizobacteria are able to modulate the production of primary or secondary metabolites in plants. According to previously reported literature, the increased signals have different roles in alleviating Fe deficiency stress in plants [12,28–33]. Other identified compounds included those involved in plant primary metabolism, at 88.08 *m/z* (pyruvate), 97.83 *m/z* (glycolate), 790.06 *m/z* (coenzyme A) [34]; brassinosteroid metabolism, at 397.20 *m/z* (5-dehydroepisterol), 419.15 *m/z* (6-deoxocathasterone), and 467.07 *m/z* (6-alpha-hydroxycastasterone) [35–38]; compounds with antioxidant roles in plants, at 266.20 *m/z* (thiamine), 366.17 *m/z* (galactinol), and 933.53 *m/z* (notoginsenoside R1) [39–41], and some flavonoids with antioxidant capacity, which act as chemotactic signals for symbiotic nitrogen-fixing bacteria of legumes, at 275.13 *m/z* (fustin), 287.19 *m/z* (kaempferol), and 291.15 *m/z* (formononetin) [10,42–45] (Table 1).

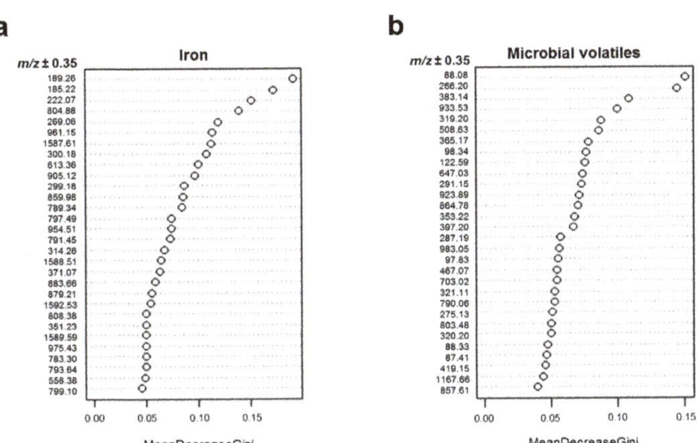


**Figure 4.** Ion importance ranking obtained by Random Forest model for differentiating between sample treatments with DLI-ESI mass spectra data. The thirty most important ions are shown for discriminating between Fe growth conditions (**a**) and the effect of microbial volatiles (**b**). Ntrees = 500, OOB error = 0% for Fe growth conditions and 43.75% for volatiles.

**Table 1.** The most important compounds detected in *Medicago truncatula* seedlings by DLI-ESI-MS and putatively identified by the SpiderMass software, which were able to differentiate between samples grown under Fe-sufficient and -deficient growth conditions.

| *m/z* | Monoisotopic Mass (Da) | Ionization Mode | Compound Name | Function |
|---|---|---|---|---|
| 189.26 | 188.20 | $[M + H]^+$ | Norspermine | Stress |
| 185.22 | 184.01 | $[M + H]^+$ | 1-Deoxy-L-glycero-tetrulose 4-phosphate | Riboflavin biosynthesis |
| 222.07 | 199.03 | $[M + Na]^+$ | L-Histidinol-phosphate | Histidine biosynthesis |
| 269.06 | 246.09 | $[M + Na]^+$ | (+)-Marmesin | Stress |
| 300.18 | 299.15 | $[M + H]^+$ | (S)-N-methylcoclaurine | Stress |
| 613.36 | 612.22 | $[M + H]^+$ | Protochlorophyllide a | Chlorophyll biosynthesis |
| 299.18 | 276.11 | $[M + Na]^+$ | 5-Amino-6-(D-ribitylamino) uracil | Riboflavin biosynthesis |
| 859.98 | 837.16 | $[M + Na]^+$ | Butanoyl-CoA | Fatty acid beta oxidation |
| 797.49 | 774.53 | $[M + Na]^+$ | 1-18:3-2-18:3-Monogalactosyldiacylglycerol | Chloroplast membrane lipid |
| 314.26 | 313.24 | $[M + H]^+$ | 9,10-Epoxy-18-hydroxystearate | Cutin biosynthesis |
| 371.07 | 348.12 | $[M + Na]^+$ | Chelerythrine | Stress |
| 808.38 | 785.16 | $[M + Na]^+$ | Dioleoylphosphatidylcholine | Membranes lipid |
| 351.23 | 328.17 | $[M + Na]^+$ | Crocetin | Stress |

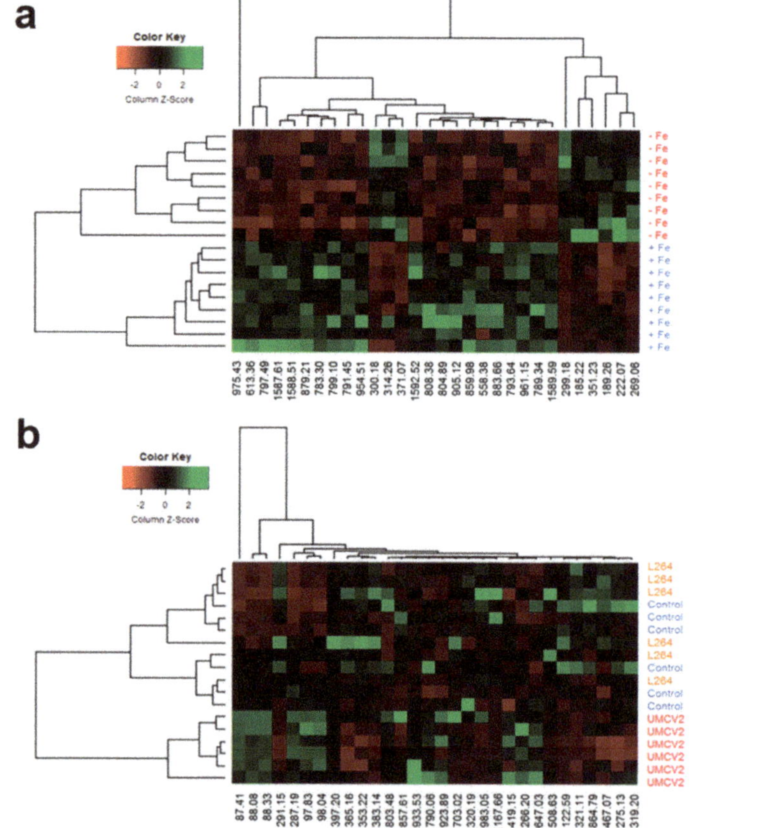

**Figure 5.** Metabolomic heatmap generated with the 30 most important ions detected by the Random Forest model for Fe availability (**a**) and bacterial volatiles (**b**). Heatmap combined with an analysis of cluster hierarchical using Euclidean distance between experimental units and Ward's algorithm for classification by ion (along x-axis) and by treatment (along *y*-axis).

**Table 2.** The most important compounds detected by DLI-ESI-MS in *Medicago truncatula* seedlings and putatively identified by the SpiderMass software that could differentiate between samples exposed to volatiles from L264, UMCV2 strains, and control.

| *m/z* | Monoisotopic Mass (Da) | Ionization Mode | Compound Name | Function |
|---|---|---|---|---|
| 88.08 | 87.01 | $[M + H]^+$ | Pyruvate | Energy |
| 266.20 | 265.11 | $[M + H]^+$ | Thiamine | Stress |
| 383.14 | 360.14 | $[M + Na]^+$ | 7-Deoxyloganate | Stress |
| 933.53 | 932.53 | $[M + H]^+$ | Notoginsenoside R1 | Stress |
| 319.20 | 296.31 | $[M + Na]^+$ | Phytol | Constituent of chlorophyll |
| 365.17 | 342.12 | $[M + Na]^+$ | Galactinol | Stress |
| 98.34 | 75.07 | $[M + Na]^+$ | *N*-Monomethylethanolamine | Choline biosynthesis |
| 291.15 | 268.07 | $[M + Na]^+$ | Formononetin | Stress |
| 353.22 | 352.18 | $[M + H]^+$ | 16-Hydroxytabersonine | Indole alkaloid biosynthesis |
| 397.20 | 396.34 | $[M + H]^+$ | 5-Dehydroepisterol | Brassinosteroid biosynthesis |
| 287.19 | 286.05 | $[M + H]^+$ | Kaempferol | Stress |
| 97.83 | 75.01 | $[M + Na]^+$ | Glycolate | Photorespiration |
| 467.07 | 466.37 | $[M + H]^+$ | 6-Hydroxycastasterone | Brassinosteroid biosynthesis |
| 321.11 | 320.09 | $[M + H]^+$ | 4-Coumaroylshikimate | Flavonoid and phenylpropanoid biosynthesis |
| 790.06 | 767.12 | $[M + Na]^+$ | Coenzyme A | Fatty acid beta oxidation |
| 275.13 | 274.08 | $[M + H]^+$ | Fustin | Stress |
| 320.20 | 297.24 | $[M + Na]^+$ | 18-Hydroxyoleate | Cutin, suberin and wax biosynthesis |
| 88.33 | 87.07 | $[M + H]^+$ | 4-Aminobutanal | Stress |
| 87.41 | 86.07 | $[M + H]^+$ | 3-Pentanone | Stress |
| 419.15 | 418.38 | $[M + H]^+$ | 6-Deoxocathasterone | Brassinosteroid biosynthesis |

Although DLI-ESI-MS provides the possible composition of the compounds with minimal sample preparation, the exact metabolite identification is limited by the lack of fragmentation data or device accuracy (~0.3 Da) [6]. Therefore, mass spectrometry coupled with a separation technique such as GC, can provide identification with a high level of confidence based on the comparison of the retention time with the appropriate standard, and in addition, it allows calculation of the concentration of the compounds in the sample. In our study, we observed variations in the intensity of many ions in the mass spectra obtained from plants treated with UMCV2. Three of the *m/z* ions were putatively identified as components of the BRs biosynthesis (Table 2); we observed that the signal 419.15 *m/z*, which is a direct precursor of brassinolide [37] mainly increased in plants inoculated with the UMCV2 strain. Thus, we decided to confirm by GC-SIM-MS whether the volatiles emitted by *A. agilis* UMCV2, stimulate the production of brassinolide in *M. truncatula*, as it is the most bioactive form of BRs in plants. For this, we acetylated the molecule to change the analyte properties, which improved the identification capability of brassinolide (Figure 6a–c).

BRs are endogenous plant hormones that are essential for plant growth and development. Additionally, BRs are involved in sensing and responding to mineral deficiency stress. A factorial analysis showed that Fe deficiency stress, as well as microbial volatile factors significantly increased the content of epibrassinolide in the plant ($p = 0.0039$ and $p = 0.0036$, respectively); and the interaction of both factors was also statistically significant ($p = 0.0153$). Plants grown in Fe-deficient conditions and inoculated with UMCV2 showed a ten-fold higher accumulation of brassinolide relative to controls (Figure 6d). This result suggests that *A. agilis* induce the synthesis of brassinolide in the plant as a part of the mechanism for Fe stress tolerance; thus, the quantification of this phytohormone in the plants may serve as a reference of the beneficial effects of rhizobacterium to plants.

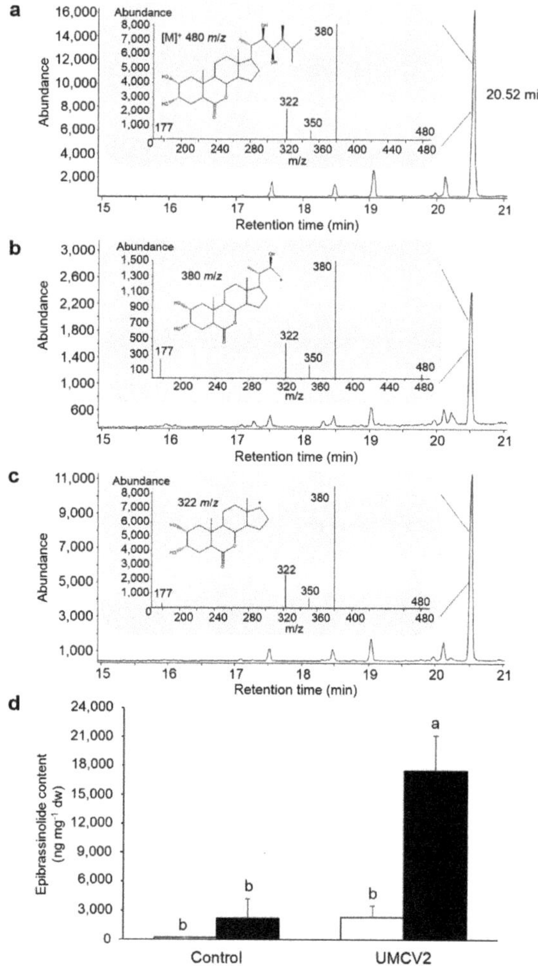

**Figure 6.** Identification of brassinolide in *Medicago truncatula* by GC-MS. (**a**) Total ion chromatogram of the epibrassinolide standard, indicating the retention time of the phytohormone and the electron impact mass spectrum during the SIM analysis. (**b**) Total ion chromatogram and mass spectrum obtained from the control plants grown under conditions of iron (Fe) deficiency. (**c**) Total ion chromatogram and mass spectrum obtained from plants grown under conditions of Fe deficiency and exposed to volatiles from *A. agilis* UMCV2. (**d**) Brassinolide content in plants grown under Fe-sufficient (white bars) and -deficient (black bars) growth conditions. Data shown are means ± standard error (n = 3). Different letters indicate significant differences ($p \leq 0.05$) among treatments determined with two-way ANOVA followed by a Tukey's test.

To summarize the information obtained in this study, a hierarchical cluster analysis was constructed for each treatment (Figure 5a,b). The metabolic heat maps based on DLI-ESI-MS data visually displayed differences between samples in the intensity of the selected *m/z* ions. Remarkably, two sets of ions were identified that accumulated under conditions of Fe deficiency, one of them was composed of the following ions 299.18, 185.22, 351.23, 189.26, 222.07, and 269.06 *m/z*, and the other was set by 300.18, 314.26, and 371.07 *m/z* whereas the remaining ions on the heat map diminished (Figure 5a); therefore, these ions could be considered as biomarkers of the plants' Fe nutritional status.

Conversely, differences were observed in the metabolite profiling of plants exposed to microbial volatiles. Uninoculated plants and those inoculated with the L264 strain shared the same conglomerate, verifying that the volatiles from the commensal bacterium did not have an effect on the overall profile of plant metabolites produced in response to inoculation. Ions obtained from plants inoculated with the UMCV2 strain formed a separate conglomerate, and ions 87.41, 88.08, and 88.33 *m/z* presented the strongest differences, suggesting an important role during the interaction between *M. truncatula* and *A. agilis* UMCV2 (Figure 5b).

## 3. Discussion

Fe is an important metal in photosynthesis. Additionally, Fe is used ubiquitously in oxidation-reduction processes [13]. These important roles make Fe an essential micronutrient for plant fitness. As a consequence, Fe deficiency is a major constraint for agricultural quality and production, eventually affecting human health via the food chain [16]. Fe scarcity leads to the activation of sophisticated mechanisms to maintain cellular Fe homeostasis. Legumes are classified as Strategy I plants, which undergo biochemical changes to increase the capacity for Fe uptake via the roots and Fe solubility in the soil [11]. In this study, we used DLI-ESI-MS and putatively identified several compounds in plants that are commonly accepted to be associated with biochemical responses and adaptation strategies under Fe deficiency conditions. Furthermore, this analytical tool showed sensitivity in discriminating plants grown under Fe-sufficient or -deficient conditions, providing evidence that the plants used in the study were metabolically stressed due to a lack of Fe. For example, we found that the mass spectra from Fe-deficient plants presented decreased signals for protochlorophyllide a (613.36 *m/z*) compared with control plants. The lower concentration of chlorophylls caused leaf yellowing, which is an important visible symptom of Fe deficiency in plants. In addition, the signal for 1-18:3-2-18:3-monogalactosyldiacylglycerol (797.49 *m/z*) also decreased. Monogalactosyldiacylglycerol is a major lipid component of chloroplast membranes and acts directly in several important plastid roles, particularly during photosynthesis [26,27]. Fe deficiency decreased the signals of other lipids including, phospholipid dioleoylphosphatidylcholine (808.38 *m/z*), which is a component of cell membranes, and 9,10-epoxy-18-hydroxystearate (314.26 *m/z*), which is involved in cutin biosynthesis [46,47].

In addition, we found that six signals were increased due to Fe deficiency which, according to the heat map, could be used as biomarkers to distinguish between Fe treatments. Two of these are precursors for riboflavin biosynthesis, 1-deoxy-L-glycero-tetrulose 4-phosphate (185.22 *m/z*) and 5-amino-6-(D-ribitylamino) uracil (299.18 *m/z*), which are subsequently transformed into FAD. Accumulation of riboflavin was observed in *M. truncatula* plants grown under Fe-deficient conditions, with or without $CaCO_3$ as a source of alkaline pH stress. The root protein profile showed the de novo accumulation of 6,7-dimethyl-8-ribityllumazine synthase (DMRLs) and GTP cyclohydrolase II (GTPcII); these proteins are involved in riboflavin biosynthesis, suggesting that the riboflavin biosynthetic pathway is upregulated under conditions of Fe deficiency [11]. Since flavin compounds are exported and accumulate in Fe-deficient roots, different roles have been proposed for riboflavin, including as an electron donor either for enzymatic Fe (III) reduction, as a cofactor, or as a metal chelator [8,11].

Increased signaling was also observed for compounds of different families such as polyamines, coumarins, and terpenes; all of which have different roles in the adaptation of plants to the environment and for overcoming stress conditions. One signal corresponded to norspermine (189.26 *m/z*), a polyamine previously identified in *Medicago* plants [48], which accumulated in response to Fe deficiency, inducing ferric-chelate reductase activity and the expression of genes related to Fe uptake [49]. The coumarin (+)-marmesin (269.06 *m/z*) has antioxidant properties [50], as well as the tetraterpenoid crocetin (351.23 *m/z*) [14], which may protect plants from damage induced by oxidative stress in response to Fe deficiency [9]. Finally, the signal for the putative ion identified as L-histidinol phosphate (222.07 *m/z*) also increased. This compound is a precursor of histidine; however, its role in plants under stress caused by Fe deficiency has not been fully explored. The chemical properties of the imidazole side group allow this amino acid to participate in acid-base catalysis, and in the co-ordination of metal ions [51].

PGPR application has become an increasingly common practice as part of an agricultural strategy to alleviate plant abiotic stresses in the field. The use of PGPR will help to address the challenges of producing food for a growing human population in a sustainable and environmentally friendly manner. From this perspective, we have studied the effects of volatiles from the rhizobacterium *A. agilis* UMCV2 on the growth and development of *M. truncatula* in Fe-sufficient and -deficient growth media [23]. In a previous study, we found that the UMCV2 strain induces iron acquisition mechanisms in this Strategy I plant, including rhizosphere acidification, ferric chelate reductase activity, and Fe content in plants. In the present study, using DLI-ESI-MS and the RF model, we found that volatiles from the UMCV2 strain favor the accumulation of flavonoids in leaves under conditions of Fe sufficiency and deficiency, particularly kaempferol (287.19 *m/z*). Legumes are a source of flavonoids and have a beneficial effect on human health [52]; however in plants, flavonoids have diverse roles; for example, flavonoids reduce Fe (III) to Fe (II), reduce the production of reactive oxygen species (ROS), quench ROS, have antifungal activity, chelate ions of transition metals, and quench cascades of free-radical reactions in lipid peroxidation. Besides, due to their low redox potential, they can also reduce potent free radicals (superoxides, alkyl radicals, hydroxyl radicals) [45]. Last, they are involved in plant-microbe interactions signaling, in particular, in symbiotic bacteria stimulating root colonization [10,42–44].

Besides kaempferol, five other signals, 87.41, 88.08, 88.33, 97.83, and 98.04 *m/z*, which correspond to 3-pentanone, pyruvate, 4-aminobutanal, glycolate, and *N*-monomethylethanolamine, respectively, were grouped in the same conglomerate in the heat map, indicating that these compounds are metabolite markers that are specifically produced in response to the presence of the UMCV2 strain, compared with uninoculated plants and those treated with the commensal rhizobacteria L264. These compounds have previously been reported in plants subjected to different kinds of abiotic stresses, acting as signaling molecules that regulate many cellular processes, such as plant growth/development and acclimation responses to stress [12,28–33]. Thus, these results suggest that a complex network of signaling events is activated during the interaction of *M. truncatula* with *A. agilis* UMCV2 via the emission of volatile compounds; this stimulates iron acquisition mechanisms and mediates cellular activity to alleviate plant stress.

Finally, of the 30 most important ions shown in the RF model, we identified three compounds (5-dehydroepisterol, 6-deoxocathasterone, and 6-hydroxycastasterone) related to BRs synthesis [35–38]. These phytohormones regulate the growth and development of plants, and their involvement in the detection and response to Fe deficiency in plants has only recently emerged [15,17,53]. The exogenous application of BRs to stressed plants induces stress-tolerance mechanisms. Thus, it would be timely to conduct a detailed study to ascertain whether microbial volatiles can modulate BRs signaling pathways in plants grown under conditions of Fe sufficiency and deficiency, since we found that volatiles from *A. agilis* UMCV2 promote the growth and the synthesis of BRs in plants grown under Fe-sufficient and -deficient growth conditions.

In summary, our findings show the usefulness of DLI-ESI-MS for studying the metabolic disturbances induced by Fe deficiency in plants; and also, it provided an integrated view of the cellular processes that occur following inoculation with PGPR and different metabolite markers could be identified as possible subjects for further studies. The combination of both mass spectrometry techniques allowed us to show that plants effectively sense the volatiles emitted by *A. agilis* UMCV2 and reconfigure their metabolic networks accordingly. It is probable that multiple mechanisms, including brassinosteroid production, are activated during plant-microbe interactions, either simultaneously or in succession to ameliorate plant stress. Currently, we are investigating the role of volatiles from *A. agilis* UMCV2 in the protection against oxidative stress and the production of certain flavonoids and BRs to mediate Fe stress responses.

## 4. Materials and Methods

### 4.1. Biological Material and Growth Conditions

In this study, seeds of *M. truncatula* ecotype Jemalong A17 were scarified with 2 mL of concentrated sulfuric acid for 8 min and then rinsed with five washes of sterile deionized water to remove excess acid [54]. Later, seeds were superficially disinfected with 12% sodium hypochlorite for 2 min and rinsed five times with sterile deionized water. Seeds were placed on 0.6% agar plates (Phytotechnology, Shawnee Mission, KS, USA) with 0.6% sucrose and vernalized at 4 °C for 3 days. Germination was performed in a Percival growth chamber with a photoperiod of 16 h light/8 h dark, a luminous intensity of 6100 lx and a constant temperature of 22 °C.

After 3 days of germination, three seedlings were transferred to each 170 mL glass flask with 35 mL of Hoagland medium and 0.6% agar. The Hoagland base medium was supplemented with the following salts: 1020 ppm $KNO_3$, 492 ppm $Ca(NO_3)_2 \times 4H_2O$, 230 ppm $NH_4H_2(PO_4)$, 490 ppm $MgSO_4 \times 7H_2O$, 2.80 ppm $H_3BO_3$, 1.81 ppm $MnCl_2 \times 2H_2O$, 0.08 ppm $CuSO_4 \times 5H_2O$, 0.22 ppm $ZnSO_4 \times 5H_2O$ and 0.09 ppm $Na_2MoO_4 \times H_2O$. For the Fe-sufficient treatment, Hoagland medium was supplemented with 20 µM $FeSO4$, and for the Fe-deficient treatment, no source of Fe was added.

The UMCV2 strain (CECT-7743, Spanish Type Culture Collection, Valencia, Spain) was grown on nutrient agar (3 g $L^{-1}$ of meat extract, 5 g $L^{-1}$ peptone, and 1.5% bacteriological agar) at 22 °C. Also, we used the commensal rhizobacterium *Bacillus* sp. L264 as a control [55,56]. The L264 strain was maintained under similar conditions to UMCV2.

### 4.2. Plant-Bacteria Interaction through the Emission of Volatiles

A system with separate compartments was used (Figure 1a–f). Two days after transferring the plants to Hoagland media, 20 µL of each rhizobacteria (0.05 D.O.$_{595nm}$) was inoculated into a glass vial with 2 mL nutritive agar medium. For the treatment of the uninoculated plants, 20 µL of water was added instead of bacterial inoculum. Then, seedlings were allowed to grow in the growth chamber under the controlled conditions of light and temperature mentioned above. After 10 days, the seedlings were carefully removed from the medium, and the trifoliate leaves were harvested, immediately frozen with liquid nitrogen, and maintained at −80 °C. Other plants were dried to a constant weight at 68 °C for 7 days. Dry weight was analyzed using a factorial design, comprising of two factors (Fe availability with two levels and bacterial volatiles with three levels), followed by Tukey's test ($p \leq 0.05$).

### 4.3. Metabolite Extraction from M. truncatula Leaves

Frozen leaves were lyophilized and 3 mg of dry tissue was ground in a Mixer Mill (MM 400-Retsch, Verder Scientific GmbH & Co. KG; Haan, Germany) at 30 Hz for 30 s. The extraction was carried out with 500 µL 75% methanol grade HPLC acidified with 0.5% formic acid, and samples were then sonicated for 30 min, centrifuged at 10,000 rpm for 10 min at 4 °C, and filtered with a 4 mm syringe filter sterile through a 0.22 µm pore size hydrophilic nylon membrane. The samples were directly injected into the DLI-ESI mass spectrometer without further pre-treatment.

### 4.4. Non-Targeted Metabolic Profiling by DLI-ESI Mass Spectrometry

The samples were analyzed on a SQ-Detector 2 spectrometer (ESI/APCI/ESCi, multimode, Waters, Milford, MA, USA) with the fabricant software MassLynx 4.1. The measurements were made with electrospray ionization in positive mode, a capillary voltage of 3 kV, a cone voltage of 30 V and an extractor voltage of 3 V, source temperature of 135 °C, desolvation temperature and flow of 250 °C and 250 L $h^{-1}$, respectively, and cone gas flow of 50 L $h^{-1}$. The RF lens was set to 2.5 V. In the analyzer section, a resolution LM and HM of 10 and 14.6, respectively, and an energy ion of −0.1 were used. The samples were injected with a flow rate of 10 µL $min^{-1}$. The spectra were collected within the range 50–2000 *m/z*, the duration of the run was 1 min, and one scan was obtained per second.

The spectra obtained were converted from the raw extension to mzXML using MSConverte 3.0 from the ProteoWizard Library open-source initiative (https://proteowizard.sourceforge.net). The software mMass version 5.5.0 [57] was used to subtract the noise from the spectra, normalize the base peak, and obtain an average mass spectrum. Using the R language (Version 3.4.1 https://www.rstudio.com/) and the MALDIquant package [58], a database was obtained in text format with the ions present in the mass spectra. The database was used for statistical analyses.

The chemical profiles of the leaves were compared using the hierarchical cluster analysis (HCA) approach and by generating heatmaps with the ion intensities. To determine the contribution of each ion, the Rattle packet [59] was used in the R interface to generate a Random Forest (RF) model for each variable (iron and bacterial volatiles). The RF algorithm consisted of training, validation, and test steps using $m/z$ ions. To obtain the most important ions in the study, 500 decision trees were created using 70% of the samples; the remaining samples were used for validation (15%) and testing (15%). The importance of the ion variable was determined by measuring the mean decrease by the Gini index.

### 4.5. Identification of Significant Ions

The most important ions, according to the RF algorithm, were putatively identified using SpiderMass software [60] and a *M. truncatula* metabolite database (PlantCyc database, https://www.plantcyc.org/) with a tolerance of ± 0.35 $m/z$.

### 4.6. Brassinosteroids Determination

The presence of BRs in the samples was confirmed by GC-MS (Agilent, Foster City, CA, USA) analysis. Each treatment consisted in three composed samples of three plants. The extracts were evaporated to dryness under a stream of nitrogen in a reaction vial. Then, they were treated with acetic anhydride (1.5 mL) and dichloromethane (1 mL), and heated at 75 °C for 90 min. Acetylation decreases the boiling point of the phytohormone, improves the thermal stability in the GC injection port and allows a better chromatographic separation. After cooling, the acetylated sample was diluted with chloroform (2 mL) and washed with deionized water (4 mL) three times. The organic phase was recovered, dried over anhydrous $Na_2SO_4$, evaporated and re-dissolved in 50 μL chloroform for GC-selected ion monitoring-MS analysis (GC-SIM-MS). The molecular ion of the acetylated compound at 652 $m/z$ was very weak and sometimes not observed. Thus, the fragmented ions used for the SIM-MS method were 480, 380, 350, 322, and 177 $m/z$. These ions have previously been reported as characteristic ions for the structural determination of brassinolide rings, which have been analyzed by electron impact MS detector [61]. In addition, the phytohormone was further confirmed by comparing the retention time in the extract to a pure epibrassinolide standard (SIGMA-ALDRICH, Saint Louis, MO, USA, catalog no. E1641). The standard was also acetylated and to estimate the amount of the compound in the sample, we constructed a calibration curve ($R^2 = 1$).

The phytohormone (2 μL) was analyzed using an Agilent 6850 Series II gas chromatograph equipped with an Agilent MS detector (model 5973) (Agilent) and a 5% phenyl methyl silicone capillary column (HP-5 MS) (30 m × 0.25 mm I.D., 0.25 mm film thickness). The operating conditions were 1 mL min$^{-1}$ of helium as the carrier gas, 300 °C as the detection temperature, and 270 °C as the injection temperature. The column was held for 3 min at 180 °C and programmed at 5 °C min$^{-1}$ to reach a final temperature of 300 °C for 12 min. The ions were monitored after electron impact ionization (70 eV).

Brassinolide concentration was analyzed using a factorial design, which comprised of two factors (Fe-rich and -deficient media, and the presence and absence of volatiles from *A. agilis* UMCV2), followed by Tukey's test ($p \leq 0.05$).

**Author Contributions:** Conceptualization, L.M.-R., E.V.-C. and R.W.; Methodology, L.M.-R., E.V.-C. and R.W.; Validation, L.M.-R. and I.F.-C., Formal analysis, I.F.-C., A.R.-O., M.T.C.-R. and M.I.C.E.-A.; Investigation, I.F.-C., A.R.-O., M.T.C.-R. and M.I.C.E.-A., Resources, L.M.-R., E.V.-C. and R.W.; Writing—Original Draft Preparation, I.F.-C., A.R.-O. and L.M.-R.; Writing—Review & Editing, L.M.-R., E.V.-C. and R.W.; Supervision, L.M.-R.; Project Administration, L.M.-R. and R.W. Funding Acquisition, L.M.-R. and R.W.

**Funding:** This research was funded by the Fronteras project 2015-2/814, the bilateral grant CONACyT-DFG 2016/277850, and the Consejo de la Investigación Científica (UMSNH) (grant number 2.24). I. Flores-Cortez is indebted to CONACyT for providing a PhD fellowship (grant number 164395).

**Conflicts of Interest:** The authors declare that they have no conflict of interest.

## References


1. Analytical Methods Committee AMCTB No. 81. A "periodic table" of mass spectrometry instrumentation and acronyms. *Anal. Methods* **2017**, *9*, 5086–5090.
2. García-Flores, M.; Juárez-Colunga, S.; García-Casarrubias, A.; Trachsel, S.; Winkler, R.; Tiessen, A. Metabolic profiling of plant extracts using direct-injection electrospray ionization mass spectrometry allows for high-throughput phenotypic characterization according to genetic and environmental effects. *J. Agric. Food Chem.* **2015**, *63*, 1042–1052. [CrossRef] [PubMed]
3. Gamboa-Becerra, R.; Montero-Vargas, J.; Martínez-Jarquín, S.; Gálvez-Ponce, E.; Moreno-Pedraza, A.; Winkler, R. Rapid classification of coffee products by data mining models from direct electrospray and plasma-based mass spectrometry analyses. *Food Anal. Methods* **2016**, *10*, 1359–1368. [CrossRef]
4. González-Domínguez, R.; Sayago, A.; Fernández-Recamales, A. High-throughput mass-spectrometry based-metabolomics to characterize metabolite fingerprints associated with Alzheimer's disease pathogenesis. *Metabolites* **2018**, *8*, 52. [CrossRef] [PubMed]
5. Montero-Vargas, J.; Casarrubias-Castillo, K.; Martínez-Gallardo, N.; Ordaz-Ortiz, J.; Délano-Frier, J.; Winkler, R. Modulation of steroidal glycoalkaloid biosynthesis in tomato (*Solanum lycopersicum*) by jasmonic acid. *Plant. Sci.* **2018**, *277*, 155–165. [CrossRef] [PubMed]
6. García-Flores, M.; Juárez-Colunga, S.; Montero-Vargas, J.M.; López-Arciniega, J.A.I.; Chagolla, A.; Tiessen, A.; Winkler, R. Evaluating the physiological state of maize (Zea mays L.) plants by direct-injection electrospray mass spectrometry (DIESI-MS). *Mol. BioSyst.* **2012**, *8*, 1658–1660. [CrossRef] [PubMed]
7. Rendón-Anaya, M.; Montero-Vargas, J.M.; Saburido-Álvarez, S.; Vlasova, A.; Capella-Gutierrez, S.; Ordaz-Ortiz, J.J.; Aguilar, O.M.; Vianello-Brondani, R.P.; Santalla, M.; Delaye, L.; et al. Genomic history of the origin and domestication of common bean unveils its closest sister species. *Genome Biol.* **2017**, *18*, 60. [CrossRef]
8. González-Vallejo, E.B.; Susín, S.; Abadía, A.; Abadía, J. Changes in sugar beet leaf plasma membrane Fe(III)-chelate reductase activities mediated by Fe-deficiency, assay buffer composition, anaerobiosis and the presence of flavins. *Protoplasma* **1998**, *205*, 163–168. [CrossRef]
9. Salama, Z.; El-Beltagi, H.; El-Hariri, D.M. Effect of Fe deficiency on antioxidant system in leaves of three flax cultivars. *Not. Bot. Hort. Agrobot. Cluj.* **2009**, *37*, 122–128.
10. Jin, C.W.; Li, G.X.; Yu, X.H.; Zheng, S.J. Plant Fe status affects the composition of siderophore-secreting microbes in the rhizosphere. *Ann. Bot.* **2010**, *105*, 835–841. [CrossRef]
11. Rodríguez-Celma, J.; Lattanzio, G.; Grusak, M.A.; Abadía, A.; Abadía, J.; López-Millán, A.F. Root responses of *Medicago truncatula* plants grown in two different iron deficiency conditions: Changes in root protein profile and riboflavin biosynthesis. *J. Proteome Res.* **2011**, *10*, 2590–2601. [CrossRef] [PubMed]
12. Fariduddin, Q.; Varshney, P.; Yusuf, M.; Ahmad, A. Polyamines: Potent modulators of plant responses to stress. *J. Plant. Interact.* **2013**, *8*, 1–16. [CrossRef]
13. Ravet, K.; Pilon, M. Copper and iron homeostasis in plants: The challenges of oxidative stress. *Antioxid Redox Signal.* **2013**, *19*, 919–932. [CrossRef]
14. Baba, S.A.; Malik, A.H.; Wani, Z.A.; Mohiuddin, T.; Shah, Z.; Abbas, N.; Ashraf, N. Phytochemical analysis and antioxidant activity of different tissue types of *Crocus sativus* and oxidative stress alleviating potential of saffron extract in plants, bacteria, and yeast. *S. Afr. J. Bot.* **2015**, *99*, 80–87. [CrossRef]
15. Wang, B.; Li, G.; Zhang, W.H. Brassinosteroids are involved in Fe homeostasis in rice (*Oryza sativa* L.). *J. Exp. Bot.* **2015**, *66*, 2749–2761. [CrossRef] [PubMed]
16. Li, W.; Lan, P. The understanding of the plant iron deficiency responses in strategy I plants and the role of ethylene in this process by omic approaches. *Front. Plant. Sci.* **2017**, *8*, 40. [CrossRef] [PubMed]
17. Lima, M.D.R.; Barros Junior, U.O.; Batista, B.L.; da Silva Lobato, A.K. Brassinosteroids mitigate iron deficiency improving nutritional status and photochemical efficiency in *Eucalyptus urophylla* plants. *Trees* **2018**, *32*, 1681–1694. [CrossRef]

18. Pérez-Montaño, F.; Alías-Villegas, C.; Bellogín, R.A.; del Cerro, P.; Espuny, M.R.; Jiménez-Guerrero, I.; López-Baena, F.J.; Ollero, F.J.; Cubo, T. Plant growth promotion in cereal and leguminous agricultural important plants: From microorganism capacities to crop production. *Microbiol. Res.* **2014**, *169*, 325–336. [CrossRef]

19. Ryu, C.M.; Farag, M.A.; Hu, C.H.; Reddy, M.S.; Wei, H.X.; Pare, P.W.; Kloepper, J.W. Bacterial volatiles promote growth in *Arabidopsis*. *Proc. Natl. Acad. Sci.* **2003**, *100*, 4927–4932. [CrossRef]

20. Zhang, H.; Sun, Y.; Xie, X.; Kim, M.S.; Dowd, S.E.; Paré, P.W. A soil bacterium regulates plant acquisition of iron via deficiency-inducible mechanisms. *Plant. J.* **2009**, *58*, 568–577. [CrossRef]

21. Valencia-Cantero, E.; Hernández-Calderón, E.; Velázquez-Becerra, C.; López-Meza, J.E.; Alfaro-Cuevas, R.; López-Bucio, J. Role of dissimilatory fermentative iron-reducing bacteria in Fe uptake by common bean (*Phaseolus vulgaris* L.) plants grown in alkaline soil. *Plant. Soil.* **2007**, *291*, 263–273. [CrossRef]

22. Velázquez-Becerra, C.; Macías-Rodríguez, L.I.; López-Bucio, J.; Altamirano-Hernández, J.; Flores-Cortez, I.; Valencia-Cantero, E. A volatile organic compound analysis from *Arthrobacter agilis* identifies dimethylhexadecylamine, an amino-containing lipid modulating bacterial growth and *Medicago sativa* morphogenesis in vitro. *Plant. Soil.* **2011**, *339*, 329–340. [CrossRef]

23. Orozco-Mosqueda, M.C.; Velázquez-Becerra, C.; Macías-Rodríguez, L.I.; Santoyo, G.; Flores-Cortez, I.; Alfaro-Cuevas, R.; Valencia-Cantero, E. *Arthrobacter agilis* UMCV2 induces iron acquisition in *Medicago truncatula* (strategy I plant) in vitro via dimethylhexadecylamine emission. *Plant. Soil.* **2013**, *362*, 51–66. [CrossRef]

24. Castulo-Rubio, D.Y.; Alejandre-Ramírez, N.A.; Orozco-Mosqueda, M.C.; Santoyo, G.; Macías-Rodríguez, L.I.; Valencia-Cantero, E. Volatile organic compounds produced by the rhizobacterium *Arthrobacter agilis* UMCV2 modulate *Sorghum bicolor* (strategy II plant) morphogenesis and *SbFRO1* transcription in vitro. *J. Plant. Growth Regul.* **2015**, *34*, 611–623. [CrossRef]

25. Montejano-Ramírez, V.; Martínez-Cámara, R.; García-Pineda, E.; Valencia-Cantero, E. Rhizobacterium *Arthrobacter agilis* UMCV2 increases organ-specific expression of *FRO* genes in conjunction with genes associated with the systemic resistance pathways of *Medicago truncatula*. *Acta Physiol. Plant.* **2018**, *40*, 138. [CrossRef]

26. Shimojima, M.; Ohta, H. Critical regulation of galactolipid synthesis controls membrane differentiation and remodeling in distinct plant organs and following environmental changes. *Prog. Lipid Res.* **2011**, *50*, 258–266. [CrossRef]

27. Aronsson, H.; Schottler, M.A.; Kelly, A.A.; Sundqvist, C.; Dormann, P.; Karim, S.; Jarvis, P. Monogalactosyldiacylglycerol deficiency in *Arabidopsis* affects pigment composition in the prolamellar body and impairs thylakoid membrane energization and photoprotection in leaves. *Plant. Physiol.* **2008**, *148*, 580–592. [CrossRef]

28. McNeil, S.D.; Nuccio, M.L.; Ziemak, M.J.; Hanson, A.D. Enhanced synthesis of choline and glycine betaine in transgenic tobacco plants that overexpress phosphoethanolamine N-methyltransferase. *Proc. Natl. Acad. Sci.* **2001**, *98*, 10001–10005. [CrossRef]

29. Mohamed, A.A.; Aly, A.A. Iron Deficiency Stimulated Some Enzymes Activity, Lipid Peroxidation and free radicals production in *Borage officinalis* induced in vitro. *Int. J. Agri. Biol.* **2004**, *6*, 179–184.

30. Sperotto, R.A.; Boff, T.; Duarte, G.L.; Fett, J.P. Increased senescence-associated gene expression and lipid peroxidation induced by iron deficiency in rice roots. *Plant. Cell Rep.* **2008**, *27*, 183–195. [CrossRef]

31. Vigani, G.; Zocchi, G. The fate and the role of mitochondria in Fe-deficient roots of Strategy I plants. *Plant. Signal. Behav.* **2009**, *4*, 375–379. [CrossRef] [PubMed]

32. Bitrián, M.; Zarza, X.; Altabella, T.; Tiburcio, A.F.; Alcázar, R. Polyamines under abiotic stress: Metabolic crossroads and hormonal crosstalks in plants. *Metabolites* **2012**, *2*, 516–528. [CrossRef]

33. Voss, I.; Sunil, B.; Scheibe, R.; Raghavendra, A.S. Emerging concept for the role of photorespiration as an important part of abiotic stress response. *Plant. Biol.* **2013**, *15*, 713–722. [CrossRef] [PubMed]

34. Leonardi, R.; Zhang, Y.M.; Rock, C.O.; Jackowski, S. Coenzyme A: Back in action. *Prog Lipid Res.* **2005**, *44*, 125–153. [CrossRef] [PubMed]

35. Fujioka, S.; Yokota, T. Biosynthesis and metabolism of brassinosteroids. *Annu Rev. Plant. Biol.* **2003**, *54*, 137–164. [CrossRef] [PubMed]

36. Bishop, G.J.; Nomura, T.; Yokota, T.; Harrison, K.; Noguchi, T.; Fujioka, S.; Takatsuto, S.; Jones, J.D.G.; Kamiya, Y. The tomato DWARF enzyme catalyses C-6 oxidation in brassinosteroid biosynthesis. *Proc. Natl. Acad. Sci.* **1999**, *96*, 1761–1766. [CrossRef] [PubMed]

37. Nomura, T.; Sato, T.; Bishop, G.J.; Kamiya, Y.; Takatsuto, S.; Yokota, T. Acumulation of 6-deoxocathasterone and 6-deoxocatasterone in *Arabidopsis*, pea and tomato is suggestive of common rate limiting steps in brassinosteroid biosynthesis. *Phytochemistry* **2001**, *57*, 171–178. [CrossRef]

38. Clouse, S.D. Brassinosteroids. *Arab. Book* **2002**, *1*, 1–23. [CrossRef] [PubMed]

39. Goyer, A. Thiamine in plants: Aspects of its metabolism and functions. *Phytochemistry* **2010**, *71*, 1615–1624. [CrossRef]

40. Nishizawa, A.; Yabuta, Y.; Shigeoka, S. Galactinol and raffinose constitute a novel function to protect plants from oxidative damage. *Plant. Physiol.* **2008**, *147*, 1251–1263. [CrossRef]

41. He, N.W.; Zhao, Y.; Guo, L.; Shang, J.; Yang, X.B. Antioxidant, antiproliferative, and pro-apoptotic activities of a saponin extract derived from the roots of *Panax notoginseng* (Burk.) F.H. Chen. *J. Med. Food.* **2012**, *15*, 350–359. [CrossRef] [PubMed]

42. Wasson, A.P.; Pellerone, F.I.; Mathesius, U. Silencing the flavonoid pathway in *Medicago truncatula* inhibits root nodule formation and prevents auxin transport regulation by *Rhizobia*. *Plant. Cell* **2006**, *18*, 1617–1629. [CrossRef] [PubMed]

43. Maj, D.; Wielbo, J.; Marek-Kozaczu, M.; Skorupska, A. Response to flavonoids as a factor influencing competitiveness and symbiotic activity of *Rhizobium leguminosarum*. *Microbiol. Res.* **2010**, *165*, 50–60. [CrossRef] [PubMed]

44. Abdel-Lateif, K.; Bogusz, D.; Hocher, V. The role of flavonoids in the establishment of plant roots endosymbioses with arbuscular mycorrhiza fungi, rhizobia and *Frankia* bacteria. *Plant. Signal. Behav.* **2012**, *7*, 636–641. [CrossRef] [PubMed]

45. Mierziak, J.; Kostyn, K.; Kulma, A. Flavonoids as important molecules of plant interactions with the environment. *Molecules* **2014**, *19*, 16240–16265. [CrossRef] [PubMed]

46. Meyerowitz, E.M.; Somerville, C.R. *Arabidopsis*; Cold Spring Harbor Laboratory Press: Cold Spring Harbor, NY, USA, 1994; p. 1270.

47. Pollard, M.; Beisson, F.; Li, Y.; Ohlrogge, J.B. Building lipid barriers: Biosynthesis of cutin and suberin. *Trends Plant. Sci.* **2008**, *13*, 236–246. [CrossRef] [PubMed]

48. Rodriguez-Garay, B.; Phillips, G.C.; Kuehn, G.D. Detection of norspermidine and norspermine in *Medicago sativa* L. (alfalfa). *Plant. Physiol.* **1989**, *89*, 525–529. [CrossRef] [PubMed]

49. Zhu, X.F.; Wang, B.; Song, W.F.; Zheng, S.J.; Shen, R.F. Putrescine alleviates iron deficiency via NO-dependent reutilization of root cell-wall Fe in *Arabidopsis*. *Plant. Physiol.* **2016**, *170*, 558–567. [CrossRef] [PubMed]

50. Bourgaud, F.; Hehn, A.; Larbat, R.; Doerper, S.; Gontier, E.; Kellner, S.; Matern, U. Biosynthesis of coumarins in plants: A major pathway still to be unravelled for cytochrome P450 enzymes. *Phytochem. Rev.* **2006**, *5*, 293–308. [CrossRef]

51. Ingle, R.A. Histidine biosynthesis. *Arab. Book* **2011**, *9*, 1–9. [CrossRef]

52. Velazquez, E.; Silva, L.R.; Peix, A. Legumes: A healthy and ecological source of flavonoids. *Curr. Nutr. Food Sci.* **2010**, *6*, 109–144. [CrossRef]

53. Wang, B.; Li, Y.; Zhang, W.H. Brassinosteroids are involved in response of cucumber (*Cucumis sativus*) to iron deficiency. *Ann. Bot.* **2012**, *110*, 681–688. [CrossRef] [PubMed]

54. Boisson-Dernier, A.; Andriankaja, A.; Chabaud, M.; Niebel, A.; Journet, E.P.; Barker, D.G.; de Carvalho-Niebel, F. *MtENOD11* Gene activation during rhizobial infection and mycorrhizal arbuscule development requires a common AT-rich-containing regulatory sequence. *MPMI* **2005**, *18*, 1269–1276. [CrossRef] [PubMed]

55. Gutiérrez-Luna, F.M.; López-Bucio, J.; Altamirano-Hernández, J.; Valencia-Cantero, E.; Reyes-de la Cruz, H.; Macías-Rodríguez, L. Plant growth-promoting rhizobacteria modulate root-system architecture in *Arabidopsis thaliana* through volatile organic compound emission. *Symbiosis* **2010**, *51*, 75–83. [CrossRef]

56. Hernández-Calderón, E.; Aviles-Garcia, M.E.; Castulo-Rubio, D.Y.; Macías-Rodríguez, L.; Montejano-Ramírez, V.; Santoyo, G.; López-Bucio, J.; Valencia-Cantero, E. Volatile compounds from beneficial or pathogenic bacteria differentially regulate root exudation, transcription of iron transporters, and defense signaling pathways in *Sorghum bicolor*. *Plant. Mol. Biol.* **2018**, *96*, 291–304. [CrossRef] [PubMed]

57. Strohalm, M.; Hassman, M.; Košata, B.; Kodíček, M. mMass data miner: An open source alternatative for mass spectrometric data analysis. *Rapid Commun. Mass Spectrom.* **2008**, *22*, 905–908. [CrossRef] [PubMed]
58. Gibb, S.; Strimmer, K. MALDIquant: A versatile R package for the analysis of mass spectrometry data. *Bioinformatics* **2012**, *28*, 2270–2271. [CrossRef]
59. Williams, G.J. *Data Mining with Rattle and R: The Art of Excavating Data for Knowledge Discovery*; Springer: New York, NY, USA, 2011; p. 374.
60. Winkler, R. SpiderMass: Semantic database creation and tripartite metabolite identification strategy. *J. Mass Spectrom.* **2015**, *50*, 538–541. [CrossRef]
61. Ikekawa, N.; Takatsuto, S. Microanalysis of brassinosteroids in plants by gas chromatography/mass spectrometry. *J. Mass Spectrom. Soc. Jpn.* **1984**, *32*, 55–70. [CrossRef]

**Sample Availability:** Samples of the compounds are not available from the authors.

Article

# GC-MS Analysis of the Composition of the Extracts and Essential Oil from *Myristica fragrans* Seeds Using Magnesium Aluminometasilicate as Excipient

Inga Matulyte [1], Mindaugas Marksa [2], Liudas Ivanauskas [2], Zenona Kalvėnienė [1,3], Robertas Lazauskas [4] and Jurga Bernatoniene [1,3,*]

1   Department of Drug Technology and Social Pharmacy, Medical Academy, Lithuanian University of Health Sciences, Kaunas LT-50161, Lithuania; inga.matulyte@lsmu.lt (I.M.); zenona.kalveniene@lsmuni.lt (Z.K.)
2   Department of Analytical and Toxicological Chemistry, Medical Academy, Lithuanian University of Health Sciences, Kaunas LT-50161, Lithuania; mindaugas.m.lsmu@gmail.com (M.M.); Liudas.ivanauskas@lsmuni.lt (L.I.)
3   Institute of Pharmaceutical Technologies, Medical Academy, Lithuanian University of Health Sciences, Kaunas LT-50161, Lithuania
4   Institute of Physiology and Pharmacology, Medical Academy, Lithuanian University of Health Sciences, Kaunas LT-50161, Lithuania; Robertas.lazauskas@lsmuni.lt
*   Correspondence: jurga.bernatoniene@lsmuni.lt

Academic Editor: Igor Jerković
Received: 13 February 2019; Accepted: 14 March 2019; Published: 18 March 2019

**Abstract:** *Myristica fragrans* (f. Myristicaceae) seeds are better known as a spice, but their chemical compounds may have a pharmacological effect. The yield of their composition of extracts and essential oils differs due to different methodologies. The aim of this study was to evaluate an excipient material—magnesium aluminometasilicate—and to determine its influence on the qualitative composition of nutmeg extracts and essential oils. Furthermore, we wanted to compare the yield of essential oil. The extracts were prepared by maceration (M) and ultrasound bath-assisted extraction (UAE), and the essential oil—by hydrodistillation (HD). Conventional methods (UAE, HD) were modified with magnesium aluminometasilicate. The samples were analyzed by gas chromatography-mass spectrometry (GC-MS) method. From 16 to 19 chemical compounds were obtained using UAE with magnesium aluminometasilicate, while only 8 to 13 compounds were obtained using UAE without an excipient. Using our conditions and plant material, for the first time eight new chemical compounds in nutmeg essential oil were identified. Two of these compounds (γ-amorphene and cis-α-bergamotene) were obtained with the use of excipient, the other six (β-copaene, bergamotene, citronellyl decanoate, cubebol, cubenene, orthodene) by conventional hydrodistillation. Magnesium aluminometasilicate significantly increased the quantity of sabinene (from 6.53% to 61.42%) and limonene (from 0% to 5.62%) in essential oil. The yield of the essential oil from nutmeg seeds was significantly higher using magnesium aluminometasilicate; it increased from 5.25 ± 0.04% to 10.43 ± 0.09%.

**Keywords:** *Myristica fragrans*; nutmeg; essential oil; extract; magnesium aluminometasilicate; hydrodistillation

---

## 1. Introduction

Since ancient times, nutmeg (*Myristica fragrans*, f. Myristicaceae) has been used as a spice. Its aphrodisiac effect is often mentioned [1]. In traditional medicine, nutmeg is used to treat rheumatism, pain, nausea, stomach cramps, and other illnesses [2]. A lot of research is being done in order to test the pharmacological effects of nutmeg. Scientific sources say that nutmeg has antibacterial, insecticidal, antioxidant, anticancer, antidepressant, hepatoprotective, and many other effects [3–6]. The most common studies about essential oils are of nutmeg, which is rich in terpenes, phenols, various organic acids, and other compounds. Using different solvents, methods, and conditions, the composition of extracts and essential oils variations were studied [3,7–9]. Nutmeg essential oil is colorless to pale-yellow with a specific odor. The essential oil yield may be more than 10% [10] and varies by 5–15% [11]. *Myristica fragrans* essential oil has predominantly monoterpene hydrocarbons (α-thujene, β-pinene, sabinene), sesquiterpene hydrocarbons (D-germacrene, trans-β-bergamotene), monoterpene alcohols (linalool, α-terpineol, *cis*-p-menth-2-en-1-ol, terpinen-4-ol), esters (α-terpinyl acetate, *cis*-sabinene hydrate acetate, citronellyl acetate), aromatics (eugenol, myristicin, elemicin, methoxyeugenol), and other compounds [5,7–9,12].

Traditional methods such as maceration, percolation, infusion, decoction, and Soxhlet method are commonly used to produce extracts [8,13]. These methods are cheap but less efficient, which wastes a lot of time and raw materials [14]. One of the more recent methods for extracts is extraction in an ultrasound-assisted bath, it is an effective method, in which it is possible to choose the desired conditions (i.e., temperature, duration), and it saves on materials and time [2,15]. Ultrasound-assisted extraction (UAE) has many modifications with solvents, and extra material can be used for extraction; for example: surfactants [16], polysaccharides [17], salts [18], acids [19], and other excipients, which can increase the efficiency of extraction.

The most popular methods for extracting essential oils are distillation with water (hydrodistillation) and distillation with steam or steam and water [12,14,20]. Other methods are cold or hot-pressing, supercritical fluid extraction, enfleurage, microwave, and ultrasound-assisted methods, solvent extraction, and aqueous infusion [18,21].

Hydrodistillation is a method used for extracting essential oils. It is cheap because it mostly uses water as a solvent. The equipment is not expensive, most commonly a Clevenger-type apparatus or similar modifications are used [14,20]. In different sources, hydrodistillation is carried out for various times and solid:solvent ratios. Hydrodistillation is carried out using water, and few studies have been carried out using extra material for extraction such as sea water [22], non-ionic surfactants [23] or 5% NaCl salt solution. [14]. According to the information found, it can be said that excipient substances are rarely used for hydrodistillation.

Excipients are used for improving extraction and changing environmental conditions. For example, salts that change ion voltage, surfactants, emulsifiers (sorbitan esters (Spans®), polysorbates (Tweens®)), and pH-adjusting substances are used [9,24,25]. The magnesium aluminometasilicate will be used for the first time as an excipient for chemical compound extractions, previously it was used as an excipient only in solid dosage forms [26,27]. It is a white amorphous powder with high surface area, practically insoluble in water. The magnesium aluminometasilicate has the ability to absorb materials, which is characterized by water absorption capacity [26,28].

The purpose of this study is to evaluate excipient material—magnesium aluminometasilicate— and to determine its influence on the qualitative composition of nutmeg extracts and essential oils. Furthermore, to compare the yield of essential oil using conventional hydrodistillation and hydrodistillation with magnesium aluminometasilicate.

## 2. Results

Using the traditional maceration method (conditions are mentioned in Table 1, see Materials and Methods) myrislignan and elemicin were established at 22.59% and 13.99%, respectively, (Table 2). Maceration as a method was modified with magnesium aluminometasilicate (0.5%, 1%, and 2% excipient), but the results were not significant. The magnesium aluminometasilicate did not increase the quantity of chemical compounds, so these conditions and results were not mentioned in the research.

**Table 1.** Extraction conditions used for each experiment.

| Method | Sample Code | Temperature | Time (h) | Solvent | Solvent: Nutmeg Ratio | Magnesium Aluminometasilicate: Nutmeg Ratio |
|---|---|---|---|---|---|---|
| **Maceration** | M1 | ambient | 72 | purified water | 20:1 | - |
| **Ultrasound-Assisted Extraction** | UAE1 | 25 °C | 0.5 | ethanol 50% | 20:1 | - |
| | UAE2 | | | ethanol 70% | 20:1 | - |
| | UAE3 | | | ethanol 96% | 20:1 | - |
| | UAE4 | | | ethanol 70% + 0.5% magnesium aluminometasilicate | 20:1 | 10:1 |
| | UAE5 | | | ethanol 70% + 1% magnesium aluminometasilicate | 20:1 | 5:1 |
| | UAE6 | | | ethanol 70% + 2% magnesium aluminometasilicate | 20:1 | 2.5:1 |
| **Hydrodistillation** | HD1 | 100 °C | 4 | purified water | 20:1 | - |
| | HD2 | | | purified water + 0.5% magnesium aluminometasilicate | 20:1 | 10:1 |
| | HD3 | | | purified water + 1% magnesium aluminometasilicate | 20:1 | 5:1 |
| | HD4 | | | purified water + 2% magnesium aluminometasilicate | 20:1 | 2.5:1 |

Using other extraction techniques (i.e., UAE, HD), myrislignan and elemicin were not found. The extract (M1) has 10 different active compounds: $\alpha$-phellandrene, $\beta$-myrcene, $\gamma$-asarone, elemicin, and others (Table 2).

Analyzing the influence of ethanol concentrations on the extraction of compounds, it was found that most of the different compounds were extracted using 70% ethanol (fifteen compounds), or at least 50% ethanol (eight compounds). The higher the ethanol concentration, the better the extracts were extracted, for example with $\alpha$-pinene, 2.4%, 8.3%, and 14.67%, respectively (Table 2). In contrast, when weaker ethanol concentrations were used more isolemicin was extracted (62.24%, 23.63%). In the extract of 96% ethanol isolemicin was not found.

The data of the extracts with magnesium aluminometasilicate, using 70% ethanol as a solvent (UAE4, UAE5, UAE6, Table 2) showed that when 1% magnesium aluminometasilicate was used, the utmost quantity of compounds was identified (90.92%). Magnesium aluminometasilicate helped extract sabinene (from 29.9 to 32.69%, in extracts without magnesium aluminometasilicate sabinene was not found). The following compounds were also detected in extracts with magnesium aluminometasilicate exposure: myrcene, isoterpinolene, isogermacrene, limonene, *cis*-sabinene hydrate, $\gamma$-terpinene, $\gamma$-amorphene, and $\beta$-myrcene (Table 2).

The composition of the essential oils, which were obtained using hydrodistillation (HD1) and hydrodistillation with excipients from HD2 to HD4 (Table 1), showed that 1% of magnesium aluminometasilicate significantly improved amount of $\alpha$-pinene and sabinene compared to conventional hydrodistillation, respectively 54.95% and 106.32%. Using 0.5% and 2% of magnesium aluminometasilicate the amount of the $\alpha$-pinene and sabinene was higher than HD1 but lower than HD3 (Figure 1).

**Table 2.** Chemical composition of nutmeg extracts and essential oils.

| Compound | RI [d] | Essential Oil Compounds | | | | | | Extracts Compounds | | | | | |
|---|---|---|---|---|---|---|---|---|---|---|---|---|---|
| | | HD [e] 1 (%) | HD2 (%) | HD3 (%) | HD4 (%) | M [f] 1 (%) | UAE [g] 1 (%) | UAE2 (%) | UAE3 (%) | UAE4 (%) | UAE5 (%) | UAE6 (%) |
| 4-Carene, trans -(+)- | 911 | 7.77 | - | - | - | 3.37 | 1.19 | 6.56 | 13.22 | - | - | - |
| α-Thujene | 928 | 0.99 | 1.04 | 0.93 | 0.99 | - | 2.63 | 0.72 | 1.42 | 0.84 | 0.92 | 0.81 |
| α-Pinene | 934 | 8.27 | 10.04 | 15.05 | 11.5 | - | 2.4 | 8.3 | 14.67 | 9.65 | 8.13 | 9.94 |
| Camphene | 944 | - | 2.8 | 0.11 | 0.11 | - | - | - | - | - | - | 0.15 |
| Orthodene [b] | 949 | 0.51 | - | - | - | - | - | - | 0.59 | - | - | - |
| 3-Carene | 956 | 0.53 | - | - | - | - | - | 0.42 | 0.78 | - | - | - |
| α-Phellandrene | 957 | - | - | - | - | 13.04 | - | 1.41 | 2.38 | - | - | - |
| β-Myrcene | 958 | - | - | - | - | 4.6 | - | - | - | 9.47 | 7.26 | 9.19 |
| 2-Carene | 960 | 1.99 | - | - | - | - | - | 0.75 | 1.15 | - | - | - |
| Sabinene | 969 | 6.53 | 49.64 | 61.42 | 47.1 | - | - | - | - | 29.9 | 31.74 | 32.69 |
| β-Pinene | 970 | 26.61 | 4.03 | 4.32 | 3.84 | - | - | - | - | - | - | - |
| Myristicin | 981 | 3.26 | 2.34 | 2.15 | 2.16 | - | - | - | - | 1.79 | 1.65 | 1.96 |
| γ-Terpinene | 993 | - | 0.78 | 0.68 | 0.73 | - | - | - | - | 0.68 | 1.65 | 0.71 |
| α-Terpinene | 999 | 1.23 | 1.23 | 0.94 | 1.23 | - | - | 0.6 | 1.07 | 0.67 | 1.18 | 0.69 |
| 3,7,7-trimethylcyclohepta-1,3,5-triene | 1005 | 0.18 | 0.47 | 0.33 | 0.44 | - | - | - | - | 0.37 | - | - |
| Limonene | 1009 | - | 5.62 | 4.2 | 5.23 | - | - | - | - | 4.2 | 4.46 | 4.63 |
| Isoterpinolene | 1030 | - | 2.11 | 1.18 | 2.07 | - | - | - | - | 1.19 | 2.02 | 1.22 |
| Cis-sabinen hydrate | 1037 | 7.76 | 0.58 | 0.3 | 0.83 | - | - | - | - | 1.3 | 2.32 | 1.46 |
| γ-Terpineol | 1043 | - | 0.04 | 0.04 | 0.04 | - | - | - | 0.14 | - | - | 0.09 |
| α-Terpinolene | 1051 | - | 0.7 | 0.38 | 0.7 | - | - | - | - | - | - | 0.54 |
| Sylvestrene | 1059 | - | - | - | - | 1.57 | - | - | - | - | - | - |
| Isomethyleugenol | 1062 | 0.2 | - | - | - | 6.38 | - | - | - | - | - | - |
| Cis-p-menth-2-en-1-ol | 1076 | 2.02 | 0.38 | 0.37 | 0.38 | 0.43 | 3.27 | 5.88 | 3.25 | 1.32 | 2.42 | 1.62 |
| 4-Propenyl syringol | 1083 | - | - | - | - | - | 2.16 | 1.55 | - | - | - | - |
| Citronellyl Decanoate [b] | 1110 | 0.3 | - | - | - | - | - | 0.43 | 0.16 | - | - | - |
| 1,1-dimethyl-2-[(1E)-3-methylbuta-1,3-dienyl cyclopropane | 1116 | - | - | - | - | - | 8.4 | 29.99 | 47.32 | - | - | - |
| Bicyclogermacrene | 1125 | 0.29 | - | - | - | - | - | - | - | - | - | - |
| 4-Terpineol | 1152 | 0.74 | 0.34 | 0.03 | 0.35 | - | - | 0.77 | 0.3 | 0.47 | 0.47 | 0.47 |
| Cubebol [b] | 1174 | 0.05 | - | - | - | - | - | - | - | - | - | - |
| Cubenene [b] | 1177 | 0.07 | - | - | - | - | - | - | - | - | - | - |
| Piperitol | 1189 | 0.09 | - | - | - | - | - | - | - | - | - | - |
| Isoelemicin | 1217 | 5.98 | - | - | - | - | 62.24 | 23.-63 | - | 21.5 | 21.5 | 18.85 |
| Copaene | 1228 | - | 0.25 | 0.01 | 0.25 | - | - | - | - | - | - | - |
| β-Copaene [b] | 1233 | 0.25 | - | - | - | - | - | - | - | - | - | - |
| γ-Amorphene [b] | 1245 | - | 0.75 | 0.03 | 0.78 | - | - | - | - | 0.43 | 0.92 | 0.08 |
| Cis-α-bergamotene [b] | 1283 | - | 0.09 | 0.08 | 0.1 | - | - | - | - | - | - | - |
| Isogermacrene | 1311 | - | 0.99 | 0.01 | 1.15 | 1.61 | - | - | - | 0.86 | 2.44 | 0.82 |
| Licarin B | 1328 | - | - | - | - | - | - | 0.72 | - | 0.71 | - | - |
| Bergamotene [b] | 1339 | 0.07 | - | - | - | - | - | - | - | - | - | - |
| γ-Asarone | 1361 | 3.69 | 1.21 | 1.22 | 3.51 | 0.79 | 0.84 | 2.07 | - | 0.82 | 1.84 | 1.25 |
| Elemicin | 1542 | - | - | - | - | 13.99 | - | - | - | - | - | - |
| Myrislignan | 2946 | - | - | - | - | 22.59 | - | - | - | - | - | - |
| **Total Identified Compounds % [c]** | | 79.38 | 85.43 | 93.78 | 83.6 | 68.37 | 83.13 | 83.8 | 86.45 | 86.1 | 90.92 | 87.17 |

All data are given with an 0.05 accuracy. [a] Sample code presented in Table 1. [b] First time chemical compounds extracted from nutmeg. [c] Quantity of all of the identified compounds. [d] Retention indices using RTX-5MS column and n-alkanes ($C_9$–$C_{22}$) as references. [e] HD–hydrodistillation. [f] M–maceration. [g] UAE–ultrasound-assisted extraction.

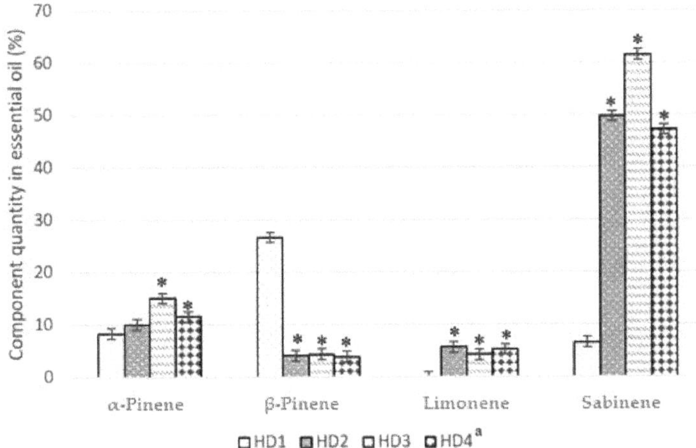



**Figure 1.** Magnesium aluminometasilicate's influence on the amount of α and β-pinene, limonene, and sabinene in the essential oils of nutmeg. Component quantity in essential oils, * $p < 0.05$ versus hydrodistillation without magnesium aluminometasilicate. [a] Sample code can be seen in Table 1.

In HD1, essential oil consisted of β-copaene, bergamotene, bicyclogermacrene, cubebol, cubenene, and piperitol, and these compounds were not identified in other extracts. Ten compounds using magnesium aluminometasilicate were obtained, and in comparison to the essential oil of HD1, their amount was not large but it was about $13.05 \pm 3.54\%$ of the quantity of all of the identified compounds in essential oil.

Comparing water and ethanol as solvents, and maceration with ultrasound-assisted extraction methods, it was found that water is better in extracting α-phellandrene (13.04%) than 70% ethanol (1.41%), but worse in γ-asarone (0.79% versus 2.07%), *cis*-p-menth-2-en-1-ol (0.43% versus 5.88%), 4-carene, trans-(+)-(3.37% versus 6.56%) (Table 2).

By using hydrodistillation from nutmeg (HD1), $0.79 \pm 0.04$ g of essential oil was obtained (Figure 2).

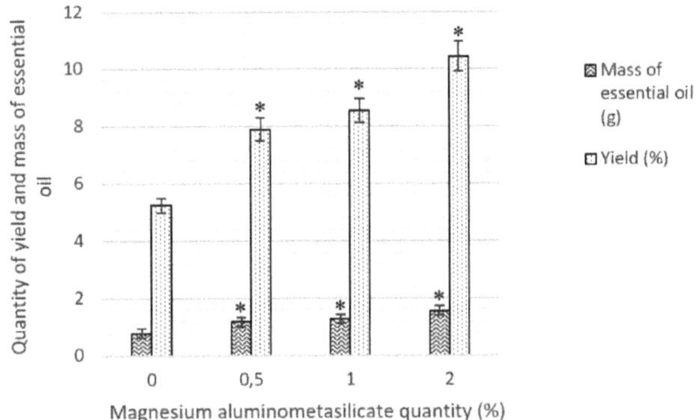



**Figure 2.** Essential oils' mass and yield from nutmeg using HD and HD with magnesium aluminometasilicate. Mass and yield of essential oil, * $p < 0.05$ versus hydrodistillation without magnesium aluminometasilicate.

At least this amount of mass was gained in all of the samples. Then we used the magnesium aluminometasilicate, and the essential oil was obtained in higher quantities—from $1.19 \pm 0.09$ g to 1.57

± 0.09 g ($p < 0.05$). The highest yield of essential oil (HD4, 10.43%) was obtained using 2% magnesium aluminometasilicate (Figure 2).

## 3. Discussion

In this study, the extracts, essential oil, and the influence of excipient were analyzed. The composition of the aqueous (M1) and ethanolic (UAE1–UAE6) extracts and essential oil (HD1–HD4) of nutmeg were determined (Table 1). Our results show that the extract which was obtained using maceration with water had 10 different chemical compounds of which the most commonly found are myrislignan (22.59%), elemicin (13.99%), and α-phellandrene (13.04%). We did not find any research done with aqueous extract of nutmeg. In ethanolic extract, which was made by using maceration, the quantity of elemicin (5.7%) was 2.45 times lower, and the quantity of α-phellandrene was 21.73 times higher [8]. Elemicin has antibacterial effects [29] and α-phellandrene, anti-inflammatory [30], so using maceration with water is more effective if a bigger effect is needed. Myrislignan has anti-inflammatory [31] and anti-cancer [32] effects. This compound was not extracted using ethanol [8].

In our study, when 96% ethanol was used (ultrasound extraction, sample UAE3), the α-pinene content was obtained at 14.67%, in other study using the same method, $6.75 \pm 0.92\%$ was obtained [8]. In one of the studies, the compound of *cis*-p-menth-2-en-1-ol was obtained at about 0.1% [8]. However, in our study using 96% ethanol the compound was obtained at 3.25%, which is 32.5 times more than in the previously mentioned study, and if we compare this with extraction using 70% ethanol, the yield of *cis*-p-menth-2-en-1-ol was obtained at 58.8 (5.88%) times more in our research (Table 2). Using nutmeg seeds material from Grenada we did not find any isoelemicin (ultrasound extraction, 96% ethanol), the other researchers found $0.8 \pm 0.46\%$ [8], but we used 50% and 70% ethanol and determined the quantity of isoelemicin, which was 62.24% and 23.63%, respectively. The excipients are used to improve the extraction of the active compounds. Surfactants used in experiments improve the extraction of certain compounds, such as Sodium dodecyle sulfate (SDS), Triton X 100, PEG 2000, Brij 35, Tween, etc. [25]. In our study we used magnesium aluminometasilicate as excipient, and our purpose was to research and determine the effect of magnesium aluminometasilicate on the extraction of active compounds. Previously, this substance has not been used in studies with extracts and essential oils. Magnesium aluminometasilicate is used as excipient in solid drug form formulations [26,27,33]. Using this material, we have relied on its mechanism of action. It adsorbs high loads of oil [34], so we predicted that it could extract more essential oil components. The results of GC-MS analysis showed the influence of the excipients on gaining different quantities of certain compounds. Magnesium aluminometasilicate significantly improved the amount of sabinene. If we compare it to the essential oils that were made by using simple hydrodistillation (Table 2), we see that 0.5% and 2% of magnesium aluminometasilicate increased sabinene by $7.41 \pm 0.2$ times (49.6% and 47.1%, traditional hydrodistillation—6.53%), the 1% of magnesium aluminometasilicate increased sabinene by 9.41 times (61.42%). These results are significant. Dupuy [7] reported 6.00–36.08% quantity of sabinene, and the nutmeg seeds were from Indonesia. Also, magnesium aluminometasilicate had influence on limonene (Figure 1) and γ-terpinene extraction. The other samples made without the excipient did not have these compounds (M1, HD1, UAE1-UAE3) (Table 2). Sabinene has anti-inflammatory, antifungal, and antioxidant effects [35,36], limonene has low-toxicity and anticancer effect [37], and γ-terpinene characterizes as an antimicrobial and antioxidant agent [38]. Magnesium aluminometasilicate's influence on α-pinene and β-pinene was different. The 1% of magnesium aluminometasilicate increased the quantity of α-pinene by 45% (from 8.27% to 15.05%) and decreased β-pinene by 84% (from 26.61% to 4.32%). Piaru [9] results showed that *Myristica fragrans* seeds's essential oil from Malaysia contained 8.5% α-pinene and 3.5% β-pinene.

After analyzing many studies with nutmeg essential oil we found a few compounds which were not mentioned in publications. Our study showed that essential oil HD1 from nutmeg (Grenada) contains β-copaene (0.25%), bergamotene (0.07%), citronellyl decanoate (0.3%), cubebol (0.05%), cubenene (0.07%), and orthodene (0.51%) (Figure 3).

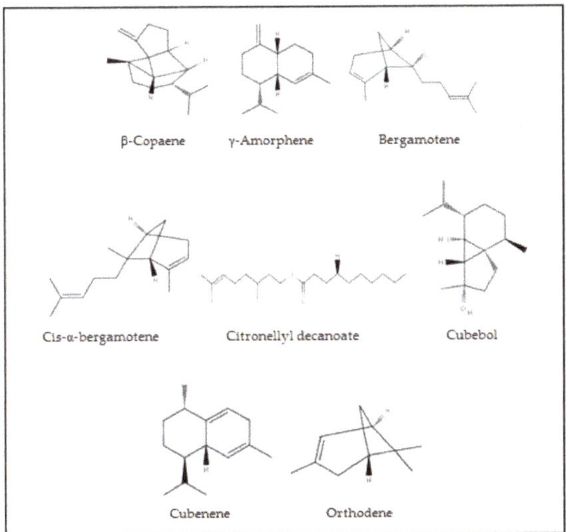

**Figure 3.** First time extracted and identified chemical compounds in essential oil of nutmeg.

In other research we found that the oleoresins of *Myristica fragrans* (Egypt) have α-copaene (0.63%) [8] and trans-α-bergamotene (0.1%) [39]. The compound of cubebol was found in *Piper Cuberica Linn.* essential oil and oleoresins [40] and *Chromolaena odorata* essential oil (8.6%) [41]. Cubenene was obtained in *Caryopteris incana* essential oil (9.7%) [42]. The γ-amorphene (0.03–0.78%) and *cis*-α-bergamotene (0.08–0.1%) were obtained by using magnesium aluminometasilicate in hydrodistillation (HD2–HD3) (Figure 3).

Volatile compound composition in extracts which have been produced using magnesium aluminometasilicate have similarities with essential oil. Seventy-one percent of the chemical compounds in the extract correspond to the composition of essential oil. However, only extracts with magnesium aluminometasilicate extracted β-myrcene and isoelemicin, and these compounds are not present in essential oil.

Our experimental data showed that the essential oil yield increased if we used magnesium aluminometasilicate as excipient which facilitated volatile compounds extraction. Essential oil yield from *Myristica fragrans* was 5.25 ± 0.04%; however, the magnesium aluminometasilicate was used for the yield of the essential oil and it was from 7.9 ± 0.09% to 10.43 ± 0.09%. Using sea water, the yield of oil-bearing rose increased statistically insignificantly by 0.045%, but the sea water significantly decreases the citronellol rate from 41.49% to 33.56%, and significantly increases geraniol (from 17.58% to 27.44%) [22]. Using non-ionic surfactants for extraction of sage essential oil it was found that Tween 20 increased the yield of oil at 1.83% and Span 80 or Span 20 had no effect [23]. Using 5% NaCl salt solution for hydrodistillation, the yield of lavender oil was 5.01 ± 0.45% in comparison with only water, where the yield was 4.55 ± 0.14% [14]. All essential oil samples with magnesium aluminometasilicate have 21 chemical compounds, although essential oil made by simple hydrodistillation has 24 compounds but that is not significantly important. The yield of nutmeg essential oil significantly increases with magnesium aluminometasilicate as excipient. The essential oil yield from nutmeg seeds (Nigeria) was 1.46% [43], from other nutmeg seeds (China) it was 2.4% [44]. Nutmeg seeds from Indonesia were hydrodistillated and the yield of essential oil was obtained at 5.84%.

## 4. Materials and Methods

### 4.1. Materials

The dried seeds of *Myristica fragrans* (nutmeg) were from Grenada. The seeds were identified by Jurga Bernatoniene, Medical Academy, Lithuania University of Health Sciences, Kaunas, Lithuania. A voucher specimen (I 18922) was placed for storage at the Herbarium of the Department of Drug Technology and Social Pharmacy, Lithuanian University of Health Sciences, Lithuania. The seeds were a brown-beige color, had a characteristic odor, and strong, bitter, and spicy flavor. Dried nutmeg seeds were ground into a powder (particles smaller than 0.5 mm) with a laboratory mill. The samples were kept in an airtight container in a dark environment and in ambient temperature.

Magnesium aluminometasilicate (Neusilin® US2, Fuji Chemical Industries Co., Ltd., Toyoma, Japan) was used as excipient. Ethanol 96% (Vilniaus degtine, Lithuania) was used as solvent for extraction. Distilled water was used throughout the experiment.

### 4.2. Maceration

The nutmeg seeds' powder was extracted using a maceration technique with water in ambient temperature while occasionally stirring. The process lasted 3 days. Powder and solvent ratio were 1:20 (w/v). The extracts were filtered with paper filter (0.22 μm pore size) and stored in refrigerator at 4 °C until further analysis. Conditions are mentioned in Table 1.

### 4.3. Ultrasound Assisted Extraction

Extracts were also prepared by using ultrasound-assisted extraction. This method was used as a simple and effective method for nutmeg extracts. The 1 g material and 20 mL solvent were used. We used different concentrations of ethanol for extraction (96%, 70%, and 50%). The flask with substance was put in an ultrasound bath (Sonorex Digitec DT 156 BH, Germany). Extracts were prepared on 25 °C temperature for 30 min. After that, the extracts were filtered with paper filter and stored in a refrigerator at 4 °C until further analysis.

Some samples were modified and made with magnesium aluminometasilicate. The extracts were made in the same conditions which were listed earlier. But for the solvent only a certain concentration of ethanol (70%) was used and the excipient was added in the mixture. The excipients' concentration was from 0.5% to 2%. Magnesium aluminometasilicate (g) was based on ethanol quantity. Extracts were filtered and stored. Conditions are mentioned in Table 1.

### 4.4. Hydrodistillation

Samples were prepared from nutmeg seeds and purified with water. The solid material was 15 g and the water was 300 mL. A Clevenger-type apparatus (European pharmacopoeia) for hydrodistillation was used. Hydrodistillation was carried out for 4 h. A colorless essential oil with specific odor was obtained. The essential oil was collected with water in airtight bottle and stored in a refrigerator at 4 °C until needed.

The hydrodistillation method was modified and magnesium aluminometasilicate as excipient was used. The amount of magnesium aluminometasilicate was based on water quantity, its concentration was from 0.5% to 2%. Fifteen grams of nutmeg seed powder were used and all of the sample ratios of solvent were the same (20:1). The essential oil was obtained and stored in the same conditions. Conditions are mentioned in Table 1.

### 4.5. Gas Chromatography-Mass Spectrometry Analysis

Gas chromatography-mass spectrometry analysis was performed on a GCMS-QP2010 system (Shimadzu, Tokyo, Japan). Twenty microliters of sample (extract or essential oil) was diluted to 1 mL with hexane (≥99%, Sigma–Aldrich, Germany). The column used was a 30 m × 0.25 mm i.d. × 0.25

µL film thickness RTX-5MS column. Flow rate of helium (99.999%, AGA Lithuania) carrier gas was set at 1.23 mL/min. The oven temperature was maintained at 40 °C for 2 min after injection and then programmed at 3 °C/min to 210 °C, at which the column was maintained for 10 min. The split ratio was 1:10. The mass detector electron ionization was 70 eV. Identification of volatile compounds was carried out using mass spectra library search (NIST 14) and compared with the mass spectral data from literature [45].

### 4.6. Statistical Analysis

Data are presented as mean $\pm$ SEM. Statistical analysis was preformed using Student's *t*-test. The results were significant when $p < 0.05$.

## 5. Conclusions

Extracts and essential oils from *Myristica fragrans* seeds (Grenada) were prepared using three different methods (M, UAE, HD). The volatile compounds were analyzed by GC-MS. The results showed the samples' quality and quantity of the chemical compounds composition. During the experiment magnesium aluminometasilicate was used for extraction in the hydrodistillation process. Previously, this excipient was used for the production of solid dosage forms. Ten chemical compounds were obtained using maceration, from eight to thirteen compounds were obtained using ultrasound-assisted extraction with different ethanol concentrations and without excipient, using UAE with magnesium aluminometasilicate, 16–19 compounds were obtained. Hydrodistillation with magnesium aluminometasilicate lead to 21 compounds, and finally, 24 compounds were obtained without the use of excipient. Although using hydrodistillation with the excipient fewer chemical compounds were obtained, some of the compounds' quantities were bigger than using hydrodistillation without excipient. Magnesium aluminometasilicate as excipient improved the extraction of $\alpha$-pinene, limonene, isoterpinolene, and sabinene, although it reduced the quantity of $\beta$-pinene. For the first time, eight chemical compounds in nutmeg essential oil were identified. Two of these compounds ($\gamma$-amorphene and cis-$\alpha$-bergamotene) were obtained by the use of magnesium aluminometasilicate.

The yield of the essential oil from nutmeg seeds was significantly higher using magnesium aluminometasilicate. The data shows that essential oil yield which was obtained by hydrodistillation with water was $5.25 \pm 0.04\%$ and it increased using hydrodistillation with water and 2% of magnesium aluminometasilicate to $10.43 \pm 0.09\%$. Using the 2% of magnesium aluminometasilicate the yield of essential oil increased by almost twice.

The use of magnesium aluminometasilicate in the extraction process improved the yield of essential oil and the number of compounds in the extracts.

**Author Contributions:** I.M., Z.K., and J.B. conceived and designed the experiments; L.I. and M.M. performed experiments with GC-MS; I.M., M.M., R.L. and J.B. analyzed the date; I.M. and J.B. wrote the paper.

**Acknowledgments:** The authors would like to thank Open Access Centre for the Advanced Pharmaceutical and Health Technologies (Lithuanian university of Health Sciences) and for the opportunity to use modern infrastructure and perform this research.

**Conflicts of Interest:** The authors declare that there are no competing interests regarding the publication of this paper.

## References

1.  Tajuddin, S.; Ahmad, A.; Latif, I.; Qasmi, A.; Amin, K.M.Y. An experimental study of sexual function improving effect of Myristica fragrans Houtt. (nutmeg). *BMC Complement. Altern. Med.* **2005**, *5*, 16–23. [CrossRef]

2.  Ziyatdinova, G.; Ziganshina, E.; Cong, P.N.; Budnikov, H. Ultrasound-assisted micellar extraction of phenolic antioxidants from spices and antioxidant properties of the extracts based on coulometric titration data. *Anal. Methods* **2016**, *8*, 7150–7157. [CrossRef]

3. Adiani, V.; Gupta, S.; Chatterjee, S.; Variyar, P.S.; Sharma, A. Activity guided characterization of antioxidant components from essential oil of Nutmeg (Myristica fragrans). *J. Food Sci. Technol.* **2015**, *52*, 221–230. [CrossRef]

4. D'Souza, S.P.; Chavannavar, S.V.; Kanchanashri, B.; Niveditha, S.B. Pharmaceutical Perspectives of Spices and Condiments as Alternative Antimicrobial Remedy. *J. Evid.-Based Complement. Altern. Med.* **2017**, *22*, 1002–1010. [CrossRef]

5. Gupta, A.D.; Bansal, V.K.; Babu, V.; Maithil, N. Chemistry, antioxidant and antimicrobial potential of nutmeg (Myristica fragrans Houtt). *J. Genet. Eng. Biotechnol.* **2013**, *11*, 25–31. [CrossRef]

6. Nagja, T.; Vimal, K.; Sanjeev, A. Myristica Fragrans: A Comprehensive Review. *Int. J. Pharm. Pharm. Sci.* **2016**, *8*, 9–12.

7. Dupuy, N.; Molinet, J.; Mehl, F.; Nanlohy, F.; Le Dréau, Y.; Kister, J. Chemometric analysis of mid infrared and gas chromatography data of Indonesian nutmeg essential oils. *Ind. Crop. Prod.* **2013**, *43*, 596–601. [CrossRef]

8. Morsy, N.F.S. A comparative study of nutmeg (Myristica fragrans Houtt.) oleoresins obtained by conventional and green extraction techniques. *J. Food Sci. Technol.* **2016**, *53*, 3770–3777. [CrossRef]

9. Piaru, S.P.; Mahmud, R.; Majid, A.M.S.; Nassar, Z.D. Antioxidant and antiangiogenic activities of the essential oils of Myristica fragrans and Morinda citrifolia. *Asian Pac. J. Trop. Med.* **2012**, *5*, 294–298.

10. Djilani, A.; Dicko, A. The Therapeutic Benefits of Essential Oils. In *Nutrition, Well-Being and Health*; Bouayed, J., Ed.; InTech: Rijeka, Croatia, 2012; pp. 155–178.

11. Barceloux, D.G. Nutmeg (Myristica fragrans Houtt.). *Dis. Mon.* **2009**, *55*, 373–379. [CrossRef] [PubMed]

12. Lanari, D.; Marcotullio, M.; Neri, A. A Design of Experiment Approach for Ionic Liquid-Based Extraction of Toxic Components-Minimized Essential Oil from Myristica fragrans Houtt. Fruits. *Molecules* **2018**, *23*, 2817. [CrossRef] [PubMed]

13. Azwanida, N.N. A Review on the Extraction Methods Use in Medicinal Plants, Principle, Strength and Limitation. *Med. Aromat. Plants* **2015**, *4*, 3–8.

14. Filly, A.; Fabiano-Tixier, A.S.; Louis, C.; Fernandez, X.; Chemat, F. Water as a green solvent combined with different techniques for extraction of essential oil from lavender flowers. *C. R. Chim.* **2016**, *19*, 707–717. [CrossRef]

15. Choi, M.P.K.; Chan, K.K.C.; Leung, H.W.; Huie, C.W. Pressurized liquid extraction of active ingredients (ginsenosides) from medicinal plants using non-ionic surfactant solutions. *J. Chromatogr. A* **2003**, *983*, 153–162. [CrossRef]

16. Feng, L.; Aun, R.; Yan-Wei, W.; Xiu-Quan, X.; Guan-Hua, C. Optimization of Surfactant-Mediated, Ultrasonic-assisted Extraction of Antioxidant Polyphenols from Rattan Tea (Ampelopsis grossedentata) Using Response Surface Methodology. *Pharmacogn. Mag.* **2017**, *13*, 446–453.

17. Zhang, X.; Ban, Q.; Wang, X.; Wang, Z. Green and Efficient PEG-Based Ultrasonic-Assisted Extraction of Polysaccharides from Tree Peony Pods and the Evaluation of Their Antioxidant Activity In Vitro. *Biomed. Res. Int.* **2018**, *2018*, 2121385. [CrossRef] [PubMed]

18. Rezaeepour, R.; Heydari, R.; Ismaili, A. Ultrasound and salt-assisted liquid-liquid extraction as an efficient method for natural product extraction. *Anal. Methods* **2015**, *7*, 3253–3259. [CrossRef]

19. Minjares-Fuentes, R.; Femenia, A.; Garau, M.C.; Meza-Velázquez, J.A.; Simal, S.; Rossello, C. Ultrasound-assisted extraction of pectins from grape pomace using citric acid: A response surface methodology approach. *Carbohydr. Polym.* **2014**, *106*, 179–189. [CrossRef]

20. Tongnuanchan, P.; Benjakul, S. Essential Oils: Extraction, Bioactivities, and Their Uses for Food Preservation. *J. Food Sci.* **2014**, *79*, 1231–1249. [CrossRef]

21. Da Porto, C.; Decorti, D.; Kikic, I. Flavour compounds of *Lavandula angustifolia* L. to use in food manufacturing: Comparison of three different extraction methods. *Food Chem.* **2009**, *112*, 1072–1078. [CrossRef]

22. Kara, N.; Erbas, D.; Baydar, H. The Effect of Seawater Used for Hydrodistillation on Essential Oil Yield and Composition of Oil-Bearing Rose ( Rosa damascena Mill.). *Int. J. Sec. Metab.* **2017**, *4*, 423–428.

23. Charchari, S.; Abdelli, M. Enhanced Extraction by Hydrodistillation of Sage (*Salvia officinalis* L.) Essential Oil Using Water Solutions of Non-ionic Surfactants. *J. Essent. Oil Bear. Plants* **2015**, *5026*, 1094–1099.

24. Hosseinzadeh, R.; Khorsandi, K.; Hemmaty, S. Study of the Effect of Surfactants on Extraction and Determination of Polyphenolic Compounds and Antioxidant Capacity of Fruits Extracts. *PLoS ONE* **2013**, *8*, e57353. [CrossRef] [PubMed]

25. Sharma, S.; Kori, S.; Parmar, A. Surfactant mediated extraction of total phenolic contents (TPC) and antioxidants from fruits juices. *Food Chem.* **2015**, *185*, 284–288. [CrossRef]

26. Krupa, A.; Majda, D.; Jachowicz, R.; Mozgawa, W. Solid-state interaction of ibuprofen and Neusilin US2. *Thermochim. Acta* **2010**, *509*, 12–17. [CrossRef]

27. Sander, C.; Holm, P. Porous Magnesium Aluminometasilicate Tablets as Carrier of a Cyclosporine Self-Emulsifying Formulation. *AAPS PharmSciTech* **2009**, *10*, 1388. [CrossRef]

28. Naiserova, M.; Kubova, K.; Vyslouzil, J.; Pavlokova, S.; Vetchy, D.; Urbanova, M.; Brus, J.; Vyslouzil, J.; Kulich, P. Investigation of Dissolution Behavior HPMC/Eudragit®/Magnesium Aluminometasilicate Oral Matrices Based on NMR Solid-State Spectroscopy and Dynamic Characteristics of Gel Layer. *AAPS PharmSciTech* **2017**, *19*, 681–692. [CrossRef] [PubMed]

29. Rossi, P.G.; Bao, L.; Luciani, A.; Panighi, J.; Desjobert, J.M.; Costa, J.; Casanova, J.; Bolla, J.M.; Berti, L. (E)-Methylisoeugenol and Elemicin: Antibacterial Components of Daucus carota L. Essential Oil against Campylobacter jejuni. *J. Agric. Food Chem.* **2007**, *55*, 7332–7336. [CrossRef]

30. Siqueira, H.D.S.; Neto, B.S.; Sousa, D.P.; Gomes, B.S.; da Silva, F.V.; Cunha, F.V.M.; Wanderley, C.W.S.; Pinheiro, G.; Candido, A.G.F.; Wong, D.V.H. α-Phellandrene, a cyclic monoterpene, attenuates inflammatory response through neutrophil migration inhibition and mast cell degranulation. *Life Sci.* **2016**, *160*, 27–33. [CrossRef]

31. Jin, H.; Zhu, Z.G.; Yu, P.J.; Wang, G.F.; Zhang, J.Y.; Li, J.R.; Ai, R.T.; Li, Z.H.; Tian, Y.X.; Zhang, W.X.; et al. Myrislignan attenuates lipopolysaccharide-induced inflammation reaction in murine macrophage cells through inhibition of NF-κB signalling pathway activation. *Phyther. Res.* **2012**, *26*, 1320–1326. [CrossRef]

32. Lu, X.; Yang, L.; Chen, J.; Zhou, J.; Tang, X.; Zhu, Y.; Qiu, H.; Shen, J. The action and mechanism of myrislignan on A549 cells in vitro and in vivo. *J. Nat. Med.* **2017**, *71*, 76–85. [CrossRef] [PubMed]

33. Gupta, M.K.; Vanwert, A.; Bogner, R.H. Formation of Physically Stable Amorphous Drugs by Milling with Neusilin. *J. Pharm. Sci.* **2003**, *92*, 536–551. [CrossRef] [PubMed]

34. Khade, S.; Pore, Y. Formulation and Evaluation Of Neusilin® Us2 Adsorbed Amorphous Solid Self-Microemulsifying Delivery System of Atorvastatin Calcium. *Asian J. Pharm. Clin. Res.* **2016**, *9*, 93–100.

35. Mihajilov-Krstev, T.; Jovanovic, B.; Jovic, J.; Ilic, B.; Miladinovic, D.; Matejic, J.; Rajkovic, J.; Dordevic, L.; Cvetkovic, V.; Zlatkovic, B. Antimicrobial, Antioxidative, and Insect Repellent Effects of Artemisia absinthium Essential Oil. *Planta Med.* **2014**, *80*, 1698–1705. [CrossRef] [PubMed]

36. Valente, J.; Zuzarte, M.; Goncalves, M.J.; Lopes, M.C.; Cavaleiro, C.; Salgueiro, L.; Cruz, M.T. Antifungal, antioxidant and anti-inflammatory activities of Oenanthe crocata L. essential oil. *Food Chem. Toxicol.* **2013**, *62*, 349–354. [CrossRef] [PubMed]

37. Sun, S. D-limonene: Safety and clinical applications. *Altern. Med. Rev.* **2007**, *12*, 259–264. [PubMed]

38. Giweli, A.; Dzamic, A.M.; Sokovic, M.; Ristic, M.S.; Marin, P.D. Antimicrobial and antioxidant activities of essential oils of satureja thymbra growing wild in libya. *Molecules* **2012**, *17*, 4836–4850. [CrossRef]

39. Mallavarapu, G.R.; Ramesh, S. Composition of essential oils of nutmeg and mace. *J. Med. Arom. Plant Sci* **1998**, *20*, 746–748.

40. Gupta, M. Pharmacological properties and traditional therapeutic uses of important indian spices: A review. *Int. J. Food Prop.* **2010**, *13*, 1092–1116. [CrossRef]

41. Joshi, R.K. Chemical Composition of the Essential Oils of Aerial Parts and Flowers of *Chromolaena odorata* (L.) R. M. King & H. Rob. from Western Ghats Region of North West Karnataka, India. *J. Essent. Oil-Bear. Plants* **2013**, *16*, 71–75.

42. Chu, S.; Liu, O.; Zhou, L.; Du, S.; Liu, Z. Chemical composition and toxic activity of essential oil of Caryopteris incana against Sitophilus zeamais. *Afr. J. Biotechnol.* **2013**, *10*, 8476–8480.

43. Ekundayo, O.; Ogunwande, I.A.; Olawore, N.O.; Adeleke, K.A. Chemical composition of essential oil of myristica fragrans houtt (nutmeg) from nigeria. *J. Essent. Oil-Bear. Plants* **2003**, *6*, 21–26.

44. Du, S.S.; Yang, K.; Wang, C.F.; You, C.X.; Geng, Z.F.; Guo, S.S.; Deng, Z.W.; Liu, Z.L. Chemical constituents and activities of the essential oil from myristica fragrans against cigarette beetle lasioderma serricorne. *Chem. Biodivers.* **2014**, *11*, 1449–1456. [CrossRef] [PubMed]
45. Adams, R.P. *Identification of Essential Oil Components by Gas Chromatography/Mass Spectrometry*; Allured Publishing Corporation: Carol Stream, IL, USA, 1995.

**Sample Availability:** Samples of the compounds are available from the authors.

Article

# Yeast Smell Like What They Eat: Analysis of Volatile Organic Compounds of *Malassezia furfur* in Growth Media Supplemented with Different Lipids

Mabel Gonzalez [1], Adriana M. Celis [2,3,*], Marcela I. Guevara-Suarez [2,3], Jorge Molina [4] and Chiara Carazzone [1,*]

[1] Laboratory of Advanced Analytical Techniques in Natural Products (LATNAP), Universidad de los Andes, Cra 1 No. 18A-12, Bogotá 111711, Cundinamarca, Colombia; mabel.c.gonzalez@uniandes.edu.co
[2] Grupo de Investigación Celular y Molecular de Microorganismos Patógenos (CeMoP), Universidad de los Andes, Cra 1 No. 18A-12, Bogotá 111711, Cundinamarca, Colombia; mi.guevara34@uniandes.edu.co
[3] Laboratorio de Micología y Fitopatología (LAMFU), Universidad de los Andes, Cra 1 No. 18A-12, Bogotá 111711, Cundinamarca, Colombia
[4] Centro de Investigaciones en Microbiología y Parasitología Tropical (CIMPAT), Universidad de los Andes, Cra 1 No. 18A-12, Bogotá 111711, Cundinamarca, Colombia; jmolina@uniandes.edu.co
* Correspondence: acelis@uniandes.edu.co (A.M.C.); c.carazzone@uniandes.edu.co (C.C.); Tel.: +57-13394949 (ext. 3757) (A.M.C.); +57-13394949 (ext. 1480) (C.C.)

Academic Editor: Igor Jerković
Received: 13 December 2018; Accepted: 21 January 2019; Published: 24 January 2019

**Abstract:** *Malassezia furfur* is part of the human skin microbiota. Its volatile organic compounds (VOCs) possibly contribute to the characteristic odour in humans, as well as to microbiota interaction. The aim of this study was to investigate how the lipid composition of the liquid medium influences the production of VOCs. Growth was performed in four media: (1) mDixon, (2) oleic acid (OA), (3) oleic acid + palmitic acid (OA+PA), and (4) palmitic acid (PA). The profiles of the VOCs were characterized by HS-SPME/GC-MS in the exponential and stationary phases. A total number of 61 VOCs was found in *M. furfur*, among which alkanes, alcohols, ketones, and furanic compounds were the most abundant. Some compounds previously reported for *Malassezia* (γ-dodecalactone, 3-methylbutan-1-ol, and hexan-1-ol) were also found. Through our experiments, using univariate and multivariate unsupervised (Hierarchical Cluster Analysis (HCA) and Principal Component Analysis (PCA)) and supervised (Projection to Latent Structures Discriminant Analysis (PLS-DA)) statistical techniques, we have proven that each tested growth medium stimulates the production of a different volatiles profile in *M. furfur*. Carbon dioxide, hexan-1-ol, pentyl acetate, isomer5 of methyldecane, dimethyl sulphide, undec-5-ene, isomer2 of methylundecane, isomer1 of methyldecane, and 2-methyltetrahydrofuran were established as differentiating compounds among treatments by all the techniques. The significance of our findings deserves future research to investigate if certain volatile profiles could be related to the beneficial or pathogenic role of this yeast.

**Keywords:** fungal VOCs; GC-MS; HS-SPME; lipids; exponential growth phase; stationary growth phase; oleic acid; palmitic acid

---

## 1. Introduction

All species in the genus *Malassezia* are lipid-dependent because of their inability to synthesize de novo $C_{14}$ or $C_{16}$ fatty acids (FAs) [1] due to the lack of a cytosolic fatty acid synthase complex (FAS) [2]. In order to survive, *Malassezia* rely on lipids encountered in host skin that are subsequently metabolized by lipases and phospholipases [3]. The 18 species of the genus belong to class Malasseziomycetes of phylum *Basidiomycota* [4]. *Malassezia* species could represent between 53% and 80% of human skin fungal microbiota, accumulating at different body sites, especially behind the ears [5] and on the scalp [6].

These organisms have been recognized as aetiological agents of pityriasis versicolor and might also be related to other skin diseases, such as seborrheic dermatitis, dandruff, folliculitis, atopic/eczema dermatitis syndrome, and psoriasis [7]. Depending on the stimulation, or occasionally the downregulation of the immune response against *Malassezia*, these yeasts could remain as a commensal or eventually become pathogenic [8]. The causes related to this interconversion are still unknown. There are different studies suggesting that a relationship exists between pathogenicity and modifications in the lipid compositions of the host [9,10] or changes in the skin microbiota [11,12]. The complexity of the human skin environment hinders the possibility of fully comprehending the host–microbe and microbe–microbe interactions [13], because the skin determines the microbiota and, at the same time, the resident microbial communities modify this environment in a dynamic process [14]. Host demographics and genetics, human behavior, local and regional environmental characteristics, and transmission events determine the human skin microbiota variability [15]. The spatial distribution of the micro-organisms is also highly variable [5]. *Malassezia* and *Cutibacterium acnes* (formerly *Propionibacterium acnes*) [16] reside in areas with a high density of sebaceous glands, whereas *Staphylococcus* and *Corynebacterium* are mostly prevalent in areas with a high temperature and humidity [17]. Any changes in these conditions may affect homeostasis and promote a pathogen or even a commensal, such as *Malassezia*, to cause disease.

Some *Malassezia* species can easily be differentiated by the types of lipids necessary to grow them in certain media [18,19]. Therefore, the *Malassezia* assimilation assay is widely used to determine the lipidic requirement of different strains [3]. For example, palmitic acid (PA) has a fungicidal activity in *Malassezia pachydermatis*, in an atypical strain of *Malassezia furfur*, and in *Malassezia sympodialis*, but not in *M. furfur* CBS 1878 [20]. The addition of oleic acid (OA) produces a fungistatic effect on *M. pachydermatis*, on atypical *M. furfur*, and on *M. furfur* CBS 1878, but on *M. sympodialis* OA produces a fungicidal effect [21]. Therefore, Dixon, mDixon, and Leeming and Notham (LNA) media are the most widely used for culturing *Malassezia*, because their more complex composition and the presence of Ox bile (obtained after purifying and drying fresh bile from oxen) offer the essential nutritional requirements for their growth [8,22]. The ability of *M. furfur* to survive in PA could not be related to the presence of a Δ9-desaturase gene OLE1, which is involved in the conversion of saturated to unsaturated acids, such as PA to palmitoleic acid, because this gene has also been found in other species/strains where PA is fungicidal [3,21].

Thus, the lipid-dependent and lipophilic lifestyle of *Malassezia* spp. could determine its ecological niche in the skin environment of human and animal hosts [20]. Understanding new aspects of the lipid metabolism of these yeasts is expected to provide new insights into their biology and ecology, and, therefore, illuminate the path to understanding interconversions between commensalism and pathogenicity [21]. As they can be easily collected, fungal volatile organic compounds (VOCs) are one of the means to assess the metabolic profiles of fungi related to certain diseases [23–29]. Furthermore, they can be used to relatively quickly monitor static or dynamic metabolic changes [27,30–32]. Other important uses of fungal VOCs include biotechnological processes [30,33,34], signalling of ecological communication between different species [31,35], and obtaining information about fungal development [36,37]. The high diversity of VOCs released by bacteria and fungi justifies the creation of a database that summarizes the taxonomic distribution of around 2000 compounds and their possible functions. The first version of the mVOC Database was created in 2014, and was renewed this year [38,39]. No volatile compounds of *Malassezia* have been reported in this database, because the data sources used have been limited to PubMed publications [39]. The first and only description of *Malassezia* volatiles, in which the authors reported the presence of 11 different VOCs, was performed in 1979, when the genus was recognized as *Pityrosporum* [40].

The human skin sebum contains a complex mixture of triglycerides, fatty acids, wax esters, sterol esters, cholesterol, and squalene. Among the fatty acids, the unsaturated ones predominate, but the sebum is generally composed of OA, PA, stearic acid, linoleic acid, sapienic acid, and palmitoleic acid [41]. OA is an irritant component that can induce dandruff in dandruff-susceptible people [10,41]. Among animals, PA is a fatty acid found in the human sebum in a relatively high amount [21]. Having this in mind, the aim of this study is to investigate how the lipid composition of a liquid medium influences the production of VOCs released by *M. furfur*, the only *Malassezia* species known to survive in the PA medium. To accomplish this, four different media for growing the strain CBS 1878 were prepared: (1) mDixon, (2) minimal medium (MM) supplemented with oleic acid (OA), (3) MM with oleic + palmitic acid (OA+PA), and (4) MM with palmitic acid (PA). The VOCs released by the yeast grown in these media were analyzed using headspace solid-phase microextraction (HS-SPME) and gas chromatography-mass spectrometry (GC-MS). The mDixon medium was selected as the reference for in vitro growing conditions. The other media serve to identify the VOCs released under restricted nutritional conditions when an unsaturated acid (OA), a saturated acid (PA), or a combination of both is present in the medium. In addition, samplings of VOCs were performed for each medium in the exponential and stationary phase. Finally, results from eight experimental treatments (mDixon in both growing phases: De, Ds; oleic acid in both growing phases: Oe, Os; oleic + palmitic acid in both growing phases: OPe, OPs; and, finally, palmitic acid in both growing phases: Pe, Ps) were obtained.

Previous reports on other species have demonstrated that changes in the microbial growth conditions produce qualitative and quantitative alterations in the VOCs released by an organism [28,42,43]. Based on these findings, we hypothesized the existence of a chemical differentiation in the VOCs profiles released by *M. furfur* CBS 1878 due to the different lipid composition of the medium and to the phase of growth. Univariate comparisons and unsupervised and supervised multivariate analyses have been applied to measure the metabolic differentiation and its significance.

## 2. Results and Discussion

### 2.1. Profiling of VOCs from M. furfur

HS-SPME/CG-MS analysis was applied to determine the profiles of VOCs released by *M. furfur* CBS 1878 growing in four different media: mDixon, OA, OA+PA, and PA in both growing phases. With the objective of analyzing exclusively VOCs released by the organism, the following criteria were established for the inclusion of a compound originating from the micro-organisms in the succeeding matrices: (1) the VOC was absent in any replicate of all medium control analyses (for more details, see the Materials and Methods section); and (2) the VOC was found in more than one replicate of the yeast. Qualitative differences between the chromatographic profiles of the eight experimental treatments (Figure 1) show a combination of both: the VOCs released by *M. furfur* and the VOCs released from growth media.

HS-SPME/CG-MS analysis of *M. furfur* allowed for the annotation of 61 different VOCs (Table 1), among which alkanes, alcohols, ketones, and furanic compounds were the most prevalent. Carbon dioxide, hexan-1-ol, octane, and nonane showed the highest peak areas. The lack of terpenes, which are common VOCs found in fungi [44–46], such as *Candida albicans* [44] and *Aspergillus fumigatus* [47], is remarkable. Average peak areas, standard deviations, and the number of replicates in which each VOC was found are summarized in Table 1. A tentative annotation of 27 compounds was carried out by comparison of mass spectra with databases and experimental retention index (RI) deviations below 30 units from the theoretical RI. In opposition, 30 compounds were annotated to specific moieties, or at least to a chemical class, based only on mass spectra comparisons. Among these, most of the alkanes could not be exhaustively identified because their electron ionization (EI) fragmentation patterns are very similar. Four compounds were described as unknowns.

The presence of γ-dodecalactone, 3-methylbutan-1-ol, and hexan-1-ol, previously described as signature VOCs of *Malassezia*, was reported [40]. The γ-dodecalactone was detected in the mDixon and the OA medium in both growth phases, while the hexan-1-ol was found in all treatments, with the exception of the PA medium. In contrast to the first analysis of Labows et al. (1979), who found many different gamma lactones in the volatiles profile of *Malassezia*, we detected the presence of only one of these gamma lactones. These differences could be related firstly to the genetic differences among the tested strains, because, in 1979, they used the strains ATCC 24,047, 12,078, and 14,521 of *M. furfur* [40]; secondly, to the growing conditions, because the authors used solid media and Tween 80; or, finally, to the sampling conditions, because they absorbed the compounds in tenax, instead of using SPME (solid-phase microextracion) fibers [40]. Although mDixon also contains OA, the dietary importance of this unsaturated acid in the biosynthesis of γ-dodecalactone in *M. furfur* remains unknown. If the presence of OA controlled this pathway, the emission of γ-dodecalactone would have been observable also in the OA+PA medium, unless PA was an inhibitory supplement. In other fungi, OA is indeed a suitable substrate for this biosynthesis, undergoing hydroxylation, β-oxidation, isomerization, and lactonization reactions [33]. Taking into account the rest of the VOCs, it is clear that the DVB/CAR/PDMS-SPME fiber allowed for the sampling of a wider range of polarities than the ones analysed by Labows et al. using tenax in 1979. This resulted in the description of VOCs not previously reported for *Malassezia* and even for fungi, according to the data summarized in the mVOC 2.0 Database. The 16 compounds previously reported in other fungi are marked with * in Table 1.

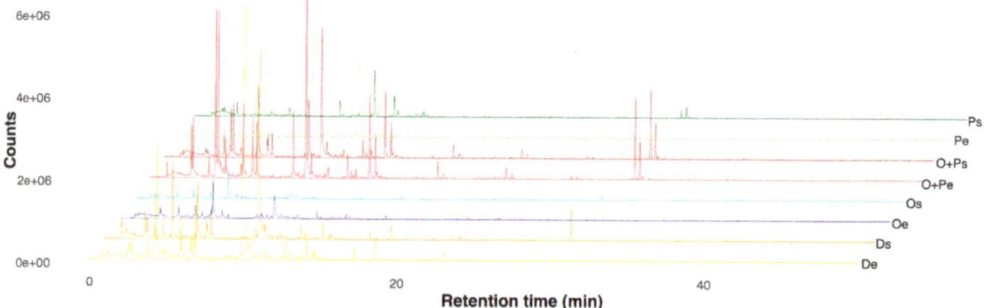


**Figure 1.** Chromatograms of *Malassezia furfur* in exponential and stationary phases of growth and in eight treatments with different fatty acid (FA) content. De = Dixon in Exponential phase, Ds = Dixon in Stationary phase, Oe = Oleic acid in Exponential phase, Os = Oleic acid in Stationary phase, OPe = Oleic acid + Palmitic acid in Exponential phase, OPs = Oleic acid + Palmitic acid in Stationary phase, Pe = Palmitic acid in Exponential phase, Ps = Palmitic acid in Stationary phase. The Y-axis denotes chromatographic intensity of signals in an absolute scale after being detected in the mass spectrometer.

From all of the volatiles profiles, 25 compounds were detected in the five replicates of all media, but the absence/presence of most of the compounds was highly variable. The maximum number of VOCs (27) was reported for Ds, followed by 23 compounds detected in Oe. Subsequently, 20 compounds were encountered in OPe, and, finally, just eight VOCs were revealed in Ps. PA medium produced a drastic reduction in the number of VOCs and even in the chemiodiversity, limited almost exclusively to alkanes. The comparison between the VOCs obtained from *M. furfur* CBS 1878 in the four media discussed, and the ones released by the strains ATCC 24,047, 12,078, and 14,521 of *M. furfur*, *M. pachydermatis*, and *Malassezia globose* grown in different media [40,48], demonstrates the importance of the lipid substrate for both the growing of micro-organisms and the releasing of VOCs. The interaction between the two variables growth and diversity of volatiles in *M. furfur* is still unknown because growing (determined by the CFU (colony forming units)) and volatile sampling (through HS-SPME/GC-MS) could not be performed simultaneously on each replicate in order to maintain the headspace conditions without air contamination. Anyways, we were able to collect indirect evidence that the lack of growth (the static patterns observed for the growth curves in PA and OA+PA media in Figure S1) does not correlate with the quantity of released VOCs. While the number of detected VOCs shows that PA more greatly reduces the diversity of VOCs, in the other growth-static medium (OA+PA) *M. furfur* released more than 15 volatiles (Table 1).

**Table 1.** Volatiles profiles of *M. furfur* in four media and two growth phases.

| Compound | Crit. | RI | RI Exp | Treatment Ret. Time | De | | | Ds | | | Oe | | | Os | | | Pe | | | Ps | | | OPe | | | OPs | | |
|---|---|---|---|---|---|---|---|---|---|---|---|---|---|---|---|---|---|---|---|---|---|---|---|---|---|---|---|---|
| | | | | | X | SD | N | X | SD | N | X | SD | N | X | SD | N | X | SD | N | X | SD | N | X | SD | N | X | SD | N |
| Carbon dioxide * | a | | | 1.3 | 7.7 | 4.2 | 5 | 14.6 | 15.8 | 5 | 9.8 | 6.6 | 5 | 8.4 | 6.2 | 5 | 61.5 | 21.4 | 5 | 45.3 | 22.0 | 5 | 2.5 | 0.4 | 5 | 2.4 | 1.7 | 5 |
| Pentane * | a | 500 | | 1.6 | | | | 1.1 | | 1 | | | | | | | 2.0 | 0.3 | 2 | | | | 0.1 | 0.1 | 2 | | | |
| Dimethyl sulphide * | a | 526 | | 1.7 | 0.1 | 0.1 | 2 | 0.3 | 0.1 | 5 | | | | | | | | | | | | | | | | | | |
| 2-methylpent-1-ene | a | 588 | | 2.0 | | | | 0.2 | 0.2 | 3 | 0.3 | | 1 | | | | | | | | | | | | | | | |
| 2-methylpropan-1-ol * | a,b | 629 | 624 | 2.2 | 0.1 | | 1 | 0.2 | 0.1 | 5 | | | | | | | | | | | | | | | | | | |
| Butan-1-ol * | a,b | 663 | 658 | 2.6 | 0.5 | 0.2 | 5 | 0.4 | 0.1 | 5 | 1.0 | 0.3 | 5 | 0.8 | 0.3 | 5 | | | | | | | 0.1 | 0.1 | 5 | 2.1 | 3.5 | 5 |
| 2-methyltetrahydrofuran | a,b | 678 | 665 | 2.6 | | | | | | | | | | | | | | | | | | | 0.1 | 0.1 | 5 | 0.3 | 0.0 | 5 |
| Isomer of pentan-2-one | a | | 678 | 2.8 | | | | 0.6 | 0.3 | 4 | | | | | | | | | | | | | | | | | | |
| Pentan-2-one * | a,b | 674 | 684 | 2.8 | 0.8 | 0.2 | 3 | 6.2 | 2.0 | 5 | 0.2 | 0.1 | 4 | 3.6 | 5.6 | 5 | | | | | | | | | | | | |
| Hept-2-ene | a,b | est: 725 | 702 | 3.0 | | | | | | | | | | | | | | | | | | | 0.6 | 1.0 | 4 | 2.4 | | 1 |
| Hept-3-ene | a,b | 730 | 713 | 3.2 | | | | | | | | | | | | | | | | | | | 0.1 | 0.1 | 5 | | | |
| 2,5-dimethylhexane | a | est: 688 | 723 | 3.4 | | | | | | | | | | | | | 2.4 | 0.4 | 3 | | | | | | | | | |
| 3-methylbutan-1-ol * | a,b | 746 | 730 | 3.5 | 0.9 | 0.2 | 5 | 3.0 | 0.5 | 5 | | | | | | | | | | 11.9 | | 1 | | | | | | |
| 2-methylbutan-1-ol * | a,b | 745 | 734 | 3.5 | 0.1 | 0.0 | 5 | 0.2 | 0.1 | 2 | | | | | | | | | | | | | | | | | | |
| 3-methylpentan-2-one * | a,b | 752 | 748 | 3.8 | | | | 0.1 | 0.0 | 4 | | | | | | | | | | | | | | | | | | |
| Octane * | a,b | 800 | 800 | 4.6 | 49.3 | 11.4 | 5 | 20.1 | 10.9 | 5 | 2.8 | 1.4 | 4 | 1.6 | 1.3 | 5 | 11.6 | 10.2 | 5 | 4.8 | | 1 | 51.9 | 2.7 | 5 | 15.3 | 3.7 | 5 |
| Oct-2-ene * | a,b | 810 | 806 | 4.8 | | | | | | | | | | | | | | | | | | | 0.2 | | 1 | 0.2 | 0.1 | 3 |
| 2-Furanmethanol | a,b | 858 | 852 | 5.9 | 0.3 | 0.0 | 2 | 0.4 | 0.2 | 5 | | | | | | | | | | | | | | | | | | |
| Hexan-1-ol * | a,b | 870 | 865 | 6.3 | 17.9 | 7.2 | 5 | 32.9 | 11.8 | 5 | 72.9 | 12.6 | 5 | 80.5 | 6.3 | 5 | | | | | | | 28.6 | 3.2 | 5 | 60.0 | 21.4 | 5 |
| Nonane | a,b | 900 | 900 | 7.2 | 19.7 | 7.2 | 5 | 15.6 | 8.1 | 5 | 0.8 | | 1 | 0.7 | 0.3 | 2 | 9.3 | 1.3 | 5 | 11.9 | 10.6 | 5 | 14.4 | 1.0 | 5 | 14.5 | 12.8 | 5 |
| Pentyl acetate * | a,b | est: 924 | 912 | 7.6 | | | | | | | | | | | | | | | | | | | 0.1 | 0.0 | 5 | 0.2 | 0.1 | 5 |
| Unknown bp: 56 | | | 925 | 8.0 | | | | | | | | | | | | | | | | | | | 0.1 | 0.0 | 4 | 0.1 | 0.0 | 2 |
| Unknown compound 1 with Nitrogen and C=O | a | | 955 | 9.1 | | | | | | | | | | | | | | | | | | | 0.0 | 0.0 | 3 | | | |
| Hept-6-en-1-ol | a,b | est: 950 | 959 | 9.3 | | | | | | | | | | | | | | | | | | | 0.3 | 0.1 | 5 | 1.3 | 0.5 | 5 |
| Unknown1 bp: 95 | | | 980 | 10.0 | 0.1 | 0.0 | 3 | 0.1 | 0.0 | 2 | | | | | | | | | | | | | | | | | | |
| Decane | a,b | 1000 | 1000 | 10.7 | | | | | | | 0.7 | 0.1 | 3 | | | | | | | | | | | | | | | |
| Dec-4-ene | a,b | est: 1023 | 1005 | 10.9 | | | | | | | | | | | | | | | | | | | | | | 0.3 | 0.0 | 3 |
| Isomer1 of methyldecane | a | 1001 | 1038 | 12.3 | 0.5 | 0.2 | 5 | 0.4 | 0.1 | 5 | | | | | | | | | | | | | | | | | | |
| 2,6,7-trimethyldecane | a,b | est: 1048 | 1052 | 12.9 | | | | 1.9 | 0.4 | 5 | | | | | | | | | | | | | | | | | | |
| Isomer4 of methyldecane | a | | 1059 | 13.2 | 0.3 | 0.1 | 3 | 0.2 | 0.0 | 3 | | | | | | | | | | | | | | | | | | |

**Table 1.** *Cont.*

| Compound | Crit. | RI | RI Exp | Treatment | De | | | Ds | | | Oe | | | Os | | | Pe | | | Ps | | | OPe | | | OPs | | |
|---|---|---|---|---|---|---|---|---|---|---|---|---|---|---|---|---|---|---|---|---|---|---|---|---|---|---|---|---|
| | | | | Ret. Time | X | SD | N | X | SD | N | X | SD | N | X | SD | N | X | SD | N | X | SD | N | X | SD | N | X | SD | N |
| Isomer5 of methyldecane | a | | 1069 | 13.6 | 0.5 | 0.4 | 5 | 0.5 | 0.1 | 5 | | | | | | | | | | | | | | | | | | |
| (5-Ethyl-cyclopent-1-enyl)-methanol | a,b | 1073 | 1073 | 13.7 | | | | | | | 0.8 | 0.4 | 3 | 1.2 | | 1 | | | | | | | | | | | | |
| Isomer1 of Undec-1-ene | a | | 1079 | 14.0 | 1.9 | 0.9 | 3 | 1.1 | 0.1 | 3 | | | | | | | | | | | | | | | | | | |
| Isomer2 of Undec-1-ene | a | | 1083 | 14.2 | | | | | | | 8.4 | 3.8 | 3 | 2.0 | 1.2 | 3 | | | | | | | | | | | | |
| Undec-5-ene | a,b | 1092 | 1105 | 15.1 | | | | | | | | | | | | | | | | | | | 0.3 | 0.0 | 5 | 0.9 | 0.2 | 5 |
| Heptyl acetate | a,b | 1110 | 1112 | 15.4 | | | | | | | | | | | | | | | | | | | 0.1 | 0.0 | 3 | | | |
| Isomer3 of Undec-1-ene | a | | 1114 | 15.5 | | | | | | | | | | | | | | | | | | | | | | 1.5 | 0.4 | 5 |
| Isomer1 of methylundecane | a | | 1124 | 15.9 | | | | | | | | | | | | | 7.8 | 3.3 | 2 | | | | | | | | | |
| Isomer2 of methylundecane | a | | 1124 | 15.9 | 0.4 | 0.2 | 5 | 0.4 | 0.0 | 5 | | | | | | | | | | | | | | | | | | |
| Non-3-en-1-ol, (Z)- * | a,b | 1158 | 1148 | 17.0 | | | | | | | 0.2 | | 1 | | | | | | | | | | | | | 0.1 | 0.1 | 3 |
| Non-2-en-1-ol, (E)- * | a | 1120 | 1167 | 17.8 | | | | | | | | | | | | | | | | | | | 0.3 | 0.2 | 4 | | | |
| Unknown2 bp:95 | | | 1181 | 18.5 | | | | 0.3 | 0.0 | 4 | | | | | | | | | | | | | | | | | | |
| 2-Nonyloxirane | a,b | est: 1205 | 1210 | 19.7 | | | | | | | | | | | | | | | | | | | 0.2 | 0.0 | 4 | 0.4 | 0.1 | 5 |
| 4,7-dimethylundecane | a,b | est: 1185 | 1211 | 19.8 | | | | 0.3 | 0.0 | 2 | | | | | | | 5.4 | 1.3 | 4 | 11.7 | 3.8 | 2 | | | | | | |
| Isomer2 of methyldodecane | a | | 1237 | 20.9 | | | | | | | 1.2 | 0.6 | 3 | 0.8 | | 1 | | | | | | | | | | | | |
| Isomer3 of methyldodecane | a | | 1245 | 21.3 | 0.2 | 0.0 | 2 | | | | 1.8 | 0.6 | 3 | 0.9 | | 1 | | | | | | | 0.1 | 0.1 | 4 | 0.3 | 0.1 | 5 |
| Isomer of Decenol | a | | 1268 | 22.3 | | | | | | | | | | | | | | | | | | | 0.4 | 0.3 | 4 | | | |
| Isomer5 of methyldodecane | a | | 1274 | 22.6 | | | | | | | 0.6 | 0.7 | 2 | | | | | | | | | | | | | | | |
| Undecan-2-one * | a,b | 1294 | 1292 | 23.4 | | | | | | | | | | | | | | | | 16.7 | 8.7 | 4 | | | | | | |
| Isomer6 of methyldodecane | a | | 1295 | 23.5 | | | | | | | 1.0 | 0.0 | 2 | 1.5 | | 1 | | | | | | | | | | | | |
| Isomer2 of methyldecanol | a | | 1295 | 23.5 | | | | | | | | | | | | | 4.7 | 2.2 | 2 | 12.2 | 2.0 | 2 | | | | | | |
| Unknown furanic | a | | 1301 | 23.7 | | | | 0.7 | 0.5 | 2 | | | | | | | | | | | | | | | | | | |
| Isomer1 of trimethyldodecane | a | | 1302 | 23.8 | | | | | | | 2.2 | 0.7 | 3 | 1.2 | 0.2 | 2 | | | | | | | | | | | | |
| Isomer2 of trimethyldodecane | a | | 1320 | 24.6 | | | | | | | 1.0 | 0.2 | 3 | 0.7 | | 1 | | | | | | | | | | | | |
| Isomer of methylundecanol | a | | 1323 | 24.7 | | | | | | | 1.8 | 0.5 | 3 | 1.0 | | 1 | | | | | | | | | | | | |
| Isomer1 of methyltridecane | a | | 1336 | 25.3 | | | | | | | 0.5 | 0.2 | 4 | 1.6 | 1.2 | 3 | | | | | | | | | | | | |
| Isomer2 of methyltridecane | a | | 1343 | 25.6 | | | | | | | 1.2 | 0.4 | 3 | 0.8 | | 1 | | | | | | | | | | | | |
| Isomer3 of methyltridecane | a | | 1354 | 26.0 | | | | | | | 0.8 | 0.0 | 2 | | | | | | | | | | | | | | | |
| Unknown3 bp:95 | | | 1381 | 27.2 | | | | 0.3 | 0.1 | 4 | | | | | | | | | | | | | | | | | | |
| Isomer of methyltetradecane | a | | 1496 | 31.9 | | | | | | | 1.2 | 0.4 | 3 | | | | | | | | | | | | | | | |
| γ-Dodecalactone | a,b | 1682 | 1679 | 37.5 | 0.3 | 0.2 | 3 | 0.4 | 0.1 | 4 | 0.6 | | 1 | 0.9 | 0.1 | 3 | | | | | | | | | | | | |

Average area percentage for each sample is represented as X, standard deviation is represented as SD, and number of replicates in which a volatile organic compound (VOC) was found as N. RI = Retention index published in literature. RI Exp. = Retention index relative to n-alkanes ($C_8 - C_{20}$) on a BP-5 column (30 m × 0.25 mm × 0.25 μm). We used two annotation criteria: *a* = Comparison of experimental mass spectra with the NIST14 database and *b* = experimental RI deviations below 30 units from the theoretical RI. Compounds with * represent VOCs reported previously for other Fungi in the mVOC 2.0 Database. In the table, only the isomers annotated and absent in the medium control analysis were reported (i.e., released by *M. furfur*).

### 2.2. VOCs Profile Differentiation in Four Growth Media in the Exponential and Stationary Phase

Unpublished data by Celis et al. (2017) showed that *M. furfur* CBS 1878 is capable of growing in modified Dixon (mDixon) and OA media, whereas it can only survive in OA+PA and PA media (Figure S1), as can be verified by the absence of growth (exponential phase) in these last two media [21]. When the growth medium is supplemented with a mixture of PA and OA, the toxic effect of PA is, at least partly, relieved and *M. furfur* is able to grow (Figure S1). In this research, using a univariate (UVA) and a multivariate (MVA) statistical analysis, we proved that the growth media and phases produced differences in the volatile profiles released by *M. furfur*.

We demonstrated that, for 17 VOCs, peak areas significantly change when the medium-growth phase is modified (ANOVA, 2.37 < F < 560.26, $P < 0.05$) (Kruskal–Wallis test, 14.34 < K < 34.60, $P < 0.05$) (Table S1). We reported VOCs with $P < 0.05$ in a parametric and a non-parametric test, because more than half of the compounds (34 of 61) did not have a normal distribution (Shapiro–Wilk test, 0.59 < W < 0.91, $P < 0.05$). Figure 2 summarizes the semi-quantitative differences in the total peak areas between each treatment for these 17 VOCs. The peak areas of carbon dioxide released by *M. furfur* growing in mDixon were higher than in the other treatments, evidencing an increased respiration in this medium due to an increase in growth (Figure S1). Octane and nonane were released in the mDixon and OA+PA media with semi-quantitative differences in their areas. Hexan-1-ol, butan-1-ol, and the Isomer3 of methyldodecane were released in all media, except in PA. In contrast, the rest of the VOCs were released by *M. furfur* almost exclusively in the mDixon or in the OA+PA media (Figure 2). Differences among exponential and stationary phases were also observed, being some compounds increased in the stationary phase, such as dimethyl sulphide, and others decreased, such as octane. The presence of dimethyl sulphide, and other volatile organic sulphur compounds (VOSC), has been previously reported in bacteria [49], ascomycota yeasts [50,51], and basidiomycota yeast [52]; however, to the best of our knowledge, this is the first report of this compound in the genus *Malassezia*. Another commensal microfungi, *Candida albicans*, shared the presence of dimethyl sulphide, 2-methylpropan-1-ol, pentan-2-one, and 3-methylbutan-1-ol with *M. furfur* [53]; however, the ecological functions of these VOCs for both remain unknown. The presence/absence of some specific VOCs in *M. furfur* highlights the importance of analyzing the different volatiles profiles of a single species in different environments, especially when the objective of the research is to track biomarker compounds for chemotaxonomic, biotechnological [31], or therapeutic purposes [46]. Chemical differentiation in the VOCs released should be studied whenever possible. Previous reports using UVA showed that the emission of some VOCs in other fungi, such as *Ceratocystis sp.*, *Lentinus lepideus*, *Aspergillus versicolor*, *Aspergillus fumigatus*, *Penicillium commune*, *Cladosporium cladosporioides*, *Paecilomyces variotii*, *Phialophora fastigiata Trichoderma spp.*, and *Kluyveromyces marxianus*, depends highly on the differences in the nutritional supplement of the growth medium [25,42,43,54–56].

Using Hierarchical Cluster Analysis (HCA), we discovered that the five replicates from *M. furfur* growing in the mDixon and OA+PA in the exponential and stationary phase have a signature profile that is easily differentiated from the profiles of other treatments; however, the grouping is not related to a specific chemical class of compounds (Figure 3). This aggrupation validates the reproducibility of the volatiles profiles, although some VOCs showed an evident high variance of transformed and scaled peak areas in the heatmap. In addition, the topology demonstrates that a dynamic transformation of VOCs production occurs when the yeast moves from the exponential to the stationary phase, even for the OA+PA medium in which growth remains static (Figure 3 and Figure S1). In the PA medium, higher semi-quantitative similarities between samples of different growth phases were found, which explain the lack of clustering of the five replicates of the exponential and stationary phase. Nevertheless, volatiles profiles from the PA medium are clustered together, demonstrating the differentiation among the VOCs. In contrast, the OA medium does not show a distinctive volatiles profile because of the higher variability between the VOCs found in each replicate in both growth phases. Some of these replicates only had a couple of VOCs with low-intensity peak areas and were clustered with the mDixon. The other group of replicates were clustered as the outgroup of the Dixon, OA, and OA+PA group, having more VOCs with higher peak areas. VOCs from the chemical groups detected in *M. furfur* can be related to main microbial biochemical pathways. Therefore, it can be suggested that furanic compounds, such as lactones, methylketones, esters, alkenes, and maybe alcohols, could be obtained through the catabolism of fatty acids; the alcohols also could be synthesized from pyruvate pathways intermediaries, alkanes probably could be obtained from poliketides generated in the acetate pathway, and sulphur and nitrogen compounds might be produced from amino acids [57,58].

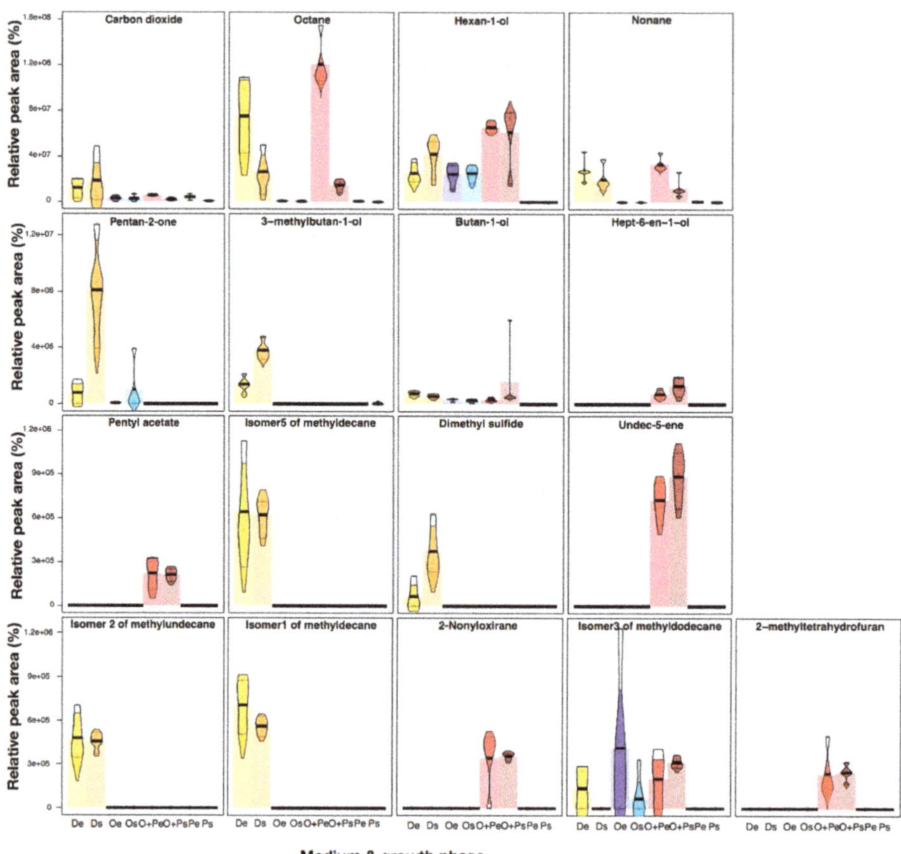

**Figure 2.** Pirate-plots representing univariate statistical differences in the absolute peak areas of 17 VOCs in which the grouping variable Medium-Phase significantly affects the release ($P < 0.05$ in ANOVA and a Kruskal—Wallis test). Variation of data is represented by Bayesian High-Density Intervals after running 10,000 iterations. The bars show the median values for each treatment. The first line of plots is y-scaled from 0 to $16 \times 10^7$; in the following line, the y-axis was scaled from 0 to $1.2 \times 10^7$, while the other two lines are scaled from 0 to $1.2 \times 10^6$.

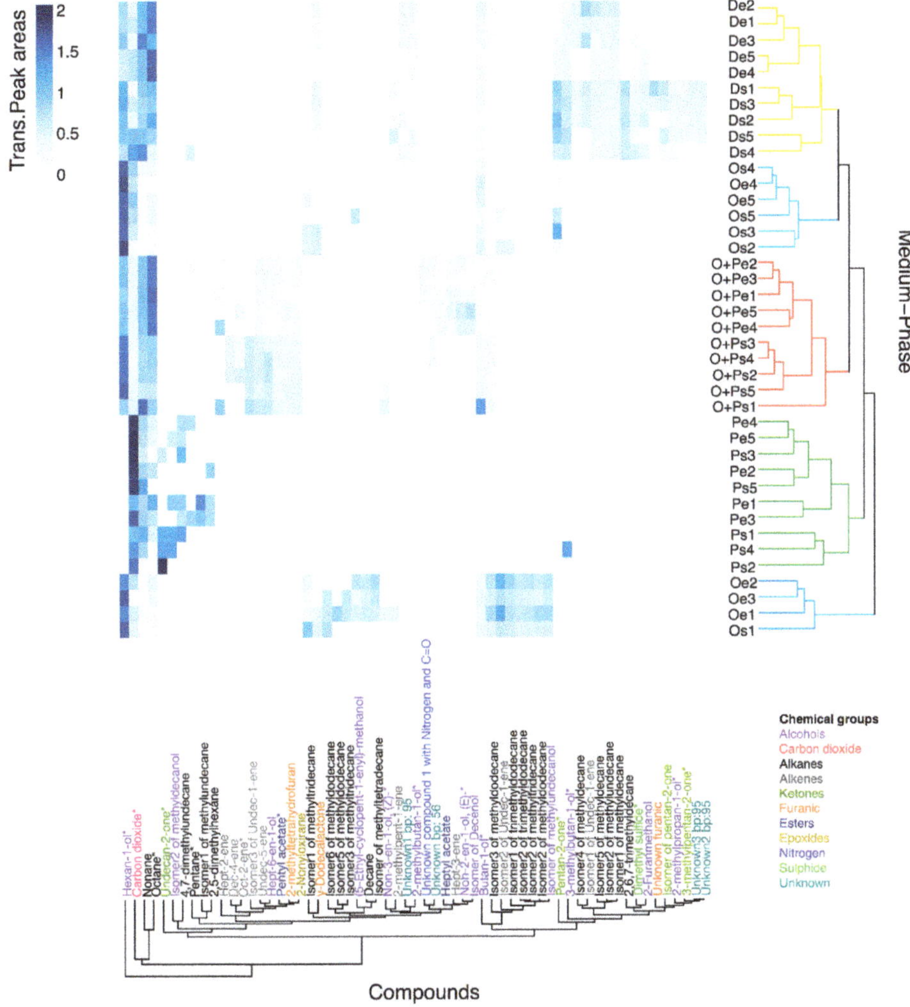

**Figure 3.** A heatmap of Hellinger transformed and Pareto-scaled peak areas from 61 tentative annotated volatiles found in an *M. furfur* analysis using HS-SPME/GC-MS. Dendrograms represent similarities between samples (objects) and VOCs (variables) using Euclidean distances as a measure. Clusters of medium-phase samples were coloured according to the four growth media, while the compounds were coloured according to the chemical group. Compounds with * represent VOCs reported previously to be released by other Fungi in the mVOC 2.0 Database.

A second multivariate non-supervised approximation was performed using Principal Component Analysis (PCA) to summarize the total variation of the volatiles profiles of *M. furfur* in each medium and growth phase, reducing the redundancy between correlated VOCs (Figures 4 and S2). The first three principal components (PCs) represented 56.35% of the explained variance from the 61 volatiles annotated in the original matrix. In comparison to the analysis obtained with the HCA, all media and medium-phases were differentiated in a three-dimensional space, including the samples obtained from the OA medium, which were grouped together in the PCA space, despite having the higher variation represented by bigger ellipses. In order to determine whether the medium or the growth phase or both were variables affecting the volatiles profiles, all three were considered as grouping variables in the PCA analysis. Taking into consideration the total multivariate volatiles profile, the differentiation is better explained through the interaction medium-phase (PERMANOVA, R = 0.6945, F = 10.393, $P$ < 0.001). Furthermore, when considered separately as independent variables, medium (PERMANOVA, R = 0.5457, F = 14.415, $P$ < 0.001) and phase (PERMANOVA, R = 0.0516, F = 2.068, $P$ = 0.043) were both also statistically supported as volatiles profile differentiation factors. PC1, which discriminates samples from both mDixon treatments, is associated with an increment in dimethyl sulphide, 2-methylpropan-1-ol, isomer1 of methyldecane, isomer5 of methyldecane, and the isomer2 of methylundecane. PC2, which discriminates samples from the OA+PA medium in the lower values, is associated with an increment in the isomer2 of undec-1-ene, the isomer2 of methyldodecane, the isomer2 of trimethyldodecane, the isomer of methylundecanol, and the isomer2 of methyltridecane. PC3, which discriminates samples from PA medium in the lower values, is associated with an increment in 2-methyltetrahydrofuran, hexan-1-ol, pentyl acetate, and undec-5-ene and a decrease in carbon dioxide. Half of the aforementioned compounds (9) also stood out in the UVA (Table S1 and Figure 2), but the PCA allows for the interpretation of the direction of differentiation of each compound according to the loading values on each axis. To the best of our knowledge, many studies have used MVA to evaluate differentiation of the metabolites profiles in bacteria and fungi to discriminate species, even for chemotaxonomic purposes [59–62], but this is the first report where MVA analysis demonstrates fungi chemical differentiation in the VOCs profiles that is caused by changes in both medium or/and phase of growth. There are some examples where the effect of one of these variables on the volatiles profile has been tested in bacteria [63–65]; however, the studies in fungi are less frequent [66].

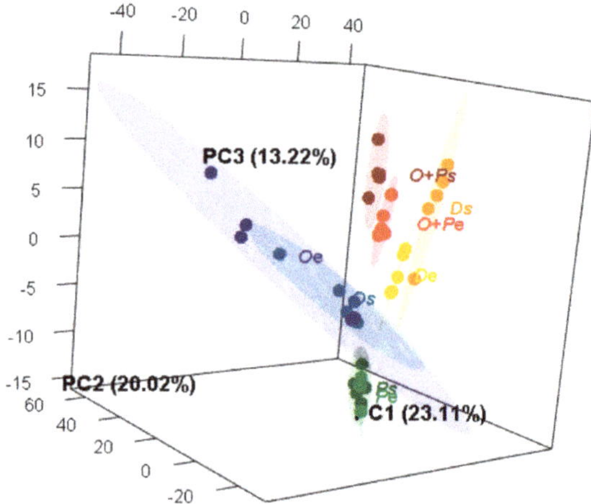



**Figure 4.** Three-dimensional Principal Component Analysis (PCA) of the volatiles profiles of *M. furfur* growing in eight different experimental treatments. The ellipses in this plot are confidence intervals of 95% using a normal distribution. For more details about the data distribution in the three axes, see the three-dimensional animation in Figure S2.

A third multivariate supervised approximation was performed using Projection to Latent Structures Discriminant Analysis (PLS-DA, Figure 5a). The quality of the model was evaluated by the R2X, R2Y, and Q2 metrics. R2X and R2Y support the goodness of the fit, while Q2 > 0.5 reflects the predictability of the model as significant (R2X = 0.686, R2Y = 0.624, Q2 = 0.505). This analysis proves that the volatiles profiles from mDixon and OA+PA are differentiated, as well as the exponential and stationary phases of these two media. To ensure that a model based on a subset of the data can perform equally well on other subsets, we used cross-validation in seven groups and 1000 permutations. Despite the differences in the number of VOCs found between OA and PA media, as discussed before, Figure 5a shows a considerable overlap of these two media in both growth phases. This evidence suggests that some metabolic changes produced in *M. furfur* in both restricted dietary conditions could be similar, in comparison with the metabolic performance in OA+PA or mDixon. Our findings using this supervised technique demonstrate that the absence of both unsaturated oleic acid and saturated palmitic acid produces major changes in the fungal VOCs. The same was observed for the VOCs produced by *Saccharomyces cerevisiae* during the fermentation of different beverages, when linoleic acid is added at different concentrations [67]. The changes observed with *S. cerevisiae* deeply affect food quality in biotechnological and industrial processes, whereas, for *M. furfur*, these modifications help define the interactions between the yeast and other micro-organisms and the host. Surprisingly, the VOCs profile of the OA+PA medium is different from the sum of the ones obtained by the OA and PA media individually (Table 1, Figures 4 and 5a). This has been previously observed in *Trichoderma spp.*, where the volatiles profile found by combining three different amino acids in the growth medium was different from the sum of the profiles obtained when each amino acid was supplied individually [42].

The variable importance on projection (VIP) was used to assess the significance of the variables in the differentiation between treatments. Usually, values higher than 1 are considered significant. Twenty-seven VOCs from *M. furfur* followed this requirement (Figure 5b). Compounds found exclusively in one treatment or medium show the higher VIP values. For example, the hept-3-ene, which was found in the OPe treatment with a value of 1.58, is the compound that better differentiates the four media and the two growth phases, followed by isomer3 of undec-1-ene (VIP = 1.49), a signature compound from *M. furfur* growing in OPs. Among these 27 compounds, 12 were also signalled as statistically significant using UVA (Table S1 and Figure 2), findings that validate the interpretation of these two types of approximations. Nevertheless, the VIP values from PLS-DA offer more information because they discriminate the experimental treatments by organizing the VOCs in order of importance. On the other hand, this is not possible for the UVA, where each VOC is analyzed separately and not correlated to other compounds of the profile. Nine of the 27 VIP compounds had the higher loadings in the PCA analysis and were also highlighted as relevant in the UVA (carbon dioxide, hexan-1-ol, pentyl acetate, the isomer5 of methyldecane, dimethyl sulphide, undec-5-ene, isomer2 of methylundecane, isomer1 of methyldecane, and 2-methyltetrahydrofuran), whereas the PCA and PLS-DA analysis shared 10 out of these 27 VIP compounds. In addition to the abovementioned nine compounds, 2-methylpropan-1-ol was an important differentiation factor in both analyses. These findings evidence the differences of the multivariate models built under supervised and unsupervised conditions and the limitations of UVA. While the construction of the PCA model does not maximize distances between treatments, the PLS-DA constructs the multivariate components as a function of this response variable (in this case, the medium and phase of growth) [68]. This could explain why the VOCs with the higher VIP (for example, hept-3-ene), found exclusively in OPe, were not recognized as significant in the UVA or the PCA.

The ecological meaning of the differentiation produced by the absence of PA and OA on the volatiles profiles of *M. furfur* need to be further explored. PA is particularly abundant in human sebum, in comparison to other animals [21], so spatial differences in the PA distribution on skin could lead to spatial changes in the VOCs released by *M. furfur* or other species from human microbiota [10]. Therefore, a homeostasis needs to be established between human skin and microbiota through the production of metabolites. An enzyme of *Malassezia* that is able to attenuate the formation of biofilms of *Staphylococcus aureus*, an opportunistic bacterium from human skin microbiota [12], is already known; however, volatile metabolites released by these micro-organisms could have similar effects. VOCs functioning as quorum-sensing molecules [36,44] or antibiotics [58,69] have been reported in other bacteria and fungi; however, the ecological function of the VOCs released by *M. furfur* should be studied in the future. These studies show how any alteration in the human skin could be responsible for a disequilibrium in the homeostasis of the total microbiota and may be related to the interconversion of commensal to pathogenic states in *M. furfur*, as well as in other opportunistic species.

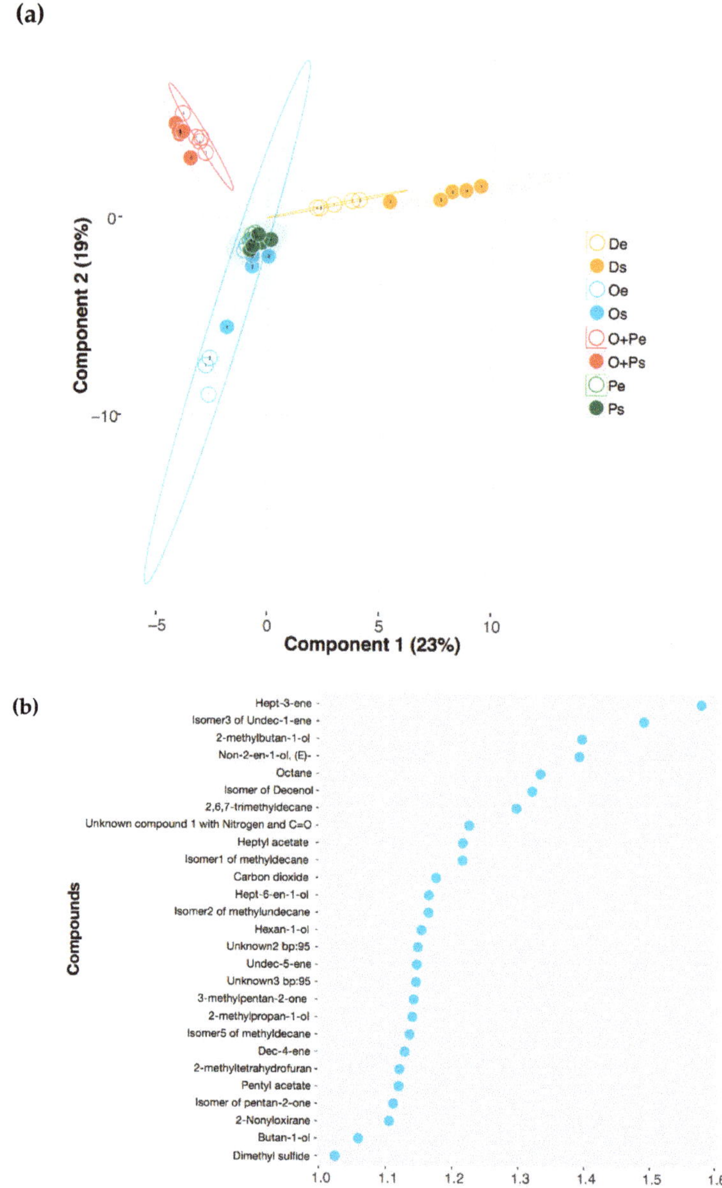

**Figure 5.** (**a**) Projection to Latent Structures Discriminant Analysis (PLS-DA) of volatiles profiles of *M. furfur* growing in eight different experimental treatments. The ellipses in this plot are confidence intervals of 95% using a *t*-distribution. (**b**) Variable importance on projection (VIP) plot showing the most important VOCs from *M. furfur* differentiated between four media and two growth phases. The VIP score of each compound was calculated as a weighted sum of the squared correlations between the PLS-DA components and the original variables.

## 3. Materials and Methods

### 3.1. Chemicals

Chemicals were purchased from Sigma (St Louis, MO, USA) unless otherwise indicated.

### 3.2. Medium Preparation

Modified Dixon broth (mDixon), OA medium, OA+PA medium, and PA medium were prepared at the "Grupo de Investigación Celular de Microorganismos Patógenos (CeMoP)" at Universidad de los Andes (Bogotá, Colombia) one day previous to carrying out the experiments. The prepared mDixon contained 36 g/L malt extract (Oxoid), 10 g/L peptone (BD), 20 g/L Ox bile, 2 mL/L glycerol, 2 mL/L oleic acid, and 10 mL/L Tween 40 (Table S2). The OA, OA+PA, and PA media were prepared using a minimal medium (MM) that contained 10 mL K-buffer pH 7.0 (200 g/L of $K_2HPO_4$, 145 g/L of $KH_2PO_4$), 20 mL of M-N (30 g/L of $MgSO_4 \cdot 7H_2O$, 15 g/L of NaCl), 1 mL 1% ($w/v$) of $CaCl_2 \cdot 2H_2O$, 10 mL 20% ($w/v$) of glucose, 10 mL of 0.01% ($w/v$) $FeSO_4$, 5 mL of spore elements (100 mg/L of $ZnSO_4 \cdot 7H_2O$, 100 mg/L of $CuSO_4 \cdot 5H_2O$, 100 mg/L of $H_3BO_3$, 100 mg/L of $MnSO_4 \cdot H_2O$, 100 mg/L of $Na_2MoO_4 \cdot 2H_2O$), and 2.5 mL 20% ($w/v$) of $NH_4NO_3$. A mixture of either 4.2 mM of palmitic acid (Merck, Darmstadt, Germany) or 4.2 mM of oleic acid (Carlo Erba), or 4.2 mM oleic acid and palmitic acid 50:50 ($v/v$), was added to the MM, respectively. Mixtures containing palmitic acid were supplemented with 1% Brij-58, an emulsifier that is not metabolized [20] (Table S3).

### 3.3. Yeast Growth

The strain CBS 1878 of *M. furfur* was purchased from the Fungal Biodiversity Center (Westerdijk institute, Utrecht, The Netherlands). Two loops of cells from 7-day-old mDixon agar grown colonies were suspended in 5 mL sterile water. Of these, 3 mL were added to 27 mL of mDixon broth or to 27 mL of MM with the respective fatty acid. After 3 days of growth at 180 rpm on a horizontal shaker (Heidolph, Germany), 0.3 mL culture were used to inoculate 29.7 mL of fresh mDixon broth or MM in 150 mL crimp-top Erlenmeyers with 20 mm PTFE septa (JG Finneran, Vineland, NJ, USA). The samples were incubated at 33 °C at 180 rpm. The reference time for the exponential phase was 1.17 days ± 0.17 and for the stationary phase was 3.08 ± 0.19 days. These times were previously validated as the mid-point of exponential phase growth and the initial stationary phase, respectively, in the mDixon (Figure S1) [21].

### 3.4. HS-SPME/GC-MS Analysis

Initially, VOCs released by each growth media alone were analyzed in three experimental replicates using the same 150 mL crimp-top Erlenmeyers, PTFE septa, under the same shaking and temperature conditions described for the growth of the yeast. These samples were used as a medium control analysis, with the aim of differentiating between the VOCs from each broth and those released by *M. furfur* CBS 1878 cultivated in such medium. Afterwards, VOCs released from *M. furfur* in both exponential and stationary phases and in the four liquid media: mDixon and MM supplemented with oleic acid (OA), or oleic and palmitic acid (OA+PA), or palmitic acid (PA), were analyzed in five replicates for each experimental treatment (De, Ds, Oe, Os, OPe, OPs, Pe, Ps).

In order to extract the widest range of polarities of VOCs, a SPME fiber of Divinylbenzene/Carboxen/Poly(dimethylsiloxane) (DVB/CAR/PDMS, 50 µm /30 µm, gray) (SUPELCO, PA, USA) was used. The fiber was exposed for 20 min to the headspace of the Erlenmeyer with growth media and yeast, maintained at 33 °C, and the agitation diminished to 90 rpm during the sampling, so that the state of the fiber was not compromised. Subsequently, the desorption process was carried out in the GC HP 6890 Series equipped with an Agilent Mass Selective Detector 5973 (Agilent technologies, Palo Alto, CA, USA) at 250 °C using splitless injection. Separation was performed on a BP-5 capillary GC column (30 m × 0.25 mm × 0.25 µm, SGE, Austin, TX, USA) using helium as a carrier gas at a flow rate of 1.3 mL/min. The temperature gradient program started at 40 °C for

0.5 min, followed by an increase to 60 °C at a rate of 6 °C/min, then the temperature was raised to 150 °C at 3 °C/min, and finally to 250 °C at 10 °C/min, and this temperature was maintained for 6 min. The GC-MS filament source and the quadrupole temperature were set at 230 and 150 °C, respectively. The electron ionization (EI) source was set at 70 eV, and the mass spectrometer was operated in full scan mode applying a mass range from $m/z$ 30 to 300 at a scan rate of 2.0 scan/s. All samples, including linear alkanes, were run under the same chromatographic conditions. Linear alkanes of the series $C_8$–$C_{20}$ were used for the determination of retention indexes (RI) and later for the tentative assignment of compounds.

### 3.5. Data Analysis

To conduct the analysis of the GC-MS data, the profiles with VOCs obtained in five biological replicates for each medium-phase treatment were analyzed with the MSD ChemStation D.02.00.275 (Agilent technologies), and automatic integration using a threshold of 14 units was performed. Only those peaks absent in any of the three replicates of each medium control were considered to be volatiles released by *M. furfur*. Tentative annotation of compounds was done using NIST MS search 2.0 with the NIST 14 database and comparing experimental RI with the RI of compounds reported in the literature from the same (or an equivalent) column. Exported data files of chromatograms in .csv format were used for the subsequent plotting of chromatograms. Automatic integration results were carefully reviewed, and peak areas were used to construct the matrix in which tentatively annotated volatile compounds are reported as columns/variables and estimated peak areas as rows/observations. As grouping variables, we included the four media (mDixon, OA, OA+PA, and PA), both growth phases (exponential and stationary), and the combination medium-phase of the eight experimental treatments (De, Ds, Oe, Os, OPe, OPs, Pe, and Ps). Each peak area was transformed using a Hellinger transformation adequate for matrices with many zeros [70], and afterwards Pareto scaling was applied, which is useful for chromatographic analysis on a GC-MS platform [71].

In order to investigate the hypothesis that supervised techniques could afford more informative models than unsupervised analysis for the differential data, and thus unravel variables that influence class separation, each data set was analyzed by three techniques: Hierarchical Cluster Analysis (HCA), Principal Component Analysis (PCA), and Projection to Latent Structures Discriminant Analysis (PLS-DA). The data were scaled to have unit variance and were zero-centred. HCA is a statistical tool to pool samples based on similarities by measuring distances between all possible pairs of samples and plotting a topology for objects (e.g., samples for each treatment) and variables (e.g., VOCs). PCA was performed to reduce the redundancy between correlated VOCs. To support this analysis, PERMANOVA with 999 permutations was used using Euclidean distances in order to determine which grouping variable (medium, growth phase, or both) better explained the volatiles profiles differentiation. The PLS-DA model intended to find components or latent variables that correlate variation from a VOCs matrix to a response variable Y. The eight experimental medium-phase treatments (De, Ds, Oe, Os, OPe, OPs, Pe, Ps) were used as the Y response variable of the model. PLS-DA was cross-validated using seven groups (Q2 >0.5 was considered significant [72]) and performing 1000 random permutations of the VOCs dataset to evaluate the Q2 value. All statistical tests were performed using R studio software (http://www.rstudio.org/) and the stats packages HybridMTest (http://bioconductor.org/packages/HybridMTest/), vegan (https://CRAN.R-project.org/package=vegan), MetabolAnalyze (https://CRAN.R-project.org/package=MetabolAnalyze), psych (http://CRAN.R-project.org/package=psych), ropls (http://bioconductor.org/packages/ropls/) and ggplot2 (https://CRAN.R-project.org/package=ggplot2).

## 4. Conclusions

The combination of UVA and MVA results provided strong evidence to support the view that changes in the lipid composition of the growing media in *M. furfur* CBS 1878 produce qualitative and semi-quantitative differences in their volatiles profiles. Additionally, the growth phase also affects the type and intensity of the released VOCs, despite the yeast growth not being equal in the tested media. In the UVA and MVA, several VOCs were differentiated as meaningful factors, including carbon dioxide, hexan-1-ol, pentyl acetate, isomer5 of methyldecane, dimethyl sulphide, undec-5-ene, isomer2 of methylundecane, isomer1 of methyldecane, and 2-methyltetrahydrofuran. The differences observed using in vitro conditions in a controlled environment suggest how the presence of saturated (as PA) or unsaturated (as OA) fatty acids in the skin changes the VOCs profile in one species of the most important fungi genus from human skin microbiota. This species is particularly relevant because it is the only one known to be able to survive in PA, the most abundant saturated fatty acid in human sebum. Some VOCs are produced only when OA, PA, or a combination of both is present, but the complexity of the physiological and biochemical changes produced by the stressful nutritional conditions in these three restricted media should be studied further. The mDixon medium, with a more complex lipid composition, promotes the most diverse chemical profile, whereas the PA medium restricts the chemiodiversity and abundance of the VOCs produced. Our results demonstrate how variable are the volatile profiles of micro-organisms grown in different media. This study's findings are especially important for chemotaxonomic purposes or for studying the ecological function of scents. In the current research, the variation on media allowed us to describe VOCs not previously reported for *Malassezia* or even for fungi. The ecological significance of our findings deserves future in vivo research to analyze the variability of lipid composition from skins of different humans, and how these variations induce changes in the volatiles profiles released by *M. furfur*. These modifications on human bouquets will determine many ecological interactions where VOCs play an important role, such as microbe–microbe, microbe–host, and microbe–insect interactions. Another important direction of research is the comparison of the VOCs profiles when *M. furfur* behaves as a commensal versus pathogenic, with the aim of better understanding the virulence of this micro-organism or the beneficial effect and how it changes according to all of the dynamic complex transformations of the microbiota.

**Supplementary Materials:** The following are available online, Figure S1: Growth curves of *M. furfur* in the mDixon, oleic acid (OA), palmitic acid (PA), and OA+PA medium and their confidence intervals of 99% using a loess method. Figure S2: Animation of rotational distribution of data in the PCA analysis of the volatiles profiles of *M. furfur* growing in eight different experimental treatments. Table S1: Significant statistical univariate parameters found on 17 VOCs from *M. furfur* and compounds highlighted by PCA and PLS-DA. Table S2: Chemical composition of mDixon medium. Table S3: Chemical composition of oleic acid (OA), palmitic acid (PA), and OA+PA media. The components of the minimal medium (MM) are listed in Materials and Methods.

**Author Contributions:** Methodology, M.G., M.I.G.-S., A.M.C. and C.C.; Software, Formal Analysis, and Writing (Original Draft Preparation), M.G.; Validation and Investigation, M.G. and A.M.C.; Data Curation, A.M.C. and C.C.; Writing (Review and Editing), all authors; Visualization, M.G.; Conceptualization, Resources, Supervision, Project Administration, and Funding Acquisition, A.M.C., J.M. and C.C.

**Funding:** This research was funded by the FAPA project of Chiara Carazzone from the Faculty of Science at Universidad de los Andes, by the grants No. 120465741393 and 120456934423 from Departamento Administrativo de Ciencia, Tecnología e Innovación (COLCIENCIAS), by the Netherlands fellowship program NFP-phd.14/99, and by COLCIENCIAS, announcement N. 757 Doctorados Nacionales.

**Acknowledgments:** We acknowledge Diana Marcela Tabares for her contributions during the preliminary analysis of VOCs.



## References

1. Shifrine, M.; Marr, A.G. The Requirement of fatty acids by Pityrosporum ovale. *J. Gen. Microbiol.* **1963**, *32*, 263–270. [CrossRef] [PubMed]
2. Xu, J.; Saunders, C.W.; Hu, P.; Grant, R.A.; Boekhout, T.; Kuramae, E.E.; Kronstad, J.W.; DeAngelis, Y.M.; Reeder, N.L.; Johnstone, K.R.; et al. Dandruff-associated Malassezia genomes reveal convergent and divergent virulence traits shared with plant and human fungal pathogens. *Proc. Natl. Acad. Sci. USA* **2007**, *104*, 18730–18735. [CrossRef] [PubMed]

3. Wu, G.; Zhao, H.; Li, C.; Rajapakse, M.P.; Wong, W.C.; Xu, J.; Saunders, C.W.; Reeder, N.L.; Reilman, R.A.; Scheynius, A.; et al. Genus-Wide comparative genomics of Malassezia delineates its phylogeny, physiology, and niche adaptation on human skin. *PLoS Genet.* **2015**, *11*, 1–26. [CrossRef] [PubMed]

4. Lorch, J.M.; Palmer, J.M.; Vanderwolf, K.J.; Schmidt, K.Z.; Verant, M.L.; Weller, T.J.; Blehert, D.S. Malassezia vespertilionis sp. nov.: A new cold-tolerant species of yeast isolated from bats. *Persoonia—Mol. Phylogeny Evol. Fungi* **2018**, *41*, 56–70. [CrossRef]

5. Gao, Z.; Perez-Perez, G.I.; Chen, Y.; Blaser, M.J. Quantitation of major human cutaneous bacterial and fungal populations. *J. Clin. Microbiol.* **2010**, *48*, 3575–3581. [CrossRef] [PubMed]

6. Ashbee, H.R.; Evans, E.G.V. Immunology of diseases associated with Malassezia species. *Clin. Microbiol. Rev.* **2002**, *15*, 21–57. [CrossRef] [PubMed]

7. Mayser, P.A.; Lang, S.K. Pathogenicity of Malassezia yeasts. In *The Mycota Human and Animal Relationships*; Brakhage, A., Zipfel, P., Eds.; Springer: Berlin/Heidelberg, Germany, 2008; pp. 115–141, ISBN 978-3-540-87406-5.

8. Boekhout, T.; Mayser, P.; Guého-Kellermann, E.; Velegraki, A. *Malassezia and the Skin. Science and Clinical Practice*; Boekhout, T., Mayser, P., Guého-Kellermann, E., Velegraki, A., Eds.; Springer: Berlin/Heidelberg, Germany, 2010; ISBN 978-3-642-03615-6.

9. Kesavan, S.; Holland, K.T.; Ingham, E. The effects of lipid extraction on the immunomodulatory activity of Malassezia species in vitro. *Med. Mycol.* **2000**, *38*, 239–247. [CrossRef]

10. DeAngelis, Y.M.; Gemmer, C.M.; Kaczvinsky, J.R.; Kenneally, D.C.; Schwartz, J.R.; Dawson, T.L. Three etiologic facets of dandruff and seborrheic dermatitis: Malassezia fungi, sebaceous lipids, and individual sensitivity. *J. Investig. Dermatol. Symp. Proc.* **2005**, *10*, 295–297. [CrossRef]

11. Bjerre, R.D.; Bandier, J.; Skov, L.; Engstrand, L.; Johansen, J.D. The role of the skin microbiome in atopic dermatitis: A systematic review. *Br. J. Dermatol.* **2017**, *177*, 1272–1278. [CrossRef]

12. Li, H.; Goh, B.N.; Teh, W.K.; Jiang, Z.; Goh, J.P.Z.; Goh, A.; Wu, G.; Hoon, S.S.; Raida, M.; Camattari, A.; et al. Skin commensal Malassezia globosa secreted protease attenuates Staphylococcus aureus biofilm formation. *J. Investig. Dermatol.* **2018**, *138*, 1137–1145. [CrossRef]

13. Cogen, A.L.; Nizet, V.; Gallo, R.L. Skin microbiota: A source of disease or defence? *Br. J. Dermatol.* **2008**, *158*, 442–455. [CrossRef] [PubMed]

14. Bouslimani, A.; Porto, C.; Rath, C.M.; Wang, M.; Guo, Y.; Gonzalez, A.; Berg-Lyon, D.; Ackermann, G.; Moeller Christensen, G.J.; Nakatsuji, T.; et al. Molecular cartography of the human skin surface in 3D. *Proc. Natl. Acad. Sci. USA* **2015**, *112*, E2120–E2129. [CrossRef] [PubMed]

15. Rosenthal, M.; Goldberg, D.; Aiello, A.; Larson, E.; Foxman, B. Skin microbiota: Microbial community structure and its potential association with health and disease. *Infect. Genet. Evol.* **2011**, *11*, 839–848. [CrossRef] [PubMed]

16. Dréno, B.; Pécastaings, S.; Corvec, S.; Veraldi, S.; Khammari, A.; Roques, C. Cutibacterium acnes (Propionibacterium acnes) and acne vulgaris: A brief look at the latest updates. *J. Eur. Acad. Dermatol. Venereol.* **2018**, *32*, 5–14. [CrossRef] [PubMed]

17. Findley, K.; Oh, J.; Yang, J.; Conlan, S.; Deming, C.; Meyer, J.A.; Schoenfeld, D.; Nomicos, E.; Park, M.; Becker, J.; et al. Topographic diversity of fungal and bacterial communities in human skin. *Nature* **2013**, *498*, 367–370. [CrossRef] [PubMed]

18. Mayser, P.; Haze, P.; Papavassilis, C.; Pickel, M.; Gruender, K.; Gueho, E. Differentiation of Malassezia species: Selectivity of Cremophor EL, castor oil and ricinoleic acid for M. furfur. *Br. J. Dermatol.* **1997**, *137*, 208–213. [CrossRef]

19. Wilde, P.F.; Stewart, P.S. A study of the fatty acid metabolism of the yeast Pityrosporum ovale. *Biochem. J.* **1968**, *108*, 225–231. [CrossRef]

20. Triana, S.; de Cock, H.; Ohm, R.A.; Danies, G.; Wösten, H.A.B.; Restrepo, S.; González Barrios, A.F.; Celis, A. Lipid metabolic versatility in Malassezia spp. yeasts studied through metabolic modeling. *Front. Microbiol.* **2017**, *8*, 1–18. [CrossRef]

21. Kaneko, T.; Makimura, K.; Onozaki, M.; Ueda, K.; Yamada, Y.; Nishiyama, Y.; Yamaguchi, H. Vital growth factors of Malassezia species on modified CHROMagar Candida. *Med. Mycol.* **2005**, *43*, 699–704. [CrossRef]

22. Celis, A.M. Unraveling Lipid Metabolism in Lipid-Dependent Pathogenic Malassezia Yeasts. Ph.D. Thesis, Utrecht University, Utrecht, The Netherlands, 2017.

23. Scotter, J.M.; Langford, V.S.; Wilson, P.F.; McEwan, M.J.; Chambers, S.T. Real-time detection of common microbial volatile organic compounds from medically important fungi by Selected Ion Flow Tube-Mass Spectrometry (SIFT-MS). *J. Microbiol. Methods* **2005**, *63*, 127–134. [CrossRef]
24. Bazemore, R.A.; Feng, J.; Cseke, L.; Podila, G.K. Biomedically important pathogenic fungi detection with volatile biomarkers. *J. Breath Res.* **2012**, *6*. [CrossRef] [PubMed]
25. Heddergott, C.; Latgé, J.P.; Calvo, A.M. The volatome of Aspergillus fumigatus. *Eukaryot. Cell* **2014**, *13*, 1014–1025. [CrossRef] [PubMed]
26. Koo, S.; Thomas, H.R.; Daniels, S.D.; Lynch, R.C.; Fortier, S.M.; Shea, M.M.; Rearden, P.; Comolli, J.C.; Baden, L.R.; Marty, F.M. A breath fungal secondary metabolite signature to diagnose invasive aspergillosis. *Clin. Infect. Dis.* **2014**, *59*, 1733–1740. [CrossRef] [PubMed]
27. Zhang, Q.; Zhou, L.; Chen, H.; Wang, C.-Z.; Xia, Z.; Yuan, C.-S. Solid-phase microextraction technology for in vitro and in vivo metabolite analysis. *Trends Analyt. Chem.* **2016**, *80*, 57–65. [CrossRef] [PubMed]
28. Chung, H.; Lee, N.; Seo, J.A.; Kim, Y.S. Comparative analysis of nonvolatile and volatile metabolites in Lichtheimia ramosa cultivated in different growth media. *Biosci. Biotechnol. Biochem.* **2017**, *81*, 565–572. [CrossRef] [PubMed]
29. Rees, C.A.; Burklund, A.; Stefanuto, P.H.; Schwartzman, J.D.; Hill, J.E. Comprehensive volatile metabolic fingerprinting of bacterial and fungal pathogen groups. *J. Breath Res.* **2018**, *12*. [CrossRef] [PubMed]
30. Morath, S.U.; Hung, R.; Bennett, J.W. Fungal volatile organic compounds: A review with emphasis on their biotechnological potential. *Fungal Biol. Rev.* **2012**, *26*, 73–83. [CrossRef]
31. Hung, R.; Lee, S.; Bennett, J.W. Fungal volatile organic compounds and their role in ecosystems. *Appl. Microbiol. Biotechnol.* **2015**, *99*, 3395–3405. [CrossRef] [PubMed]
32. Tejero Rioseras, A.; Garcia Gomez, D.; Ebert, B.E.; Blank, L.M.; Ibáñez, A.J.; Sinues, P.M.L. Comprehensive real-time analysis of the yeast Volatilome. *Sci. Rep.* **2017**, *7*, 1–9. [CrossRef]
33. Romero-Guido, C.; Belo, I.; Ta, T.M.N.; Cao-Hoang, L.; Alchihab, M.; Gomes, N.; Thonart, P.; Teixeira, J.A.; Destain, J.; Waché, Y. Biochemistry of lactone formation in yeast and fungi and its utilisation for the production of flavour and fragrance compounds. *Appl. Microbiol. Biotechnol.* **2011**, *89*, 535–547. [CrossRef]
34. Braga, A.; Belo, I. Biotechnological production of γ-decalactone, a peach like aroma, by Yarrowia lipolytica. *World J. Microbiol. Biotechnol.* **2016**, *32*, 1–8. [CrossRef] [PubMed]
35. Wheatley, R.E. The consequences of volatile organic compound mediated bacterial and fungal interactions. *Antonie Leeuwenhoek Int. J. Gen. Mol. Microbiol.* **2002**, *81*, 357–364. [CrossRef]
36. Hornby, J.M.; Jensen, E.C.; Lisec, A.D.; Tasto, J.J.; Jahnke, B.; Shoemaker, R.; Dussault, P.; Nickerson, K.W. Quorum sensing in the dimorphic fungus Candida albicans is mediated by farnesol. *Appl. Environ. Microbiol.* **2001**, *67*, 2982–2992. [CrossRef] [PubMed]
37. Bennett, J.W.; Hung, R.; Lee, S.; Padhi, S. Fungal and bacterial Volatile Organic Compounds: An overview and their role as ecological signaling agents. In *Fungal Associations*; Springer: Berlin/Heidelberg, Germany, 2012; Volume 9, pp. 373–393, ISBN 9783642308260.
38. Lemfack, M.C.; Nickel, J.; Dunkel, M.; Preissner, R.; Piechulla, B. mVOC: A database of microbial volatiles. *Nucleic Acids Res.* **2014**, *42*, D744–D748. [CrossRef]
39. Lemfack, M.C.; Gohlke, B.O.; Toguem, S.M.T.; Preissner, S.; Piechulla, B.; Preissner, R. MVOC 2.0: A database of microbial volatiles. *Nucleic Acids Res.* **2018**, *46*, D1261–D1265. [CrossRef]
40. Labows, J.N.; McGinley, K.J.; Leyden, J.J.; Webster, G.F. Characteristic γ-lactone odor production of the genus Pityrosporum. *Appl. Environ. Microbiol.* **1979**, *38*, 412–415.
41. Ro, B.I.; Dawson, T.L. The role of sebaceous gland activity and scalp microfloral metabolism in the etiology of seborrheic dermatitis and dandruff. *J. Investig. Dermatol. Symp. Proc.* **2005**, *10*, 194–197. [CrossRef]
42. Bruce, A.; Wheatley, R.E.; Humphris, S.N.; Hackett, C.A.; Florence, M.E.J. Production of volatile organic compounds by Trichoderma in media containing different amino acids and their effect on selected wood decay fungi. *Holzforschung* **2000**, *54*, 481–486. [CrossRef]
43. Sunesson, A.L.; Vaes, W.H.J.; Nilsson, C.A.; Blomquist, G.; Andersson, B.; Carlson, R. Identification of volatile metabolites from five fungal species cultivated on two media. *Appl. Environ. Microbiol.* **1995**, *61*, 2911–2918.
44. Langford, M.L.; Atkin, A.L.; Nickerson, K.W. Cellular interactions of farnesol, a quorum-sensing molecule produced by Candida albicans. *Future Microbiol.* **2009**, *4*, 1353–1362. [CrossRef]
45. Kramer, R.; Abraham, W.R. Volatile sesquiterpenes from fungi: What are they good for? *Phytochem. Rev.* **2012**, *11*, 15–37. [CrossRef]

46. Thalavitiya Acharige, M.J.; Koshy, S.S.; Koo, S. The use of microbial metabolites for the diagnosis of infectious diseases. In *Advanced Techniques in Diagnostic Microbiology*; Springer International Publishing: Cham, Switzerland, 2018; pp. 261–272, ISBN 9783319339009.

47. Lin, H.C.; Chooi, Y.H.; Dhingra, S.; Xu, W.; Calvo, A.M.; Tang, Y. The fumagillin biosynthetic gene cluster in Aspergillus fumigatus encodes a cryptic terpene cyclase involved in the formation of β-trans-bergamotene. *J. Am. Chem. Soc.* **2013**, *135*, 4616–4619. [CrossRef] [PubMed]

48. Ashbee, H.R. Update on the genus Malassezia. *Med. Mycol.* **2007**, *45*, 287–303. [CrossRef] [PubMed]

49. Weimer, B.; Seefeldt, K.; Dias, B. Sulfur metabolism in bacteria associated with cheese. *Antonie Leeuwenhoek Int. J. Gen. Mol. Microbiol.* **1999**, *76*, 247–261. [CrossRef]

50. Schöller, C.E.G.; Gürtler, H.; Pedersen, R.; Molin, S.; Wilkins, K. Volatile metabolites from Actinomycetes. *J. Agric. Food Chem.* **2002**, *50*, 2615–2621. [CrossRef] [PubMed]

51. Agrawal, R. Flavors and aromas. In *Fungal Biotechnology In Agricultural, Food and Environmental Applications*; Arora, D., Ed.; Marcel Dekker Inc.: New York, NY, USA, 2004; pp. 281–289, ISBN 3527403736.

52. Buzzini, P.; Romano, S.; Turchetti, B.; Vaughan, A.; Pagnoni, U.M.; Davoli, P. Production of volatile organic sulfur compounds (VOSCs) by basidiomycetous yeasts. *FEMS Yeast Res.* **2005**, *5*, 379–385. [CrossRef]

53. Filipiak, W.; Sponring, A.; Filipiak, A.; Baur, M.; Ager, C.; Wiesenhofer, H.; Margesin, R.; Nagl, M.; Troppmair, J.; Amann, A. Volatile Organic Compounds (VOCs) released by pathogenic microorganisms in vitro: Potential breath biomarkers for early-stage diagnosis of disease. In *Volatile Biomarkers*; Elsevier: Amsterdam, The Netherlands, 2013; pp. 463–512, ISBN 9780444626134.

54. Sprecher, E. Influence of strain specificity and culture conditions on terpene production by fungi. *Planta Med.* **1982**, *44*, 41–43. [CrossRef]

55. Bjurman, J.; Kristensson, J. Volatile production by Aspergillus versicolor as a possible cause of odor in houses affected by fungi. *Mycopathologia* **1992**, *118*, 173–178. [CrossRef]

56. Gethins, L.; Guneser, O.; Demirkol, A.; Rea, M.C.; Stanton, C.; Ross, R.P.; Karagul Yuceer, Y.; Morrissey, J.P. Influence of Carbon and Nitrogen source on production of volatile fragrance and flavour metabolites by the yeast Kluyveromyces marxianus. *Yeast* **2015**, *32*, 67–76. [CrossRef]

57. Korpi, A.; Järnberg, J.; Pasanen, A.L. Microbial volatile organic compounds. *Crit. Rev. Toxicol.* **2009**, *39*, 139–193. [CrossRef]

58. Schmidt, R.; Cordovez, V.; De Boer, W.; Raaijmakers, J.; Garbeva, P. Volatile affairs in microbial interactions. *ISME J.* **2015**, *9*, 2329–2335. [CrossRef] [PubMed]

59. Frisvad, J.C.; Andersen, B.; Thrane, U. The use of secondary metabolite profiling in chemotaxonomy of filamentous fungi. *Mycol. Res.* **2008**, *112*, 231–240. [CrossRef] [PubMed]

60. Thorn, R.M.S.; Reynolds, D.M.; Greenman, J. Multivariate analysis of bacterial volatile compound profiles for discrimination between selected species and strains in vitro. *J. Microbiol. Methods* **2011**, *84*, 258–264. [CrossRef] [PubMed]

61. Bos, L.D.J.; Sterk, P.J.; Schultz, M.J. Volatile Metabolites of Pathogens: A Systematic Review. *PLoS Pathog.* **2013**, *9*, 1–8. [CrossRef] [PubMed]

62. Boots, A.W.; Smolinska, A.; Van Berkel, J.J.B.N.; Fijten, R.R.R.; Stobberingh, E.E.; Boumans, M.L.L.; Moonen, E.J.; Wouters, E.F.M.; Dallinga, J.W.; Van Schooten, F.J. Identification of microorganisms based on headspace analysis of volatile organic compounds by gas chromatography-mass spectrometry. *J. Breath Res.* **2014**, *8*. [CrossRef] [PubMed]

63. Rees, C.A.; Smolinska, A.; Hill, J.E. The volatile metabolome of Klebsiella pneumoniae in human blood. *J. Breath Res.* **2016**, *10*, 27101. [CrossRef] [PubMed]

64. Tabares, M.; Ortiz, M.; Gonzalez, M.; Carazzone, C.; Vives Florez, M.J.; Molina, J. Behavioral responses of Rhodnius prolixus to volatile organic compounds released in vitro by bacteria isolated from human facial skin. *PLoS Negl. Trop. Dis.* **2018**, *12*, 1–16. [CrossRef]

65. Kladsomboon, S.; Thippakorn, C.; Seesaard, T. Development of organic-inorganic hybrid optical gas sensors for the non-invasive monitoring of pathogenic bacteria. *Sensors* **2018**, *18*, 3189. [CrossRef]

66. Sun, D.; She, J.; Gower, J.; Stokes, C.; Windham, G.; Baird, R.; Mlsna, T. Effects of growth parameters on the analysis of Aspergillus flavus volatile metabolites. *Separations* **2016**, *3*, 13. [CrossRef]

67. Casu, F.; Pinu, F.R.; Fedrizzi, B.; Greenwood, D.R.; Villas-Boas, S.G. The effect of linoleic acid on the Sauvignon blanc fermentation by different wine yeast strains. *FEMS Yeast Res.* **2016**, *16*, fow050. [CrossRef]

68. Gromski, P.S.; Muhamadali, H.; Ellis, D.I.; Xu, Y.; Correa, E.; Turner, M.L.; Goodacre, R. A tutorial review: Metabolomics and partial least squares-discriminant analysis—A marriage of convenience or a shotgun wedding. *Anal. Chim. Acta* **2015**, *879*, 10–23. [CrossRef]
69. Dennis, C.; Webster, J. Antagonistic properties of species-groups of Trichoderma. II. Production of volatile antibiotics. *Trans. Br. Mycol. Soc.* **1971**, *57*, 41-IN4. [CrossRef]
70. Legendre, P.; Gallagher, E.D. Ecologically meaningful transformations for ordination of species data. *Oecologia* **2001**, *129*, 271–280. [CrossRef] [PubMed]
71. Van den Berg, R.A.; Hoefsloot, H.C.J.; Westerhuis, J.A.; Smilde, A.K.; van der Werf, M.J. Centering, scaling, and transformations: Improving the biological information content of metabolomics data. *BMC Genom.* **2006**, *7*, 142. [CrossRef] [PubMed]
72. Triba, M.N.; Le Moyec, L.; Amathieu, R.; Goossens, C.; Bouchemal, N.; Nahon, P.; Rutledge, D.N.; Savarin, P. PLS/OPLS models in metabolomics: The impact of permutation of dataset rows on the K-fold cross-validation quality parameters. *Mol. Biosyst.* **2015**, *11*, 13–19. [CrossRef] [PubMed]


**Sample Availability:** Samples of *M. furfur* CBS 1878 are currently available at the "Grupo de Investigación Celular de Micoorganismos Patógenos (CeMoP)" at Universidad de los Andes.

 *molecules*

*Article*

# Volatilomic Analysis of Four Edible Flowers from *Agastache* Genus

Basma Najar [1], Ilaria Marchioni [2,3,*], Barbara Ruffoni [3], Andrea Copetta [3], Laura Pistelli [2,4,*] and Luisa Pistelli [1,4]

[1]  Dipartimento di Farmacia, Università di Pisa, Via Bonanno 6, 56126 Pisa, Italy; basmanajar@hotmail.fr (B.N.); luisa.pistelli@unipi.it (L.P.)
[2]  Dipartimento di Scienze Agrarie, Alimentari e Agro-alimentari, Università di Pisa, Via del Borghetto 80, 56124 Pisa, Italy
[3]  CREA Centro di ricerca Orticoltura e Florovivaismo, Corso Inglesi 508, 18038 Sanremo, IM, Italy; barbara.ruffoni@crea.gov.it (B.R.); andrea.copetta@crea.gov.it (A.C.)
[4]  Centro Interdipartimentale di Ricerca "Nutraceutica e Alimentazione per la Salute" (NUTRAFOOD), Università di Pisa, Via del Borghetto 80, 5614 Pisa, Italy
*  Correspondence: i.marchioni@studenti.unipi.it (I.M.); laura.pistelli@unipi.it (L.P.)

Received: 31 October 2019; Accepted: 5 December 2019; Published: 6 December 2019

**Abstract:** Volatilomes emitted from edible flowers of two species of *Agastache* (*A. aurantiaca* (A.Gray) Lint & Epling, and *A. mexicana* (Kunth) Lint & Epling) and from two hybrids (*Agastache* 'Arcado Pink' and *Agastache* 'Blue Boa') were investigated using a solid-phase microextraction technique as well as the extraction of its essential oils. Oxygenated monoterpenes were almost always the predominant class (>85%) of volatile organic compounds (VOCs) in each sample of *A. aurantiaca*, *A.* 'Blue Boa' and *A. mexicana*, with the exception of *A.* 'Arcado Pink' (38.6%). Pulegone was the main compound in *A. aurantiaca* (76.7%) and *A.* 'Blue Boa' (82.4%), while geranyl acetate (37.5%) followed by geraniol (16%) and geranial (17%) were the principal ones in *A. mexicana*. The essential oil composition showed the same behavior as the VOCs both for the main class as well as the major constituent (pulegone) with the same exception for *A. mexicana*. Total soluble sugars, secondary metabolites (polyphenols, flavonoids and anthocyanins) and antioxidant activity were also investigated to emphasize the nutraceutical properties of these edible flowers.

**Keywords:** VOCs; *A.* 'Arcado Pink'; *A. aurantica*; *A.* 'Blue Boa'; *A. mexicana*; essential oil; GC-MS; edible flowers; antioxidant activity; secondary metabolites

---

## 1. Introduction

In the last few decades, the analysis of volatilome has received big attention. In fact, plant volatilome, the aromatic complex of essential oils (EOs) and volatile organic compounds (VOCs), defined as "volatile chemiodiversity" [1], is one of the main traits that enhances our understanding and provides new insights into the physiological processes of plant growth, defense and productivity [2]. Almost all plants are able to emit VOCs, and they are released from leaves, flowers and fruits into the atmosphere, and from roots into the soil [3].

Gas chromatography (GC) in combination with mass spectrometry (GC-MS) is one of the elective techniques for volatilomics [4], the study of the volatilome. Depending on the biological problem as well as the plant material being investigated, analytical procedures for the analysis of volatiles require preparation methods for the extraction and pre-concentration of target compounds [5].

In agreement with the present guidelines aimed on the reduction of the use of organic solvent, both the traditional methods, such hydro-distillation and headspace, were shown to be the main methods used to evaluate volatiles from the plant. Since hydrodistillation was limited to plants producing essential oils, the extraction of VOCs was captured efficiently in the "head space," an enclosed volume of air surrounding the plant [6], and the use of the Solid Phase Microextraction (SPME) was shown to be the most commonly used method [5], which was a very simple and efficient technique [7] as well as an immediate one-site analysis of biogenic VOC [8]. In fact, this technique, based on the absorption-adsorption of the analytes into a coated fiber, has gained popularity in many fields of analytical chemistry, mainly in food and flavor analysis [9–11].

Looking for healthier and functional new foods, people have adopted an interest in edible flowers as creative and innovative ingredients whose popularity and consumption have been increasing worldwide since 1980 [12] due to their powerful and unique tastes, flavors, textures and pigments. The renewed success of edible flowers relies also in their nutritional and healthy properties [13,14]. In fact, several studies have highlighted high quantities of antioxidant molecules [14], primary metabolites [13], minerals [15–18] and vitamins [13,19,20]. Even though there are no official lists emitted by any international organizations (Food and Agriculture Organization of the United Nations (FAO), World Health Organization (WHO), and European Food Safety Authority (EFSA)), Lim listed over 80 species from about 32 families as edible flowers [21]. Two years later, Lu and his co-workers wrote about 97 families, 100 genera and 180 species, and they specified that the number of edible flowers varies in different countries [14]. In the framework of the INTERREG ALCOTRA Project on edible flowers (INTERREG ALCOTRA ANTEA N°1139, 2014–2020), the *Agastache* genus received special interest due to its good flavor, and so analyzing the chemical composition of its volatiles became relevant.

The *Agastache* genus belongs to Nepetoideae, a subfamily of Lamiaceae (Mint family), and includes 22 species known under the popular name 'giant hyssop' [22]. This aromatic genus includes ornamental herbaceous perennial plants, and is almost exclusively native to North America [22,23]. Lot of species is characterized by very pleasant fragrances, similar to anise, mint and licorice [24,25]. Nectar contributes to the sweetness of these flowers and makes them suitable as bee forage [26]. Moreover, some *Agastache* species are medicinal plants, characterized by several biological activities [22].

The evaluation of the flower aroma composition is essential, since it is the second greatest influence on consumer attitudes towards the edible flower's consumption, only preceded by curiosity [27]. In the last three-decades, the investigation on the *Agastache* volatilome was especially focused on its essential oil (EO) [22]. On the contrary, only two works reported the use of the headspace method for VOC analysis in this genus [28,29].

This work deals with the use of the cited technique for the analysis of four species of the *Agastache* genus: *Agastache* 'Blue Boa', *Agastache* 'Arcado Pink', *A. aurantiaca* var. 'Sunset Yellow' and *A. mexicana* var. 'Sangria'. These varieties and hybrids were selected due to their ornamental value, aroma and taste (Table 1). As a frame to the volatilome, some classes of antioxidant molecules were quantified, due to their nutraceutical potentials. Soluble sugars were also determined, considering the significant nectar production and the petal's sweet taste.

**Table 1.** Main botanical information and visual appearance of the examined *Agastache* flowers.

| Plant Name | | Plant Height | Leaves | Flowers | Blossoming Period |
|---|---|---|---|---|---|
| *A.* 'Arcado Pink' | | 60–70 cm | Medium, opposite, lanceolate, serrated, green/gray color | Purplish/pink in compact ears | May–October |

**Table 1.** *Cont.*

| Plant Name | Plant Height | Leaves | Flowers | Blossoming Period |
|---|---|---|---|---|
| *A. aurantiaca* (*A.* Gray) Lint & Epling, var. 'Sunset Yellow' | 35–40 cm | Small, opposite, lanceolate, serrated, green/gray color | Golden yellow in very loose ears | May–November |
| *A.*'Blue Boa' | 60–70 cm | Medium, opposite, lanceolate, serrated, green/gray color | Dark blue/purple in loose ears | June–October |
| *A. mexicana* (Kunth) Lint & Epling | 100–120 cm | Large, opposite, lanceolate, serrated, green/gray color | Purple red in very loose ears | June–November |

## 2. Results and Discussion

The results of the chemical composition of the volatilomes emitted by edible flowers of the four *Agastache* species led to the identification of 67 components (Table 2), representing at least 99% of the total volatiles. Oxygenated monoterpenes were the predominant class in each aroma profile and EO analysis. Their relative percentage ranged between 85.9% in *A. aurantica* and 90.6% in *A.* 'Blue Boa' in VOC analysis, and more than 86% in the distilled oils. An exception was noted for the *A.* 'Arcado Pink' SPME where the composition of the VOCs was split in three classes; two of them evidenced a similar percentage: sesquiterpene hydrocarbons (38.6%) and oxygenated monoterpene (37.9%) with a good amount of monoterpene hydrocarbons (23.2%).

In detail, the aroma profile of both the *A. aurantica* and *A.* 'Blue Boa' flowers showed the same main compound, pulegone, with a percentage of 77.7% and 84.0%, respectively. Pulegone prevailed also in *A.* 'Arcado Pink' (36.5%), followed by a good amount of $\beta$-caryophyllene (20.4%). *A. mexicana* highlighted a different composition, since geranyl acetate was the major constituent (37.5%) followed by both geranial and geraniol with more or less a similar amount (17% and 16%, respectively).

It is worthy to note that the EO composition of this latter species (*A. mexicana*) evidenced the same behavior as already reported for SPME analysis, and confirmed geranyl acetate (61.4%) as the principal compound, whose odor was described as fruity, fatty, sweet, citrus, floral, fresh and lemon-like [30], followed by geranial (11.0%) and geraniol (8.3%).

**Table 2.** Volatilome in four species of *Agastache*: *A.* 'Arcado Pink' (A.A.P.), *A. aurantica* (A.A.), *A.* 'Blue Boa' (A.B.B.) and *A. mexicana* (A.M.).

| | Compounds * | Class | LRI | LRI * | SPME | | | | EO | | | |
|---|---|---|---|---|---|---|---|---|---|---|---|---|
| | | | | | A.A.P. | A.A. | A.B.B. | A.M. | A.A.P. | A.A. | A.B.B. | A.M. |
| | | | | | Relative Percentage | | | | | | | |
| 1 | Furfural | nt | 818 | 836 | - | - | - | - | $0.2 \pm 0.08$ | - | - | - |
| 2 | 1-Octen-3-ol | nt | 979 | 979 | - | - | $0.1 \pm 0.07$ | - | - | $0.1 \pm 0.01$ | $0.2 \pm 0.00$ | - |
| 3 | $\beta$-Pinene | mh | 981 | 979 | $0.1 \pm 0.14$ | - | - | - | - | - | - | - |
| 4 | 3-Octanone | nt | 986 | 984 | - | $0.1 \pm 0.04$ | $0.2 \pm 0.00$ | - | - | - | - | - |
| 5 | Myrcene | mh | 991 | 991 | $3.5 \pm 0.21$ | $0.4 \pm 0.07$ | $0.3 \pm 0.21$ | $3.7 \pm 0.24$ | - | $0.1 \pm 0.01$ | - | - |
| 6 | *p*-Cymene | mh | 1025 | 1025 | - | - | $0.3 \pm 007$ | - | - | - | - | - |
| 7 | Limonene | mh | 1030 | 1029 | $17.1 \pm 1.70$ | $2.4 \pm 0.99$ | $3.6 \pm 0.85$ | $0.8 \pm 0.10$ | - | $0.5 \pm 0.10$ | $0.6 \pm 0.14$ | - |
| 8 | *cis-β*-Ocimene | mh | 1038 | 1037 | $0.6 \pm 0.05$ | - | - | $0.8 \pm 0.15$ | - | - | - | - |
| 9 | *trans-β*-Ocimene | mh | 1047 | 1050 | $1.0 \pm 0.4$ | - | - | $1.3 \pm 0.10$ | - | - | - | - |
| 10 | *p*-Menta-2,4(8)-diene | mh | 1086 | 1088 | $0.3 \pm 0.14$ | - | - | - | - | - | - | - |
| 11 | *p*-Cymenene | mh | 1090 | 1091 | - | - | $0.1 \pm 0.07$ | - | - | - | - | - |
| 12 | Linalool | om | 1099 | 1097 | - | - | - | $1.5 \pm 0.33$ | - | - | - | $0.5 \pm 0.14$ |
| 13 | *trans-p*-Mentha-2,8-dienol | om | 1116 | 1113 $ | - | - | - | - | $2.5 \pm 0.06$ | - | $0.3 \pm 0.07$ | - |
| 14 | 1,3,8-*p*-Menthatriene | mh | 1119 | 1110 | $0.1 \pm 0.04$ | - | $0.1 \pm 0.40$ | - | - | - | - | - |
| 15 | *trans-p*-Metha-2,8-dien-1-ol | om | 1123 | 1123 | - | - | - | - | $3.1 \pm 0.06$ | - | - | - |
| 16 | *p*-Mentha,1,5,8-triene | mh | 1130 | 1135 $ | $0.1 \pm 0.04$ | - | - | - | - | - | - | - |
| 17 | *Neo-allo*-ocimene | mh | 1131 | 1132 | $0.4 \pm 0.09$ | - | - | - | - | - | - | - |
| 18 | *cis-p*-Mentha-2,8-dien-1-ol | om | 1138 | 1138 | - | - | - | - | - | - | $0.2 \pm 0.07$ | - |
| 19 | Citronellal | om | 1153 | 1153 | - | - | - | - | - | - | - | $1.2 \pm 0.14$ |
| 20 | Menthone | om | 1154 | 1153 | - | $4.7 \pm 0.35$ | $0.4 \pm 0.35$ | - | $0.4 \pm 0.07$ | $1.4 \pm 0.10$ | $0.1 \pm 0.00$ | - |
| 21 | *iso*-Menthone | om | 1164 | 1163 | - | - | - | - | - | $0.1 \pm 0.01$ | $1.1 \pm 0.07$ | - |
| 22 | Menthofuran | om | 1165 | 1164 | $0.5 \pm 0.07$ | $1.5 \pm 0.64$ | $2.8 \pm 1.06$ | - | - | - | - | - |
| 23 | Borneol | om | 1169 | 1169 | - | - | - | - | - | - | - | $0.1 \pm 0.00$ |
| 24 | *iso*Pulegone | om | 1177 | 1179 $ | $1.1 \pm 0.42$ | $1.9 \pm 0.07$ | $2.6 \pm 0.92$ | $0.1 \pm 0.03$ | $1.1 \pm 0.0$ | $2.6 \pm 0.20$ | $2.1 \pm 0.21$ | - |
| 25 | $\alpha$-Terpineol | om | 1189 | 1189 | - | - | - | - | $0.7 \pm 0.08$ | $0.1 \pm 0.04$ | - | - |
| 26 | Verbenone | om | 1204 | 1205 | - | - | - | - | - | $0.1 \pm 0.10$ | - | - |
| 27 | 2,6,6-trimethyl,2-Cyclohexen-1-ol | nt | 1205 | 1205 $ | - | - | - | - | $0.4 \pm 0.09$ | - | - | - |
| 28 | *n*-Decanal | nt | 1206 | 1202 | - | - | - | $0.2 \pm 0.09$ | - | - | - | - |
| 29 | 4,7-dimethyl, Benzofuran | pp | 1220 | 1220 $ | - | - | $0.3 \pm 0.05$ | - | - | - | - | - |
| 30 | Citronellol | om | 1228 | 1226 | - | - | - | $6.2 \pm 1.48$ | - | - | - | $3.1 \pm 0.35$ |
| 31 | Pulegone | om | 1237 | 1237 | $36.5 \pm 1,27$ | $77.7 \pm 0.42$ | $84.0 \pm 5.30$ | - | $79.8 \pm 2.15$ | $92.5 \pm 0.80$ | $91.8 \pm 0.71$ | - |
| 32 | $\beta$-Citral (Neral) | om | 1240 | 1238 | - | - | - | $3.8 \pm 0.30$ | - | - | - | $2.9 \pm 0.14$ |
| 33 | 2-methyl-3-phenyl, Propanal | nt | 1244 | 1244 $ | - | - | - | - | - | $0.2 \pm 0.10$ | - | - |
| 34 | Piperitone | om | 1252 | 1253 | - | - | - | - | - | $0.2 \pm 0.11$ | - | - |
| 35 | Geraniol | om | 1255 | 1253 | - | - | - | $16.0 \pm 1.86$ | - | - | - | $8.3 \pm 0.49$ |
| 36 | $\alpha$-Citral (Geranial) | om | 1270 | 1267 | - | - | - | $17.0 \pm 1.56$ | - | - | - | $11.0 \pm 0.07$ |
| 37 | *iso*Bornyl acetate | om | 1286 | 1286 | $0.2 \pm 0.14$ | - | - | $1.2 \pm 0.14$ | - | - | - | $1.1 \pm 0.07$ |
| 38 | Carvyl acetate | om | 1336 | 1335 $ | - | - | $0.7 \pm 0.57$ | - | - | - | - | - |
| 39 | *trans*-Carvyl acetate | om | 1337 | 1342 | $0.3 \pm 0.07$ | - | - | - | - | - | $0.1 \pm 0.00$ | - |
| 40 | Piperitenone | om | 1340 | 1343 | - | $0.1 \pm 0.07$ | $0.1 \pm 0.02$ | - | $1.0 \pm 0.04$ | $0.5 \pm 0.20$ | $0.6 \pm 0.07$ | - |
| 41 | $\alpha$-Cubebene | sh | 1351 | 1351 | $0.1 \pm 0.00$ | - | - | - | - | - | - | - |

**Table 2.** *Cont.*

| | Compounds * | Class | LRI | LRI * | SPME | | | | EO | | | |
|---|---|---|---|---|---|---|---|---|---|---|---|---|
| | | | | | A.A.P. | A.A. | A.B.B. | A.M. | A.A.P. | A.A. | A.B.B. | A.M. |
| | | | | | Relative Percentage | | | | | | | |
| 42 | Citronellyl acetate | om | 1354 | 1353 | - | - | - | 2.6 ± 0.56 | - | - | - | 4.2 ± 0.07 |
| 43 | Neryl acetate | om | 1364 | 1362 | - | - | - | 1.6 ± 0.17 | - | - | - | 3.0 ± 0.00 |
| 44 | Geranyl acetate | om | 1381 | 1381 | - | - | - | 37.5 ± 3.78 | - | - | - | 61.4 ± 2.40 |
| 45 | α-Bourbonene | sh | 1384 | 1384 $ | 0.3 ± 0.07 | - | - | - | - | - | - | - |
| 46 | β-Elemene | sh | 1391 | 1391 | 0.7 ± 0.35 | - | - | - | - | - | - | - |
| 47 | Methyl eugenol | om | 1402 | 1404 | - | - | - | 2.0 ± 0.76 | - | - | - | 1.1 ± 0.00 |
| 48 | β-Caryophyllene | sh | 1419 | 1419 | 20.4 ± 1.27 | 8.9 ± 1.06 | 2.5 ± 1.34 | 2.7 ± 0.96 | 1.8 ± 0.06 | 0.6 ± 0.10 | 0.3 ± 0.00 | 0.5 ± 0.00 |
| 49 | β-copaene | sh | 1432 | 1432 | 0.6 ± 0.35 | - | - | - | - | - | - | - |
| 50 | *iso*Germacrene D | sh | 1448 | - | 0.2 ± 0.01 | - | - | - | - | - | - | - |
| 51 | *cis*-muurola-4(14),5-diene | sh | 1450 | 1467 | 0.3 ± 0.00 | - | - | - | - | - | - | - |
| 52 | α-Humulene | sh | 1454 | 1455 | 2.5 ± 0.07 | 0.9 ± 0.28 | 0.1 ± 0.07 | - | 0.3 ± 0.05 | 0.1 ± 0.08 | - | - |
| 53 | (E)-β-Farnesene | sh | 1457 | 1457 | - | 0.5 ± 0.40 | - | - | - | 0.1 ± 0.05 | - | - |
| 54 | *cis*-Muurola-4(15),5-diene | sh | 1463 | 1459 $ | 0.1 ± 0.10 | - | - | - | - | - | - | - |
| 55 | Germacrene D | sh | 1481 | 1485 | 10.9 ± 2,26 | 0.9 ± 0.57 | 1.1 ± 0.13 | - | 4.2 ± 1.04 | 0.3 ± 0.12 | 0.4 ± 0.00 | - |
| 56 | Valencene | sh | 1492 | 1496 | 0.1 ± 0.07 | - | - | - | - | - | - | - |
| 57 | Bicyclogermacrene | sh | 1500 | 1500 | 0.1 ± 0.02 | - | 0.4 ± 0.08 | - | - | - | 0.2 ± 0.07 | - |
| 58 | α-Farnesene | sh | 1508 | 1509 $ | 0.5 ± 0.05 | - | 0.1 ± 0.05 | - | - | - | - | - |
| 59 | γ-Cadinene | sh | 1513 | 1514 | 0.4 ± 0.07 | - | - | - | - | - | - | - |
| 60 | δ-Cadinene | sh | 1524 | 1523 | 0.6 ± 0.14 | - | - | - | - | - | - | - |
| 61 | α-Cadinene | sh | 1532 | 1536 | 0.1 ± 0.07 | - | - | - | - | - | - | - |
| 62 | Germacrene D-4-ol | os | 1574 | 1576 | 0.1 ± 0.04 | - | - | - | 3.9 ± 0.52 | 0.2 ± 0.12 | 0.1 ± 0.00 | - |
| 63 | Viridiflorol | os | 1591 | 1593 | - | - | - | - | - | 0.3 ± 0.00 | 1.8 ± 0.07 | - |
| 64 | Dodecyl acetate | nt | 1609 | 1610 $ | - | - | - | 0.2 ± 0.06 | - | - | - | - |
| 65 | *n*-Tetracosane | nt | 2400 | 2400 | - | - | - | - | - | - | - | 0.5 ± 0.42 |
| 66 | *n*-Pentacosane | nt | 2500 | 2500 | - | - | - | - | - | - | - | 1.1 ± 0.49 |
| | Unknown | | | | 0.2 ± 0.05 | 0.0 ± 0.00 | 0.2 ± 0.03 | 0.8 ± 0.19 | 0.6 ± 0.07 | 0.0 ± 0.0 | 0.1 ± 0.07 | 0.0 ± 0.00 |

| Class of Compounds | SPME | | | | EO | | | |
|---|---|---|---|---|---|---|---|---|
| | A.A.P. | A.A. | A.B.B. | A.M. | A.A.P. | A.A. | A.B.B. | A.M. |
| Monoterpene Hydrocarbons (mh) | 23.2 ± 1.13 | 2.8 ± 0.92 | 4.4 ± 1.34 | 6.6 ± 0.56 | - | 0.6 ± 0.10 | 0.6 ± 0.14 | - |
| Oxygenated Monoterpenes (om) | 38.6 ± 1.98 | 85.9 ± 1.41 | 90.6 ± 4.03 | 89.5 ± 4.96 | 88.6 ± 3.29 | 97.5 ± 0.21 | 96.3 ± 0.21 | 97.9 ± 0.92 |
| Sesquiterpene Hydrocarbons (sh) | 37.9 ± 2.90 | 11.2 ± 2.33 | 4.2 ± 0.97 | 2.7 ± | 6.3 ± 0.89 | 1.1 ± 0.00 | 0.9 ± 0.07 | 0.5 ± 0.00 |
| Oxygenated Sesquiterpenes (os) | 0.1 ± 0.04 | - | - | - | 3.9 ± 0.52 | 0.5 ± 0.00 | 1.9 ± 0.07 | - |
| Penylpropanoids (pp) | - | - | 0.3 ± 0.05 | - | - | - | - | - |
| Non-terpene Derivatives (nt) | - | 0.1 ± 0.04 | 0.3 ± 0.07 | 0.4 ± 0.02 | 0.6 ± 0.08 | 0.3 ± 0.07 | 0.2 ± 0.00 | 1.6 ± 0.09 |
| **Total Identified** | **99.8 ± 0.21** | **100.0 ± 0.00** | **99.8 ± 0.42** | **99.2 ± 0.35** | **99.4 ± 3.82** | **100.0 ± 0.00** | **99.9 ± 0.02** | **100.0 ± 0.00** |

Data are reported as mean values (n = 3; ± SD). * Compounds present with a% > 0.1%; LRI: Linear retention time experimentally determined; LRI*: Linear retention time reported by Adams 2007; $: Linear retention time reported by NIST 2014.

Notable amounts of pulegone, a compound with a strong pungent aromatic mint smell (IARC Monographs-108), was also detected in the EO composition of the remaining studied species with an amount raging between 79.8% in *A.* 'Arcado Pink' to 92.5% in *A. aurantiaca*. These two compounds (pulegone and geranyl acetate) may play a role in the aroma profile and can be perceived by the human olfactory system due to their high amount, even though the main compounds are not always the only ones responsible of the aroma [31].

The volatilome of the *A.* 'Blue Boa' inflorescences was already investigated by Wilson and collaborators [28]. Their results were in agreement with what was found herein concerning the class of compounds where oxygenated monoterpenes predominated, with methyl chavicol (estragol) as the major compound (about 40.0% in all the studied samples), while pulegone was the main compound found in this study. Estragol was the unique compound in flower spikes of *A. rugosa* [29] that was reported in the literature.

Flower EOs from two subspecies of *A. mexicana* were investigated by Estrada-Reyes and collaborators [32]. The composition of the *A. mexicana* subsp. mexicana was dominated by estragol (86.78%) and limonene (11.24%), while this latter constituent (limonene 9.49%) together with pulegone (80.07%) characterized the EOs of the *A. mexicana* subsp. xolocotziana. More recently, Kovalenko et al. [33] pointed out methyl eugenol (20.8%), estragol (15.1%) and linalool (12.7%) as main compounds of the same species. However, limonene, methyl eugenol and linalool were present in lesser amounts in the plants studied herein, but pulegone and estragol were completely absent [33]. Going back in the literature revision, the *A. mexicana* aroma composition found here was similar with that reported previously by Svoboda et al. [34], where no estragol was noted in the EOs of this species against an abundance in respect to pulegone (75.3%). Variability in the volatilome among the *Agastache* species was observed and might depend on the harvest time as well as the environmental conditions or cultivation methods [22].

The EO composition of Anice hyssop (*A. foeniculum*) was reported by Ivanov et al. [35], who confirmed the prevalence of estragol (93.45%) together with $\beta$-caryophyllene (1.2%). This latter constituent represented 2.8% of the total identified fraction in this work. On the other hand, pulegone (43.3%) together with isomenthone (27.2%) prevailed in the same species investigated by Kovalenko et al. [33]. This behavior was similar to the results obtained herein, even though the pulegone amount (82.4%) was 2-fold more abundant, and the isomenthone was of a percentage 13-fold lower (2.1%).

The EOs of three varieties of *A. aurantica* (golde gill) were also reported by Kovalenko et al. [33] who pointed out menthone (53% and 65%, respectively) and pulegone (26% and 25%, respectively) as predominant in the *A. aurantica* varieties 'Tango' and 'Fragrant Delight', while 'Apricot Sprite' evidenced isomenthone (46%) and pulegone (41.4%) as the main ones. The species analyzed in this work were different from all the studied varieties, since pulegone was in higher amount (76.7%), while menthone showed a lesser percentage (5.2%) and isomenthone was lacking.

Pulegone, a monoterpenes hydrocarbon, was produced in a higher percentage by different species of *Agastache* such as *A. rugosa* [36] and *A. scrophulariifolia* (45.2%) [37]. This compound has been reported as Generally Recognized as Safe (GRAS) by the Food and Drug Administration (FDA) since 1965, and it is used as a flavoring agent in food, perfumery and herbal products in the European Union [37,38].

The nutraceutical properties of the edible *Agastache* flowers were investigated, and primary (soluble sugars) and secondary metabolites (polyphenols, flavonoids and anthocyanins) as well as the antioxidant activity were determined (Table 3).

The total polyphenols content (TPC) and total flavonoids (TF) were higher in the two hybrids than in the *A. mexicana* and *A. aurantiaca* varieties. All four flowers showed higher TPC and TF values than other species belonging to *Agastache* genus (*A. foeniculum* and *A. rugosa*) [39], as well as other members of the Lamiaceae family (*Salvia splendens* and *Lamium album*) [20,40,41]. In literature, it was reported that the *Agastache* flowers contain interesting phenolic compounds and flavone glycosides, such as rosmarinic acid, acacetin and tilianin, with variation among different species and cultivars [42].

All these molecules have important biological activities (antioxidant, vasorelaxant, spasmolytic and antinociceptive activities) [43–45]. Therefore, more in-depth studies on the plants examined in this work could be interesting to suggest new potential uses of these flowers.

**Table 3.** Determination of secondary metabolites, ascorbic acid, radical scavenger activity and soluble sugars in the four edible flowers belonging to *Agastache* genus. Data are presented as means ± SE (n = 3). Different letters indicate statistically significant differences, at $p < 0.05$ (Fisher's probable least-squares difference test). Abbreviation: TPC—total polyphenols content; TF—total flavonoids; TA—total anthocyanins; GAE—gallic acid equivalents; CE—±catechin equivalents; ME—malvin equivalents.

| Parameters | *A. aurantiaca* | *A.* 'Arcado Pink' | *A.* 'Blue Boa' | *A. mexicana* |
|---|---|---|---|---|
| TPC (mg GAE/g FW) | 5.37 ± 0.32 [b] | 7.10 ± 0.25 [a] | 7.34 ± 0.19 [a] | 5.29 ± 0.20 [b] |
| TF (mg CE/g FW) | 3.04 ± 0.28 [b] | 5.57 ± 0.27 [a] | 4.96 ± 0.19 [a] | 3.52 ± 0.18 [b] |
| TA (mg ME/g FW) | 0.03 ± 0.00 [c] | 0.17 ± 0.02 [b] | 0.18 ± 0.01 [b] | 0.68 ± 0.02 [a] |
| Carotenoids (µg/g FW) | 198.57 ± 8.82 [a] | 1.95 ± 0.60 [b] | 11.71 ± 0.40 [b] | 8.69 ± 0.89 [b] |
| Reduced ascorbic acid (mg AsA/100 g FW) | 1.48 ± 0.06 [a] | 0.79 ± 0.02 [b] | 1.27 ± 0.10 [a] | 1.43 ± 0.08 [a] |
| Total ascorbic acid (mg AsA$_{TOT}$/100 g FW) | 1.87 ± 0.16 [b] | 1.12 ± 0.02 [c] | 3.06 ± 0.26 [a] | 2.11 ± 0.04 [a] |
| DPPH radical scavenging assay (IC$_{50}$ mg/mL) | 2.26 ± 0.15 [a] | 1.43 ± 0.04 [b] | 0.86 ± 0.01 [b] | 1.40 ± 0.05 [c] |
| Total soluble sugars (mg/g FW) | 104.44 ± 2.84 [a] | 56.41 ± 0.67 [c] | 75.81 ± 0.66 [b] | 57.52 ± 0.64 [c] |
| D-Glucose (mg/g FW) | 16.14 ± 0.67 [a] | 8.13 ± 0.19 [b] | 9.15 ± 0.53 [b] | 6.79 ± 0.32 [b] |
| Sucrose (mg/g FW) | 17.65 ± 0.44 [a] | 4.53 ± 0.35 [b] | 3.51 ± 0.34 [b] | 2.99 ± 0.20 [c] |
| D- Fructose (mg/g FW) | 13.25 ± 0.28 [a] | 7.00 ± 0.17 [b] | 6.49 ± 0.25 [b] | 4.92 ± 0.38 [c] |

Anthocyanins and carotenoids are the major pigments of petals, producing blue-to-red and yellow-to-red colors, respectively [46]. In this work, the total carotenoid content was relevant in *A. aurantiaca* (198.57 µg/g FW) compared to the other species under evaluation, as highlighted by the yellow colors of its flowers. On the other hand, *A.* 'Blue Boa', *A.* 'Arcado Pink' and *A. mexicana* are characterized by red-purple petals, so anthocyanins are present in these plants, especially in *A. mexicana*. Moreover, these three flowers are characterized by higher values of anthocyanins than *A. rugosa* [47]. However, the anthocyanins content is relatively low compared to other purple-red colored flowers (e.g., *Monarda dydima*) [20].

Ascorbic acid (vitamin C) is an essential nutrient for humans, as it is not able to be bio-synthesized [48]. The EU Regulation (No 1169/2011) recommended 80 mg as the daily intake for vitamin C (for adults) [49]. The analyzed species showed a very low amount of this compound with the highest value in *A.* 'Blue Boa' (3 mg/100 g FW). Therefore, to reach the recommended value, huge amounts of *Agastache* flowers should be consumed. Since few fresh flowers are generally added as ingredient in recipes, the species analyzed in this work cannot help to supply these EU daily requirements. To date, *Tagetes tenuifolia* (241.20 mg/100 g FW) [20] is considered a good source of ascorbic acid, and its flowers contain from 80- to 215-fold more ascorbic acid than the *Agastache* flowers.

Plants belonging to *Agastache* genus can be used in beekeeping [26] to produce a very interesting honey due to their high antimicrobial activity, and are potentially usable in skin products to treat infection caused by bacteria [50]. Nectar is a balanced sugar solution (mainly composed by fructose, glucose and saccharose) that also contains free amino acids, proteins, inorganic ions, lipids, organic acids, phenolic substances and terpenoids, among others [13,51]. In this work, the total content of soluble sugars was in *A. aurantiaca* > *A.* 'Blue Boa' > *A. mexicana* > *A.* 'Arcado Pink'. Moreover, *A. aurantiaca* showed a larger content of glucose, fructose and sucrose compared to the other flowers analyzed in this work. Until now, few studies had reported the total and the reducing sugar content of edible flowers. Grzeszczuk et al. [20] had quantified the percentages of total soluble sugars and sucrose in *Salvia splendens* and *Lavandula angustifolia*, from the Lamiaceae family. Compared with these data, our results highlighted the highest content of these primary metabolites in *Agastache* flowers.

### 3. Material and Methods

#### 3.1. Plant Material

All the *Agastache* plants (*Agastache* 'Blue Boa', *A. aurantiaca* 'Sunset Yellow', *A. mexicana* 'Sangria' and *A.*; Arcado Pink') were provided by the Chambre d'Agriculture des Alpes-Maritimes (CREAM, Nice, France) and were grown at Research Centre for Vegetable and Ornamental Crops (CREA, Sanremo, Imperia, Italy, GPS: 43.816887, 7.758900)—in large pots (30 cm of diameter; 9 L) in a substrate (Hochmoor–Terflor, Capriolo, Brescia, Italy) with a slow release fertilizer (Nitrophoska, Eurochem Agro, Cesano Maderno, Monza e della Brianza, Italy), and were irrigated with a nutrient solution (Ferti 3, Planta-Düngemittel, Regenstauf, Germany) every week. The pots were placed in greenhouses with an anti-insect net, and the plants were cultivated using the organic system (without pesticides). In order to avoid damage caused by pests, antagonist insects like *Aphidius colemani* (parasite), *Chrysoperla carnea*, *Adalia bipunctata*, *Phytoseiulus persimilis* and *Amblyseius swirskii* (predators) (Koppert Italia Srl., Bussolengo, Verona, Italy) were released in the greenhouse. *Bacillus thuringensis* subsp. *Kurstaki* (Serbios Srl, Badia Polesine, Rovigo, Italy) was sprayed to control the development of caterpillars. Flowers were picked during their flowering time (summer 2019).

#### 3.2. Headspace Trapping for the Floral Set-Up

Inflorescences of each species were accurately weighted (about 0.1 g) and carefully placed separately into 30 mL glass flasks without altering the tissue. The flasks were immediately sealed with aluminum foil and equilibrated at room temperature (around 24 °C) for 1 h.

#### 3.3. Sample Analysis: Isolation of VOCs

As recommended by manufacturer's instructions, a preconditioned 100 µm polydimethylsiloxane PDMS fiber was fitted to a manual sampling fiber holder (Supelco, Bellefonte, PA, USA). The fiber was than exposed to the headspace of the flask containing the samples for 30 min. By the end of the time, the fiber device was transferred into the GC-MS instrument for analysis. The process was repeated twice for each species.

#### 3.4. Extraction of Essential Oils

Fresh flowers were hydrodistilled for 2 h using a Clevenger apparatus as recommended by the European Pharmacopeia. The yield of the EOs was very low due to the very low quantity of plant material in disposal (1–2 g). Therefore, the EO was collected with the *n*-hexane grade for HPLC and was immediately injected in the GC-MS.

#### 3.5. GC-MS Analysis

Gas chromatography–electron impact mass spectrometry (GC–EIMS) analyses were performed with an Agilent 7890B gas chromatograph (Agilent Technologies Inc., Santa Clara, CA, USA) equipped with an Agilent HP-5MS (Agilent Technologies Inc., Santa Clara, CA, USA) capillary column (30 m × 0.25 mm; coating thickness 0.25 µm) and an Agilent 5977B single quadruple mass detector (Agilent Technologies Inc., Santa Clara, CA, USA). Analytical conditions were as follows: injector and transfer line temperatures 220 and 240 °C, respectively; oven temperature programmed from 60 to 240 °C at 3 °C/min; carrier gas helium at 1 mL/min; injection of 1 µL; split ratio 1:25. The acquisition parameters were as follows: full scan; scan range: 30–300 *m/z*; scan time: 1.0 s.

#### 3.6. Identification of VOCs and EO Composition

Identification of the constituents was based on a comparison of the retention times with those of the authentic samples, comparing their linear retention indices relative to the series of *n*-hydrocarbons. Computer matching was also used against commercial (NIST 14 and ADAMS 07) [52,53] and laboratory-developed mass spectra libraries built up of pure substances and components of known oils and MS literature data [53–58].

### 3.7. Determination of Secondary Metabolites, Ascorbic Acid and Radical Scavenging Activity (DPPH Assay)

Fresh flowers were used to quantify the total carotenoid [59], total polyphenolic content (TPC) (using the Folin—Ciocalteu method, according to Marchioni et al. [60]), total flavonoid (TF) [60] and anthocyanin content (TA) [60]. Radical scavenging activity was determined by the DPPH assay [61], reporting the results as $IC_{50}$ (mg/mL). Total ascorbate ($AsA_{TOT}$) and reduced ascorbate (AsA) were quantified according to the method of Kampfenkel et al. [62], and prior sample extraction is described in Degl'Innocenti et al. [63].

For each analysis, 200 mg of homogeneous samples (three biological replica) were used. The absorbance was read in a UV-1800 spectrophotometer (Shimadzu Corp., Kyoto, Japan).

### 3.8. Sugars Quantification

Fresh flowers (100 mg) were extracted as already described [64], and were thus spectrophotometrically estimated in their soluble sugars content. One mL of the samples was added to 4 mL of 0.2% (*w/v*) anthrone solution, and after 30 min of incubation at 90 °C, the absorbance was read at 620 nm. Sucrose, D-fructose and D-glucose determination was performed using a Sucrose/D-Fructose/D-Glucose Assay Kit (Megazyme International Ireland, Co. Wicklow, Ireland) following the manufacturer's instructions and prior to the extraction described in Tobias et al. [65]. For each analysis, three biological replicas were used.

### 3.9. Statistical Analysis

Data were statistically analyzed by one-way ANOVA followed by Fisher's probable least-squares difference test with a cut-off significance at $p \leq 0.05$ (StatView®, Version 5.0, SAS® Institute Corporation, Cary, NC, USA).

**Author Contributions:** Conceptualization, L.P. (Laura Pistelli); Data curation, B.N., I.M.; Formal analysis, B.N., I.M., A.C.; Funding acquisition, B.R., L.P. (Laura Pistelli); Investigation, B.N., I.M., B.R., A.C.; Methodology, L.P. (Luisa Pistelli), L.P. (Laura Pistelli) and B.N.; Project administration, B.R.; Supervision, B.R., L.P. (Luisa Pistelli), L.P. (Laura Pistelli); Writing—original draft, B.N., I.M.; Writing—review & editing, B.N., I.M., B.R., A.C., L.P. (Luisa Pistelli), L.P. (Laura Pistelli).

**Funding:** This research was funded by the INTERREG-ALCOTRA UE 2014–2020 Project "ANTEA" Attività innovative per lo sviluppo della filiera transfrontaliera del fiore edule (n. 1139), grant number CUP C12F17000080003.

**Acknowledgments:** The authors would like to thank Rosanna Dimita, and Sophie Descamps (CREAM, Nice, FR) for the free plant material supply.

**Conflicts of Interest:** The authors declare no conflict of interest.

### References

1. Kantsa, A.; Sotiropoulou, S.; Vaitis, M.; Petanidou, T. Plant Volatilome in Greece: A Review on the Properties, Prospects, and Chemogeography. *Chem. Biodivers.* **2015**, *12*, 1466–1480. [CrossRef]
2. Misra, B.B. Plant volatilome resources. *Curr. Metab.* **2016**, *4*, 148–150. [CrossRef]
3. Ahmadian, M.; Ahmadi, N.; Babaei, A.; Naghavi, M.R.; Ayyari, M. Comparison of volatile compounds at various developmental stages of tuberose (*Polianthes tuberosa* l. cv. mahallati) flower with different extraction methods. *J. Essent. Oil Res.* **2018**, *30*, 197–206. [CrossRef]
4. Bicchi, C.; Maffei, M. The Plant Volatilome: Methods of Analysis. In *High-Throughput Phenotyping in Plants: Methods and Protocols, Methods in Molecular Biology*; Normanly, J., Ed.; Humana Press: Totowa, NJ, USA, 2012; Chapter 15; Volume 918, pp. 289–310. [CrossRef]
5. Boiteux, J.; Monardez, C.; Fernández, M.; de los, Á.; Espino, M.; Pizzuolo, P.; Silva, M.F. *Larrea divaricata* volatilome and antimicrobial activity against *Monilinia fructicola*. *Microchem. J.* **2018**, *142*, 1–8. [CrossRef]

6.  Paul, I.; Goyal, R.; Bhadoria, P.S.; Mitra, A. Developing efficient methods for unravelling headspace flora volatilome in *Murraya paniculate* for understanding ecological interactions. In *Application of Biotechnology for Sustainable Development*; Springer: Singaopre, 2017; pp. 73–79.

7.  Longo, V.; Forleo, A.; Provenzano, S.P.; Coppola, L.; Zara, V.; Ferramosca, A.; Siciliano, P.; Capone, S. HS-SPME-GC-MS metabolomics approach for sperm quality evaluation by semen volatile organic compounds (VOCs) analysis. *Biomed. Phys. Eng. Express* **2019**, *5*, 015006. [CrossRef]

8.  Wong, Y.F.; Yan, D.D.; Shellie, R.A.; Sciarrone, D.; Marriott, P.J. Rapid plant volatiles screening using headspace SPME and person-portable Gas Chromatography–Mass Spectrometry. *Chromatographia* **2019**, *82*, 297–305. [CrossRef]

9.  Bueno, M.; Resconi, V.C.; Campo, M.M.; Ferreira, V.; Escudero, A. Development of a robust HS-SPME-GC-MS method for the analysis of solid food samples. Analysis of volatile compounds in fresh raw beef of differing lipid oxidation degrees. *Food Chem.* **2019**, *281*, 49–56. [CrossRef] [PubMed]

10. Jeleń, H.H.; Majcher, M.; Dziadas, M. Microextraction techniques in the analysis of food flavor compounds: A review. *Anal. Chim. Acta* **2012**, *738*, 13–26. [CrossRef]

11. Merkle, S.; Kleeberg, K.; Fritsche, J. Recent developments and applications of Solid Phase Microextraction (SPME) in food and environmental analysis—A Review. *Chromatography* **2015**, *2*, 293–381. [CrossRef]

12. Fernandes, L.; Saraiva, J.A.; Pereira, J.A.; Casal, S.; Ramalhosa, E. Post-harvest technologies applied to edible flowers: A review: Edible flowers preservation. *Food Rev. Int.* **2019**, *35*, 132–154. [CrossRef]

13. Fernandes, L.; Casal, S.; Pereira, J.A.; Saraiva, J.A.; Ramalhosa, E. Edible flowers: A review of the nutritional, antioxidant, antimicrobial properties and effects on human health. *J. Food Compos. Anal.* **2017**, *60*, 38–50. [CrossRef]

14. Lu, B.; Li, M.; Yin, R. Phytochemical content, health benefits, and toxicology of common edible flowers: A Review (2000–2015). *Crit. Rev. Food Sci. Nutr.* **2016**, *56*, S130–S148. [CrossRef] [PubMed]

15. Grzeszczuk, M.; Stefaniak, A.; Meller, E.; Wysocka, G. Mineral composition of some edible flowers. *J. Elementol.* **2018**, *23*, 151–162. [CrossRef]

16. Tokalıoğlu, Ş. Determination of trace elements in commonly consumed medicinal herbs by ICP-MS and multivariate analysis. *Food Chem.* **2012**, *134*, 2504–2508. [CrossRef]

17. Rop, O.; Mlcek, J.; Jurikova, T.; Neugebauerova, J.; Vabkova, J. Edible Flowers—A new promising source of mineral elements in human nutrition. *Molecules* **2012**, *17*, 6672–6683. [CrossRef]

18. Amorello, D.; Orecchio, S.; Pace, A.; Barreca, S. Discrimination of almonds (*Prunus dulcis*) geographical origin by minerals and fatty acids profiling. *Nat. Prod. Res.* **2016**, *30*, 2107–2110. [CrossRef]

19. Fernandes, L.; Ramalhosa, E.; Pereira, J.A.; Saraiva, J.A.; Casal, S. The unexplored potential of edible flowers lipids. *Agriculture* **2018**, *8*, 146. [CrossRef]

20. Grzeszczuk, M.; Stefaniak, A.; Pachlowska, A. Biological value of various edible flower species. *Acta Sci. Pol. Hortorum* **2016**, *15*, 109–119. [CrossRef]

21. Lim, T.K. Lavandula angustifolia. In *Edible Medicinal and Non Medicinal Plants: Volume 8, Flowers*; Springer: Dordrecht, The Netherlands, 2014; pp. 156–185. [CrossRef]

22. Zielińska, S.; Matkowski, A. Phytochemistry and bioactivity of aromatic and medicinal plants from the genus *Agastache* (Lamiaceae). *Phytochem. Rev.* **2014**, *13*, 391–416. [CrossRef]

23. Fuentes-Granados, R.; Widrlechner, M.P.; Wilson, L.A. An overview of *Agastache* research. *J. Herbs Spices Med. Plants* **1998**, *6*, 69–97. [CrossRef]

24. Husti, A.; Cantor, M.; Buta, E.; Horţ, D. Current Trends of Using Ornamental Plants in Culinary Arts. *ProEnvironment/ProMediu* **2013**, *6*, 52–58.

25. Myadelets, M.A.; Vorobyeva, T.A.; Domrachev, D.V. Composition of the essential oils of some species belonging to genus *Agastache Clayton* ex *Gronov* (Lamiaceae) cultivated under the conditions of the middle Ural. *Chem. Sustain. Dev.* **2013**, *21*, 397–401.

26. Ayres, G.S.; Widrlechner, M.P. The Genus Agastache as bee forage: A historical perspective. *Am. Bee J.* **1994**, *134*, 341–348.

27. Chen, N.-H.; Wei, S. Factors influencing consumers' attitudes towards the consumption of edible flowers. *Food Qual. Prefer.* **2017**, *56*, 93–100. [CrossRef]

28. Wilson, L.A.; Widrlechner, M.P.; Senechal, N.P. Headspace analysis of the volatile oils of Agastache. *J. Agric. Food Chem.* **1992**, *40*, 1362–1366. [CrossRef]

29. Yamani, H.; Mantri, N.; Morrison, P.D.; Pang, E. Analysis of the volatile organic compounds from leaves, flower spikes, and nectar of Australian grown *Agastache rugosa*. *BMC Complement. Altern. Med.* **2014**, *14*, 495. [CrossRef] [PubMed]

30. Elsharif, S.A.; Buettner, A. Structure-odor relationship study on geraniol, nerol, and their synthesized oxygenated derivatives. *J. Agric. Food Chem.* **2016**, *66*, 2324–2333. [CrossRef]

31. Kim, T.A.; Shin, H.J.; Baek, H.H.; Lee, H.J. Volatile flavour compounds in suspension culture of *Agastache rugosa* Kuntze (Korean mint). *J. Sci. Food Agric.* **2001**, *81*, 569–575. [CrossRef]

32. Estrada-Reyes, R.; Aguirre Hernández, E.; García-Argáez, A.; Soto Hernández, M.; Linares, E.; Bye, R.; Heinze, G.; Martínez-Vázquez, M. Comparative chemical composition of *Agastache mexicana* subsp. mexicana and *A. mexicana* subsp. xolocotziana. *Biochem. Syst. Ecol.* **2004**, *32*, 685–694. [CrossRef]

33. Kovalenko, N.A.; Supichenko, G.N.; Ahramovich, T.I.; Shutova, A.G.; Leontiev, V.N. Antibacterial activity of *Agastache aurantiaca* essential oils. *Khimiya Rastit. Syr'ya* **2018**, *2*, 63–70. [CrossRef]

34. Svoboda, K.P.; Gough, J.; Hampson, J.; Galambosi, B. Analysis of the essential oils of some *Agastache* species grown in Scotland from various seed sources. *Flavour Fragr. J.* **1995**, *10*, 139–145. [CrossRef]

35. Ivanov, I.G.; Vrancheva, R.Z.; Petkova, N.T.; Tumbarski, Y.; Dincheva, I.N.; Badjakov, I.K. Phytochemical compounds of anise hyssop (*Agastache foeniculum*) and antibacterial, antioxidant, and acetylcholinesterase inhibitory properties of its essential oil. *J. Appl. Pharm. Sci.* **2019**, *9*, 72–78. [CrossRef]

36. Mo, J.; Ma, L. Volatile oil of Herba *Agastache* in various growth periods and different parts by GC-MS. *Chin. J. Pharm.* **2011**, *4*.

37. Božović, M.; Ragno, R.; Tzakou, O. *Calamintha nepeta* (L.) Savi and its main essential oil constituent pulegone: Biological activities and chemistry. *Molecules* **2017**, *22*, 290. [CrossRef]

38. Zárybnický, T.; Matoušková, P.; Lancošová, B.; Šubrt, Z.; Skálová, L.; Boušová, I. Inter-individual variability in acute toxicity of R-pulegone and R-menthofuran in human liver slices and their influence on miRNA expression changes in comparison to acetaminophen. *Int. J. Mol. Sci.* **2018**, *19*, 1805. [CrossRef]

39. Shtereva, L.; Vassilevska-Ivanova, R.; Stancheva, I.; Geneva, M.; Stoyanova, E. Evaluation of antioxidant activity of *Agastache foeniculum* and *Agastache rugosa* extracts. *C. R. l'Académie Bulg. Sci.* **2016**, *69*, 295–302.

40. Li, A.-N.; Lia, S.; Lia, H.-B.; Xua, D.-P.; Xub, X.-R.; Chen, F. Total phenolic contents and antioxidant capacities of 51 edible and wild flowers. *J. Funct. Foods* **2014**, *6*, 319–330. [CrossRef]

41. Nowicka, P.; Wojdyło, A. Anti-Hyperglycemic and Anticholinergic Effects of Natural Antioxidant Contents in Edible Flowers. *Antioxidants* **2019**, *8*, 308. [CrossRef]

42. Park, W.T.; Kim, H.H.; Chae, S.C.; Cho, J.W.; Park, S.U. Phenylpropanoids in *Agastache foeniculum* and Its Cultivar *A. foeniculum* 'Golden Jubilee'. *Asian J. Chem.* **2014**, *26*, 4599–4601. [CrossRef]

43. Zielińska, S.; Kolniak-Ostek, J.; Dziadas, M.; Oszmiański, J.; Matkowski, A. Characterization of polyphenols in *Agastache rugosa* leaves and inflorescences by UPLC-qTOF-MS following FCPC separation. *J. Liq. Chromatogr. Relat. Technol.* **2016**, *39*, 209–219. [CrossRef]

44. Gonzalez-Trujano, M.E.; Ventura-Martinez, R.; Chavez, M.; Diaz-Reval, I.; Pellicer, F. Spasmolytic and antinociceptive activities of ursolic acid and acacetin identified in *Agastache mexicana*. *Planta Med.* **2013**, *78*, 793–796. [CrossRef] [PubMed]

45. Hernandez-Abreu, O.; Duran-Gomez, L.; Best-Brown, R.; Villalobos-Molina, R.; Rivera-Leyva, J.; Estrada-Soto, S. Validated liquid chromatographic method and analysis of content of tilianin on several extracts obtained from *Agastache mexicana* and its correlation with vasorelaxant effect. *J. Ethnopharmacol.* **2011**, *138*, 487–491. [CrossRef] [PubMed]

46. Ohmiya, A.; Tanase, K.; Hirashima, M.A.; Yamamizo, C.; Yagi, M. Analysis of carotenogenic gene expression in petals and leaves of carnation (*Dianthus caryophyllus* L.). *Plant Breed.* **2013**, *132*, 423–429. [CrossRef]

47. Park, C.H.; Yeo, H.J.; Baskar, T.B.; Park, Y.E.; Park, J.S.; Lee, S.Y.; Park, S.U. In Vitro Antioxidant and Antimicrobial Properties of Flower, Leaf, and Stem Extracts of Korean Mint. *Antioxidants* **2019**, *8*, 75. [CrossRef] [PubMed]

48. De Tullio, M.C. The Mystery of Vitamin C. *Nat. Educ.* **2010**, *3*, 48.

49. Regulation (EU) No 1169/2011 of the European Parliament and of the Council of 25 October 2011. In *Daily Reference Intakes for Vitamins and Minerals (Adults)*; Annex XII; European Union: Bruxelles, Belgium, 22 November 2011.

50. Anand, S.; Deighton, M.; Livanos, G.; Morrison, P.D.; Pang, E.C.K.; Mantri, N. Antimicrobial activity of Agastache honey and characterization of its bioactive compounds in comparison with important commercial honeys. *Front. Microbiol.* **2019**, *10*, 263. [CrossRef]
51. Mlcek, J.; Rop, O. Fresh edible flowers of ornamental plants—A new source of nutraceutical foods. *Trends Food Sci. Technol.* **2011**, *22*, 561–569. [CrossRef]
52. *NIST 14/EPA/NIH Mass Spectra Library*; I. Willy and Sous, Inc.: Hoboken, NJ, USA, 2014.
53. Adams, R.P. *Identification of Essential Oil Components by Gas Chromatography/Quadrupole Mass Spectroscopy*, 4th ed.; Allured Publishing Corporation: Carol Stream, IL, USA, 2007.
54. Davies, N.W. Gas chromatographic retention indices of monoterpenes and sesquiterpenes on methyl silicon and Carbowax 20M phases. *J. Chromatogr. A* **1990**, *503*, 1–24. [CrossRef]
55. Jennings, W.; Shibamoto, T. *Qualitative Analysis of Flavor and Fragrance Volatiles by Glass Capillary Gas Chromatography, Food/Nahrung*; Academic Press: New York, NY, USA, 1980.
56. Masada, Y. *Analysis of Essential Oils by Gas Chromatography and Mass Spectrometry*; John Wiley & Sons, Inc.: New York, NY, USA, 1976.
57. Stenhagen, E.; Abrahamsson, S.; McLafferty, F.W. *Registry of Mass Spectral Data*; Wiley & Sons: New York, NY, USA, 1974.
58. Swigar, A.A.; Silverstein, R.M. *Monoterpenes, Aldrich Chemical Company*; Aldrich Chemical Company: Milwaukee, WI, USA, 1981.
59. Lichtenthaler, H.K. Chlorophylls and carotenoids: Pigments of photosynthetic biomembranes. *Methods Enzymol.* **1987**, *148*, 350–382. [CrossRef]
60. Marchioni, I.; Pistelli La Ferri, B.; Cioni, P.L.; Copetta, A.; Pistelli Lu Ruffoni, B. Preliminary studies on edible saffron bio-residues during different post-harvest storages. *Bulg. Chem. Commun.* **2019**, *51*, 131–136.
61. Brand-Williams, W.; Cuvelier, M.E.; Berset, C. Use of a Free Radical Method to Evaluate Antioxidant Activity. *LWT Food Sci. Technol.* **1995**, *28*, 25–30. [CrossRef]
62. Kampfenkel, K.; Montagu, M.V.; Inzé, D. Extraction and determination of ascorbate e dehydroascorbate from plant tissue. *Anal. Chem.* **1995**, *225*, 165–167. [CrossRef] [PubMed]
63. Degl'innocenti, E.; Guidi, L.; Pardossi, A.; Tognoni, F. Biochemical Study of Leaf Browning in Minimally Processed Leaves of Lettuce (*Lactuca sativa* L. Var. Acephala. *J. Agric. Food Chem.* **2005**, *53*, 9980–9984. [CrossRef] [PubMed]
64. Das, B.K.; Choudhury, B.K.; Kar, M. Quantitative estimation of changes in biochemical constituents of mahua (*Madhuca indica* syn. *Bassia latifolia*) flowers during postharvest storage. *J. Food Process. Preserv.* **2010**, *34*, 831–844. [CrossRef]
65. Tobias, R.B.; Boyer, C.D.; Shannon, J.C. Alterations in Carbohydrate Intermediates in the Endosperm of Starch-Deficient Maize (*Zea mays* L.) Genotypes. *Plant Physiol.* **1992**, *99*, 146–152. [CrossRef]

**Sample Availability:** Samples of the plant material *Agastache* 'Blue Boa', *A. aurantiaca* 'Sunset Yellow', *A. mexicana* 'Sangria' and *A.*; Arcado Pink') are available at Research Centre for Vegetable and Ornamental Crops (CREA, Sanremo, Imperia, Italy, GPS: 43.816887, 7.758900.

 *molecules*

Article

# Volatile Compounds of Selected Raw and Cooked *Brassica* Vegetables

**Martyna N. Wieczorek and Henryk H. Jeleń ***

Faculty of Food Science and Nutrition, Poznań University of Life Sciences, Wojska Polskiego 31,
60-624 Poznań, Poland; martyna.wieczorek@up.poznan.pl
* Correspondence: henryk.jelen@up.poznan.pl; Tel.: + 48-61-848-72-73

Received: 20 December 2018; Accepted: 17 January 2019; Published: 22 January 2019

**Abstract:** *Brassica* vegetables are a significant component of the human diet and their popularity is systematically increasing. The interest in plants from this group is growing because of numerous reports focused on their pro-health properties. However, some consumers are not enthusiastic about these vegetables because of their specific bitter taste and sharp, sulfurous aroma. In this study, the volatile composition of 15 *Brassica* cultivars (five Brussels sprouts, four kohlrabi, three cauliflower and three broccoli), both raw and cooked, was analyzed by solid phase microextraction and comprehensive two-dimensional gas chromatography with time of flight mass spectrometry (SPME-GC×GC-ToFMS). Differences were found between the analyzed vegetables, as well as different cultivars of the same vegetable. Moreover, the influence of cooking on the composition of volatile compounds was evaluated. All the vegetables were frozen before analyses, which is why the impact of this process on the volatile organic compounds (VOCs) was included. The most abundant groups of compounds were sulfur components (including bioactive isothiocyanates), nitriles, aldehydes and alcohols. Cooking in general caused a decrease in the abundance of main volatiles. However, the amount of bioactive isothiocyanates increased in most cultivars after cooking. The effect of freezing on the volatile fraction was presented based on the Brussels sprout cultivars. Most of the changes were closely related to the activity of the lipoxygenase (LOX) pathway enzymes. These are characterized by a marked reduction in alcohol contents and an increment in aldehyde contents. Moreover, important changes were noted in the concentrations of bioactive components, e.g., isothiocyanates. This research included a large set of samples consisting of many cultivars of each analyzed vegetable, which is why it provides a considerable body of general information concerning volatiles in *Brassica* vegetables.

**Keywords:** *Brassica* vegetables; volatile compounds; cooking; GC×GC-ToFMS

---

## 1. Introduction

Broccoli, Brussels sprouts, cauliflower and kohlrabi are members of the Brassicaceae family, which is widely cultivated throughout the world. Some consumers reject *Brassica* vegetables due to their characteristic flavor, which is genetically related [1]. The consumption of Brassicaceae is high; however, it is mostly in the cooked form because of significant changes in flavor. Thermal treatment makes vegetables easily digestible and more acceptable for consumers. Nevertheless, some of them, e.g., broccoli, cauliflower or kohlrabi, are sometimes eaten raw, usually as an ingredient in salads. Thermal treatment affects the concentration of many beneficial compounds in vegetables, such as vitamins or phenolic compounds [2], while at the same time significantly changing the flavor profile of these vegetables. Many *Brassica* vegetables are sold as frozen florets or cubes, often found as ingredients in vegetable mixes, thus ensuring their supply year-round in a convenient form. Raw vegetables are rich in aromatic compounds, which usually are produced as a result of enzymatic reactions. Their storage in the frozen form can significantly alter biochemical reactions occurring in vegetables, thus influencing flavor also after cooking. However, there are practically no studies on the comparison of volatile compound profiles of frozen, fresh and cooked *Brassica* vegetables. In the case of the analyzed plants, the specific interest focused on bioactive molecules is particularly high and they have been extensively studied recently, mainly due to the presence of isothiocyanates. These are the products of enzymatic hydrolysis of glucosinolates and their beneficial influence on human health has been widely examined [3–8].

A wide array of analytical techniques have been used to analyze volatile compounds in plants, including fruit and vegetables. *Brassica* volatiles have been collected using simultaneous distillation-extraction (SDE) [9,10], sorbent (Tenax) trapping [11], solvent extraction [12], as well as solid phase microextraction (SPME) [13,14]. The latter is the most widely applied solvent-free isolation method for plant materials and food flavors, thanks to its sensitivity, preconcentration abilities, ruggedness, potential automation and ease of use [15,16].

The aim of the present study was to evaluate differences in the profiles of volatile compounds in fresh and cooked *Brassica* vegetables. Thanks to the results obtained for 15 different cultivars of kohlrabi, broccoli, cauliflower and Brussels sprouts, generalized conclusions may be drawn in terms of changes in volatile compounds in the main representatives of *Brassica* vegetables. The volatile compounds were analyzed using solid phase microextraction and comprehensive two-dimensional gas chromatography with time of flight mass spectrometry (SPME-GC×GC-ToFMS).

## 2. Results

### 2.1. Identification of Volatiles

To compare the profiles of volatiles for the investigated vegetables, the compounds were isolated using headspace analysis, where the volatiles were adsorbed using SPME fiber and then separated by comprehensive two-dimensional chromatography-mass spectrometry (HS-SPME-GC×GC-ToFMS). This approach allows for the specific separation and (tentative in this case) identification of volatile compounds, thus, it is widely used in volatilome fingerprinting and profiling [17,18]. As the peak capacity of GC×GC systems can reach 10,000, the number of identified compounds is usually very high. GC×GC-ToFMS provides plenty of data, for this reason extracting useful information is often a very challenging task. For our work, only the most abundant compounds with a relatively high signal to noise ratio (S/N > 250), high similarity and retention indexes comparable to literature data were taken into consideration and the comparison was based on the peak area values and the peak area percentage. This serves the assumed comparative purposes well; however, it has to be remembered that SPME is a partition coefficient based extraction method and the area percentage may be different from the data provided by exhaustive extraction methods.

Figure 1 shows the differentiation between various *Brassica* vegetables using principal component analysis (PCA) based on the profiles of volatile compounds. Each point represents a mean of three replicates analyzed for a given cultivar. One can observe obvious differences in the profiles of particular vegetables, both raw and cooked. However, there are also significant differences between particular cultivars of some investigated vegetables, which is especially noteworthy for raw broccoli and Brussels sprouts. Examination of the PCA graphs shows a similarity between cauliflower and broccoli volatiles in raw vegetables, despite the between-cultivar differences for broccoli, while kohlrabi and Brussels sprouts differ substantially in terms of their volatile compound profiles. When the data are compared for cooked vegetables, the similarities between cauliflower and broccoli can still be observed (close distance between clusters), whereas the other vegetables form distant clusters. Interestingly, after cooking, the differences between cultivars are much less pronounced. This would indicate the more uniform profile of the main volatiles. A detailed list of main compounds for all the examined cultivars is provided in Tables S1–S4 in the supplementary files. The compounds corresponding to the number in the PCA loading plots are presented in Table S5 in the supplementary files. As can be observed in the tables, the compounds can be classified into several main groups: Aldehydes (with alkanals, 2-alkenals and 2,4-alcadienals), alcohols (mainly unsaturated), isothiocyanates (mainly aliphatic, although also aromatic), other sulfur compounds (mainly sulfides and thiophenes) and nitriles. To a certain extent, the classification of volatile compounds into these classes is also associated with the sensory properties and acceptance of these compounds. Some of the identified aldehydes and alcohols may contribute to the "green", "sweet", "fatty", "soup" and other notes in the vegetable aroma. Isothiocyanates are products of the enzymatic hydrolysis of glucosinolates, which are widely investigated because of their beneficial effect on human health. However, they are also partially responsible for the specific pungent notes. Nitriles, which are an abundant group in all analyzed samples, are also found as degradation products of glucosinolates, instead of isothiocyatates [1]. Other sulfur components include mainly products of the cysteine pathway, which are sulfides. All the detected sulfides, i.e., disulfides, trisulfides, tetrasulfides, pentasulfides, and others have been collected under the general name of "sulfides", simply because of their labile nature and mutual transposition. To facilitate an easy visual comparison of raw and cooked cultivars, Figure 2 shows the main groups of the detected compounds in the four examined vegetables.

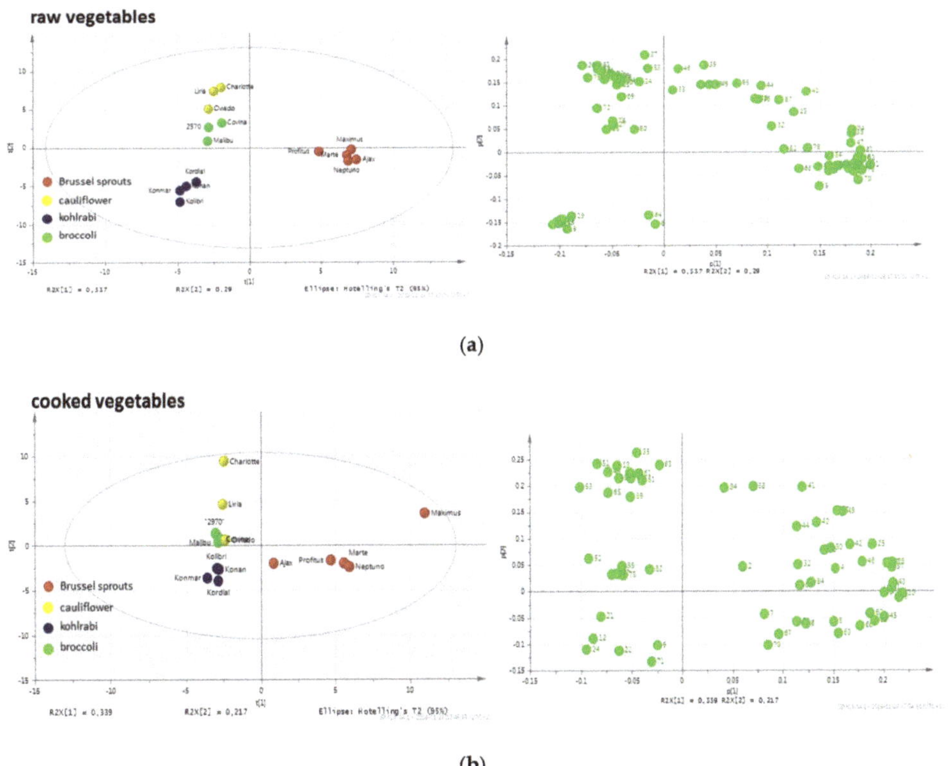


**Figure 1.** Principal component analysis (PCA) graphs of volatiles extracted from the raw (**a**) and cooked (**b**) *Brassica* vegetables investigated. Numbers on loadings plot represent compounds characteristic for analyzed vegetables. Compound numbering is uniform with data provided in the Supplementary Table S5.

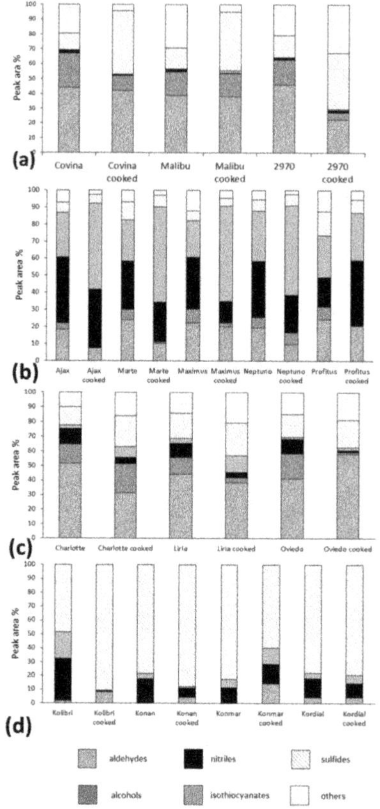



**Figure 2.** Percentage contents of isothiocyanates, alcohols, aldehydes, other sulfur compounds, nitriles, others (miscellaneous) in all analyzed samples. Broccoli (**a**): Covina, Malibu, "2970"; Brussels sprouts (**b**): Ajax, Marte, Maximus, Neptuno, Profitus; Cauliflower (**c**): Charlotte, Liria, Oviedo; Kohlrabi (**d**): Kolibri, Konan, Konmar, Kordial.

### 2.1.1. Concentrations of Isothiocyanates and Nitriles in Raw and Cooked *Brassica* Vegetables

Very small amounts of isothiocyanates were detected in all of the raw broccoli cultivars (<2% in all samples). Additionally, the nitrile levels were relatively low (<5%), which confirms very low concentrations of myrosinase hydrolysis products in the gas phase. The diversity of isothiocyanates was also very limited, barely four different isothiocyanates were identified using GC×GC. Both 2-methylbutyl isothiocyanate and isobutyl isothiocyanate were found in the cultivars Malibu and "2970", while cyclopentyl isothiocyanate and n-hexyl isothiocyanate were present in all the analyzed broccoli samples. The composition of isothiocyanates was different in cooked vegetables, as propane, 1-isothiocyanato-, butane, 1-isothiocyanato-, 1-butene, 4-isothiocyanato- and benzene, (2-isothiocyanatoethyl)- were detected in all the analyzed cooked vegetables, instead of most isothiocyanates present in raw vegetables. The most interesting result was found in the cooked cultivar "2970", where all the described isothiocyanates were present, along with a significant amount of allyl isothiocyanate (which was not present in the raw broccoli). This unexpected finding was connected with the absence of sulforaphane, one of the most explored isothiocyanates present in broccoli. Because of the health benefits of sulforaphane [6], interest in this component concentration is particularly high. Solvent assistant flavor evaporation (SAFE) extraction was performed for one broccoli cultivar to check if the sulforaphane was present in the sample. In the SAFE extract, analyzed by GC×GC-ToFMS, sulforaphane and also sulforaphane nitrile were present as glucoraphanin degradation products, although in very small concentrations. The absence of sulforaphane in the SPME extract may either be due to the labile nature of this component [19], or more probably, due to its small K value (partition coefficient sample-headspace-fiber). SAFE is an exhaustive extraction type technique, therefore other/more components can be identified.

The concentration and diversity of isothiocyanates was much greater in the cooked broccoli. The amounts were almost two times higher than in the raw vegetables. It was reported that myrosinase in the broccoli matrix is inactivated after 20 min at 60 °C [20]. No reports have been found in available literature on the stability of nitrile specific proteins. Generally, the total amount of glucosinolate degradation products decreased, especially the concentration of nitriles. Cooking caused a change in the amount of isothiocyanates, which is beneficial from the consumer's point of view.

In cauliflower, the major isothiocyanate was allyl isothiocyanate, representing 2–4% of the total volatile fraction, resulting from the high concentration of sinigrin, its precursor, in cauliflowers [12]. Allyl isothiocyanates are some of most frequently studied isothiocyanates, with many publications confirming their anticancer activity in both cultured cancer cells and animal models. Moreover, the bioavailability of this compound is very high, as almost 90% of oral intake is absorbed [21]. Other isothiocyanates were also present, although in much lower concentrations, usually less than 1%. In the Lira and Charlotte cultivars, somewhat higher amounts of isothiocyanates were recorded in the cooked vegetables compared with the raw vegetables, while in the Oviedo cultivar the situation is opposite. An abundant group in all raw cauliflower cultivars was composed of nitriles, which constituted almost 10% in all analyzed samples. It proves again, that also in raw cauliflower after tissue disruption, nitriles are formed as the main products of glucosinolate hydrolysis. It suggests that after tissue disruption, the hydrolysis of glucosinolates yields nitriles instead of bioactive isothiocyanates. The concentration of 2-methylbutyl isothiocyanate was higher in the cooked vegetables in all analyzed varieties, whereas the allyl isothiocyanate contents were higher in the raw vegetables (except for the Liria cultivar). The total amount of isothiocyanates was higher in the cooked Lira and Charlotte cultivars than in the raw vegetables (Table S2). An opposite situation was observed in the Oviedo cultivar. The nitrile concentration was significantly smaller in all analyzed cooked cauliflower samples.

The third group of analyzed plants comprised four varieties of kohlrabi: Kolibri, Kordial, Konan and Konmar. The percentage concentration of isothiocyanates in the raw vegetables was unexpectedly varied, as it was 4.16% for Konan, 18.96% in Kolibri, 6.15% Konmar and 3.71% in Kordial. The content slightly exceeded that in cauliflower. This fact was also correlated with high nitrile concentrations (more than 10% in all cases). In cooked kohlrabies, this amount increased in the Konmar and Kordial cultivars. Benzene, (2-isothiocyanatoethyl)-, which is present in uncooked vegetables at significant levels, almost disappeared in the cooked kohlrabies. The increase in isothiocyanate concentrations (Table S3) in the Konmar and Kordial cultivars was mostly caused by the increase of the n-pentyl isothiocyanate peak area in those two cultivars. The peak area (i.e., amounts) of nitriles decreased significantly in all analyzed cooked kohlrabi varieties, compared to the raw varieties.

The uncooked Brussels sprout cultivars contained mostly nitriles in their volatile fractions. Isothiocyanates constituted the second most abundant group. In the cooked plants, the isothiocyanates accounted for more than 50% in the Maximus, Marte, Ajax and Neptuno cultivars. However, the percentage contents of isothiocyanates increased, while their summary peak area changed significantly only in the Maximus cultivar. In the case of nitriles, the summary peak area decreased almost two-fold when compared with the raw varieties. Only in the Profitus cultivar did the levels of nitriles remain practically unchanged.

The important issue described in this part of the study is related with the high concentration of nitriles in some of the analyzed samples. This fact suggests that after tissue disruption in raw, defrosted vegetables, the hydrolysis of glucosinolates occurs in favor of the formation of nitriles instead of bioactive isothiocyanates. This seems to be important from a biological point of view. This is because of the health-promoting nature of isothiocyanates, which are bioactive molecules that are known for having many positive effects on human health. These isothiocyanates are found as a minority in the volatile fraction with respect to nitriles. Based on the actual in vitro data, nitriles have less beneficial health potential [19,22,23], or even harmful effects on consumers [24]. In cauliflower, the percentage of nitriles was more than 10% and the percentage of isothiocyanates was about 3% in all samples, with a similar situation observed in broccoli. In the kohlrabi cultivars, the level of isothiocyanates was slightly higher; however, the nitrile concentration was still two times higher, and in the Konan cultivar it was even four times higher than that of the isothiocyanates. As presented, the enzymatic degradation of glucosinolates leads to the formation of isothiocyanates, and in some cases, to nitriles, as the main product. The presence of modifying proteins such as the epithiospecifier protein or the nitrile-specifier proteins results in the enzymatic degradation of glucosinolates being altered in favor of nitriles [8,25]. It is important to highlight here that all vegetables were frozen before analysis; the impact of freezing on the composition of volatiles is presented in the next part of the paper. The changes in isothiocyanates and nitriles induced by cooking are presented for selected cultivars in Figure 3. It illustrates the general decrease in nitriles caused by cooking and also the different behavior of isothiocyanates in these cultivars. Initially, research on glucosinolate degradation products was mainly focused on their toxic, antinutritive and goitrogenic properties. More recently, attention has shifted to investigations concerning their beneficial effects against various diseases. Most studies are focused on different *Brassica* vegetables and the detection of glucosinolates, which are biologically non-active molecules. Still, a higher glucosinolate content does not always guarantee an increment of desirable isothiocyanates after tissue mastication. The formation of beneficial isothiocyanates depends on a variety of factors, such as the activity of myrosinase and nitrile specific proteins or domestic processing. It is worth mentioning that even if glucosinolates are not degraded, their consumption is beneficial, since they can be hydrolyzed by a healthy intestinal microbiome [26]. Studies focusing on the health-promoting effects of isothiocyanates are numerous; however, they are mostly focused on sulforaphane [27], allyl isothiocyanate [21] and benzyl isothiocyanate [28], while as presented the qualitative diversity of isothiocynates is considerable.

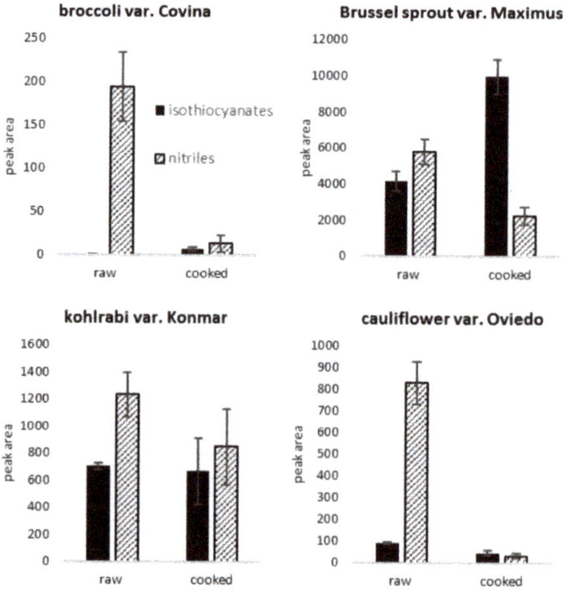


**Figure 3.** Proportions between isothiocyanates and nitriles in selected raw and cooked *Brassica* vegetables. The results are the mean of three replicates and the error bars show standard deviation. Amounts were expressed in total (sum of) peak areas.

### 2.1.2. Other Sulfur Volatiles

Derivatives from the *S*-alk(en)yl-l-cysteine pathway compounds occurred very frequently in the volatile fraction in all analyzed vegetables. In all raw kohlrabi cultivars, sulfides formed the most abundant group of volatile components. They accounted for more than 50% of all the volatiles present in the analyzed kohlrabies (Figure 2). In the Konmar cultivar, it was almost 80%. Surprisingly, a small total percentage of sulfides was noted in raw Brussels sprouts cultivars, where "other sulfides" represented less than 15% in all analyzed samples. Numerously represented sulfides (di, tri-, tetra-) are shown in the supplementary tables. The average percentage content of sulfides was approximately 10% in all analyzed *Brassicaceae*. Taking into account the extremely low odor threshold of these compounds, which are detectable at levels as low as one part per trillion by the human nose [29], their crucial role in the *Brassicaceae* characteristic flavor formation seems to be obvious. The smell of all sulfides detected here is described as extremely unpleasant [29]. The comparison of sulfide concentrations in raw and cooked vegetables was inconclusive, because in broccoli, the amount of sulfides was higher in the cooked form than in the raw form, while in kohlrabi, cauliflower and Brussels sprouts, the concentration was usually higher in the cooked vegetable. Additionally, there were some exceptions for some cultivars, where the proportions were reversed.

### 2.1.3. Aldehydes and Alcohols

All the analyzed cauliflower and broccoli cultivars contained high percentage levels of aldehydes. In broccoli, the dominating aldehydes were hexenal, 2-hexenal (more than 10% in all samples), 2,4-heptadienal and benzaldehyde. According to [29], those aldehydes contribute the "green" type aromas to the final broccoli fragrance. In the cauliflower cultivars, the content of aldehydes was even higher than in broccoli, with 2,4-heptadienal, 2-hexenal, hexanal, benzaldehyde, propanal and nonanal being dominant (found in the highest percentage levels). The presented profile of aldehydes in cauliflower was quite similar to that in broccoli. A marked decline in the diversity and percentage contents of aldehydes was observed in all kohlrabi varieties. In all of the analyzed cultivars, the level of aldehydes was below 5%. The magnitudes of the aldehyde peaks on the chromatograms were also relatively small. In comparison with cauliflower and broccoli, which contain respectively 22 and 16 different aldehydes, seven aldehydes were identified in kohlrabi after GC×GC analysis, seemingly to be a very small number. The most abundant aldehyde in all varieties was 2-butenal, with a flower-type odor [29]. In Brussels sprouts, the aldehyde percentage content was low; however, their peak areas and diversity were relatively high. Fifteen different aldehydes were identified by GC×GC, with the ratios of all aldehydes being similar and with benzaldehyde being the most abundant. Twelve different alcohols were also present in the Brussels sprouts chromatograms and their percentage contents were significantly higher than those of the aldehydes (upper 20%). Only seven different aldehydes were identified in the kohlrabi samples, at less than 5% in all samples. Despite the small amounts of aldehydes, which are mostly important for "green", "sweet" and "nice" aromas, the intensity of "raw kohlrabi" and "green" odors detected by the panelists was relatively high. Alcohols were present in trace amounts, so they were disregarded in the percentage calculations. In cooked broccoli and cauliflower, the number of aldehydes was noticeably lower in the cooked vegetables, compared to raw varieties. In contrast, the number of aldehydes in the kohlrabi was significantly greater in the cooked vegetables. The percentage contents of different aldehydes in cooked kohlrabi varied between cultivars. In Brussels sprouts, the levels of the lipoxygenase (LOX) pathway products decreased in all cooked cultivars.

Figure 4 shows the peak areas of the main groups of volatile compounds investigated in the analyzed vegetables, both raw and cooked. Total peak areas were used to indicate the differences between particular vegetables, so it can be seen that aldehydes are dominant in broccoli and cauliflower, whereas in kohlrabi, sulfides constitute the most abundant fraction. Additionally, the dominance of nitriles and isothiocyanates in raw Brussels sprouts is easily noted. The error bars indicate differences caused by different cultivars taken for comparison, therefore they are in some instances relatively high.

Similarly, the main fractions of the examined cooked vegetables can be compared, indicating an overall decrease in the amounts of volatiles caused by cooking, with some exceptions discussed above.

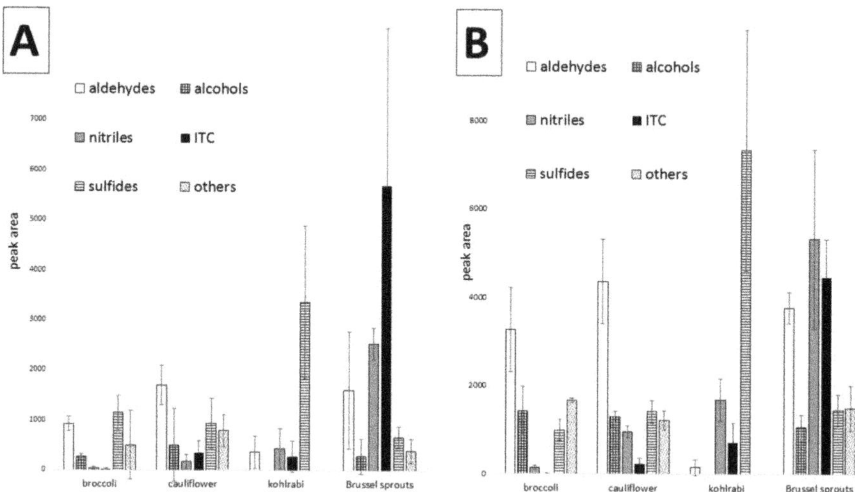


**Figure 4.** Mean values in different cultivars of analyzed raw (**A**) and cooked (**B**) vegetables. For broccoli, it is the mean value from three cultivars: Covina, Malibu and "2970", for cauliflower, from three cultivars: Charlotte, Oviedo, Liria, for kohlrabi, from four cultivars: Kolibri, Konan, Konmar, Kordial and for Brussels sprouts, from five cultivars: Ajax, Marte, Maximus, Neptuno, Profitus. The results are the mean of three replicates for each cultivar (i.e., nine analyses for broccoli and cauliflower, 15 for Brussels sprouts and 12 for kohlrabi), error bars show standard deviation. Amounts were expressed in total (sum of) peak areas.

### 2.1.4. Freezing-Thawing Effect on the Volatile Fraction

Prior to cooking, all samples were stored at −20 °C, which is why the impact of freezing needed to be evaluated. The comparison of the volatile composition in raw, fresh and frozen-thawed vegetables was performed based on Brussels sprouts. Figure 5 shows a comparison of nitrile amounts in raw (fresh) and raw frozen-thawed vegetables for all 5 Brussels sprout cultivars. Moreover, the percentage contents of aldehydes was also higher than those of alcohols in the frozen vegetables, whereas the opposite situation was observed in fresh tissue. It is worth highlighting here that in the fresh product, the concentration of isothiocyanates was more than 50% of the total value, whereas in the frozen vegetables the amount was approximately 30%. The reason for the change in the isothiocyanate levels and an increase of nitrile contents is unknown at the moment. It may be connected with cell degradation during the freezing process, leading to changes in the activity/stability of enzymes responsible for the hydrolysis of glucosinolates.

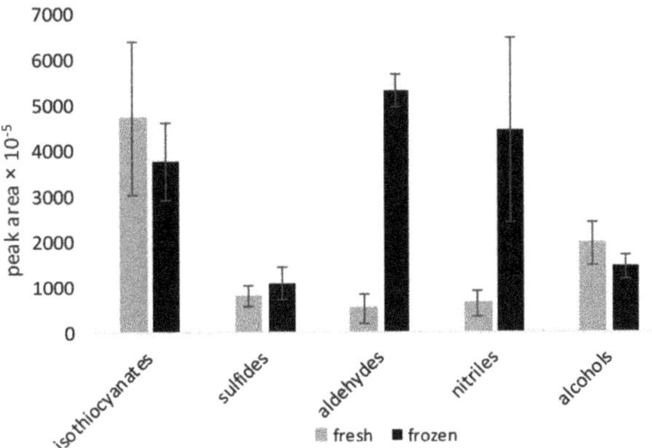



**Figure 5.** Mean values from all Brussels sprout cultivars (Ajax, Marte, Maximus, Neptuno and Profitus) for total peak areas of the main groups in raw non-frozen and frozen-thawed vegetables. The results are the mean of three replicates for each cultivar (i.e., 15 analyses), error bars show standard deviation. Amounts were expressed in total (sum of) peak areas.

## 3. Materials and Methods

### 3.1. Brassica Cultivars

Three cultivars of fresh broccoli (Covina, "2970", Malibu), three cultivars of cauliflower (Charlotte, Oviedo, Liria), four of kohlrabi (Konmar, Kolibri, Konan, Kordial) and five cultivars of Brussels sprouts (Ajax, Maximus, Profitus, Naptuno, Marte) were used for analysis. They were harvested during the same 2016 autumn season: Broccoli, cauliflower and kohlrabi were harvested in September and the Brussels sprout cultivars were harvested in December. The vegetables were delivered to the laboratory within 24 h after harvest and stored at −20 °C until analysis (approximately 4 weeks).

### 3.2. Cooking Process

All vegetables were cooked in the same way. After washing and chopping, about 200 g of plant material was placed in 1 L of boiling pure water with 7 g of salt. Cooking times differed among vegetables. It was 5 min for the broccoli, 7 min for the cauliflower and 10 min for the kohlrabi and Brussels sprouts.

### 3.3. GC×GC-ToFMS Analysis

The volatiles of the *Brassica* vegetable samples were isolated using headspace solid phase microextraction (HS-SPME) and analyzed using SPME-GC×GC-TOFMS for the comparative, semiquantitive (peak area percentage) analysis only. The SPME fiber carboxen/polydimethylsiloxane (DVB/CAR/PDMS) was used for volatile extraction (Supelco, Bellefonte, PA, USA). The vegetables were cut into approximately 0.5 cm pieces with a kitchen knife and 4 g of each was placed in a separate SPME vial. The samples were pre-incubated for 5 min at 60 °C, then the fiber was exposed for 30 min at the same temperature to adsorb volatiles. The volatile compounds isolated by SPME were desorbed in the injector port on the GC×GC–ToF-MS system (Pegasus 4D LECO, St. Joseph, MI, USA). The GC was equipped with a DB-5 column (25 m × 0.2 mm × 0.33 µm, Agilent Technologies, Santa Clara, CA, USA) and a Supelcowax 10 (1.2 m × 0.1 mm × 0.1 µm, Supelco Bellefonte, PA, USA) as the second column. The injector temperature was set at 250 °C and injection was performed in a splitless mode. The gas flow was set at 0.8 mL/min. The primary oven temperature was programed as follows: 40 °C for 1 min, then rising at 6 °C/min to 200 °C, where it then was reduced at 25 °C/min to 235 °C, where it was held for 5 min. In the secondary oven, the following programming was used: Held at

65 °C for 1 min, then rising at 6 °C/min to 225 °C, where it then decreased at 25 °C/min to 260 °C, where it was held for 5 min. The transfer line temperature was 260 °C. The modulation time was 4 s. The time-of-flight mass spectrometer was operating at a mass range of $m/z$ 38–388 and detector voltage −1700 V at 150 spectra/s. Total analysis time was 34.07 min. All analyses for any particular cultivar and preparation method were done in triplicates.

*3.4. Data Analysis*

Data were collected using the LECO ChromaTOF v.4.44 software (St. Joseph, MI, USA). Tentative identification was accomplished using the National Institute of Standards and Technology (NIST) library (version 2.0) of mass spectra. The calculations and basic statistical analysis were performed using Excel 2010 and Microsoft Office 2010. The principal component analysis (PCA) was performed by the SIMCA software, v. 14.1.0.2047 (MKS Unimetrix AB, Umea, Sweden).

## 4. Conclusions

The HS-SPME-GC×GC method was proposed for the identification of a large spectrum of volatiles from selected *Brassica* vegetables. This fast and efficient technique allowed for the determination of volatile compounds in different cultivars of broccoli, cauliflower, kohlrabi and Brussels sprouts. SPME is a method that provides a relatively fast volatile profiling of numerous samples. Despite many advantages of this technique, there are some issues which need to be mentioned here. First of all, there is no possibility to store volatile extracts (isolates), thus we face the problem of vegetable storage to analyze large sets of fresh raw vegetables. As presented, the freezing process caused some changes in the volatiles profile. Moreover, this is a non-exhaustive extraction method, so components with a low partition coefficient might be undetected, even if they are present in the sample in a relatively high concentration. Additionally, non-stable, easily reactive compounds can disintegrate, mainly due to high temperatures during thermal desorption from the fiber. The technique is very good for comparative purposes, pertinent for quantitative analysis; however, in the case of an external standard method used for quantitation, the impact of the matrix cannot be ignored, or alternatively, isotopomers of analyzed compounds could be used. Components from the following chemical groups were identified in those plants: Alcohols, aldehydes, isothiocyanates, nitriles, sulfides and others (unspecified). The proportions between the different groups were dependent on the vegetable type/species. The volatiles analyzed in raw and cooked vegetables showed changes occurring in the volatile fraction in cooked vegetables in comparison with raw ones. One of the most important differences in the volatile compounds concerned was a marked decrease in the contents of aldehydes and alcohols. The total amounts of sulfides decreased after cooking in all vegetables, except for broccoli. Isothiocyanate concentrations unexpectedly increased in most analyzed vegetables, while in the case of nitriles, a decrement was observed. Because all vegetables were frozen before analysis, the influence of freezing on the volatile fraction composition was also investigated. It was found that postharvest freezing of the *Brassica* vegetables induces biochemical changes that give rise to a significant modification of the compounds responsible for the aroma properties of those plants. Most of the changes are closely related to the activity of the LOX pathway enzymes. They are characterized by a high reduction in alcohol contents and an increase in aldehyde contents. Moreover, important changes were noted in the concentrations of bioactive components, e.g., isothiocyanates. The comparison of volatiles between fresh and thawed raw and cooked Brussels sprouts showed that after freezing, it is suggested to cook vegetables before consumption to maximize their beneficial properties.

**Supplementary Materials:** The following are available online at http://www.mdpi.com/1420-3049/24/3/391/s1, Table S1. Main volatile compounds identified using GC×GC in raw and cooked broccoli varieties; Table S2. Main volatile compounds identified using GC×GC in raw and cooked cauliflower varieties; Table S3. Main volatile compounds identified using GC×GC in raw and cooked kohlrabi varieties; Table S4. Main volatile compounds identified using GC×GC in raw and cooked Brussels sprouts varieties; Table S5. Compounds and their corresponding numbers used in PCA analysis (Figure 1).

**Author Contributions:** Conceptualization, M.N.W. and H.H.J.; Methodology, H.H.J. and M.N. Wieczorek; Software, M.N.W.; Investigation, M.N.W.; Resources, H.H.J.; Data curation, M.N.W.; Writing—original draft preparation, H.H.J. and M.N.W.; Writing—review and editing, H.H.J.; Visualization, M.N.W.; Supervision, H.H.J.; Project administration, H.H.J.; Funding acquisition, H.H.J.

**Funding:** This work was supported by the National Science Centre (Poland) under project Harmonia 7, 2015/18/M/NZ9/00372.

**Conflicts of Interest:** The authors declare no conflict of interest.

## References

1. Wieczorek, M.N.; Walczak, M.; Skrzypczak-Zielinska, M.; Jelen, H.H. Bitter taste of Brassica vegetables: The role of genetic factors, receptors, isothiocyanates, glucosinolates, and flavor context. *Crit. Rev. Food Sci. Nutr.* **2017**, 1–11. [CrossRef] [PubMed]

2. Chipurura, B.; Muchuweti, M.; Manditseraa, F. Effects of Thermal Treatment on the Phenolic Content and Antioxidant Activity of Some Vegetables. *Asian J. Clin. Nutr.* **2010**, *2*, 93–100. [CrossRef]

3. Davis, R.; Singh, K.P.; Kurzrock, R.; Shankar, S. Sulforaphane inhibits angiogenesis through activation of FOXO transcription factors. *Oncol. Rep.* **2009**, *22*, 1473–1478. [CrossRef]

4. Kong, J.S.; Yoo, S.A.; Kim, H.S.; Kim, H.A.; Yea, K.; Ryu, S.H.; Chung, Y.J.; Cho, C.S.; Kim, W.U. Inhibition of synovial hyperplasia, rheumatoid T cell activation, and experimental arthritis in mice by sulforaphane, a naturally occurring isothiocyanate. *Arthritis Rheumatol.* **2010**, *62*, 159–170. [CrossRef] [PubMed]

5. Johansson, N.L.; Pavia, C.S.; Chiao, J.W. Growth inhibition of a spectrum of bacterial and fungal pathogens by sulforaphane, an isothiocyanate product found in broccoli and other cruciferous vegetables. *Planta Med.* **2008**, *74*, 747–750. [CrossRef] [PubMed]

6. Yanaka, A.; Fahey, J.W.; Fukumoto, A.; Nakayama, M.; Inoue, S.; Zhang, S.; Tauchi, M.; Suzuki, H.; Hyodo, I.; Yamamoto, M. Dietary sulforaphane-rich broccoli sprouts reduce colonization and attenuate gastritis in Helicobacter pylori-infected mice and humans. *Cancer Prev Res.* **2009**, *2*, 353–360. [CrossRef] [PubMed]

7. Hecht, S.S. Inhibition of carcinogenesis by isothiocyanates. *Drug Metab. Rev.* **2000**, *32*, 395–411. [CrossRef] [PubMed]

8. Williams, D.J.; Critchley, C.; Pun, S.; Chaliha, M.; O'Hare, T.J. Differing mechanisms of simple nitrile formation on glucosinolate degradation in Lepidium sativum and Nasturtium officinale seeds. *Phytochemistry* **2009**, *70*, 1401–1409. [CrossRef] [PubMed]

9. Buttery, R.G.; Guadagni, D.G.; Ling, L.C.; Seifert, R.M.; Lipton, W. Additional volatile components of cabbage, broccoli, and cauliflower. *J. Agric. Food Chem.* **1976**, *24*, 829–832. [CrossRef]

10. Valette, L.; Fernandez, X.; Poulain, S.; Loiseau, A.M.; Lizzani-Cuvelier, L.; Levieil, R.; Restierc, L. Volatile constituents from Romanesco cauliflower. *Food Chem.* **2003**, *80*, 353–358. [CrossRef]

11. Jacobsson, A.; Nielsen, T.; Sjoholm, I. Influence of temperature, modified atmosphere packaging, and heat treatment on aroma compounds in broccoli. *J. Agric. Food Chem.* **2004**, *52*, 1607–1614. [CrossRef] [PubMed]

12. Engel, E.; Baty, C.; Le Corre, D.; Souchon, I.; Martin, N. Flavor-active compounds potentially implicated in cooked cauliflower acceptance. *J. Agric. Food Chem.* **2002**, *50*, 6459–6467. [CrossRef]

13. Ulrich, D.; Krumbein, A.; Schonhof, I.; Hoberg, E. Comparison of two sample preparation techniques for sniffing experiments with broccoli (Brassica oleracea var. italica Plenck). *Nahrung* **1998**, *42*, 392–394. [CrossRef]

14. De Pinho, P.G.; Valentão, P.; Gonçalves, R.F.; Sousa, C.; Andrade, P.B. Volatile composition of Brassica oleracea L. var. costata DC leaves using solid-phase microextraction and gas chromatography/ion trap mass spectrometry. *Rapid Commun. Mass Spectrom.* **2009**, *23*, 2292–2300. [CrossRef] [PubMed]

15. Jeleń, H.H.; Majcher, M.; Dziadas, M. Microextraction techniques in the analysis of food flavor compounds. A review. *Anal. Chim. Acta* **2012**, *738*, 13–26. [CrossRef] [PubMed]

16. Yang, C.; Li, D.; Wang, J. Microextraction techniques for the determination of volatile and semivolatile organic compounds from plants: A review. *Anal. Chim. Acta* **2013**, *799*, 8–22. [CrossRef] [PubMed]

17. Bojko, B.; Reyes-Garcés, N.; Bessonneau, V.; Goryński, K.; Mousavi, F.; Souza Silva, A.; Pawliszyn, J. Solid-phase microextraction in metabolomics. *Trends Anal. Chem.* **2014**, *61*, 168–180. [CrossRef]

18. Humston, E.M.; Zhang, Y.; Brabeck, G.F.; McShea, A.; Synovec, R.E. Development of a GC6GC–TOFMS method using SPME to determine volatile compounds in cacao beans. *J. Sep. Sci.* **2009**, *32*, 2289–2295. [CrossRef]
19. Matusheski, N.V.; Jeffery, E.H. Comparison of the bioactivity of two glucoraphanin hydrolysis products found in broccoli, sulforaphane and sulforaphane nitrile. *J. Agric. Food Chem.* **2001**, *49*, 5743–5749. [CrossRef]
20. Oliviero, T.; Verkerk, R.; Van Boekel, M.A.; Dekker, M. Effect of water content and temperature on inactivation kinetics of myrosinase in broccoli (Brassica oleracea var. italica). *Food Chem.* **2014**, *163*, 197–201. [CrossRef]
21. Zhang, Y. Allyl isothiocyanate as a cancer chemopreventive phytochemical. *Mol. Nutr. Food Res.* **2010**, *54*, 127–135. [CrossRef] [PubMed]
22. Jin, Y.; Wang, M.; Rosen, R.T.; Ho, C.T. Thermal degradation of sulforaphane in aqueous solution. *J. Agric. Food Chem.* **1999**, *47*, 3121–3123. [CrossRef] [PubMed]
23. Hanschen, F.S.; Herz, C.; Schlotz, N.; Kupke, F.; Bartolomé Rodríguez, M.M.; Schreiner, M.; Rohn Lamy, E. The Brassica epithionitrile 1-cyano-2,3-epithiopropane triggers cell death in human liver cancer cells in vitro. *Mol. Nutr. Food Res.* **2015**, *59*, 2178–2189. [CrossRef] [PubMed]
24. Collett, M.G.; Stegelmeier, B.L.; Tapper, B.A. Could nitrile derivatives of turnip (Brassica rapa) glucosinolates be hepato- or cholangiotoxic in cattle? *J. Agric. Food Chem.* **2014**, *62*, 7370–7375. [CrossRef] [PubMed]
25. Wittstock, U.; Burow, M. Glucosinolate breakdown in arabidopsis: Mechanism, regulation and biological significance. In *The Arabidopsis Book*; BioOne: Washington, DC, USA, 2010.
26. Oliviero, T.; Verkerk, R.; Dekker, M. Isothiocyanates from Brassica Vegetables-Effects of Processing, Cooking, Mastication, and Digestion. *Mol. Nutr. Food Res.* **2018**. [CrossRef] [PubMed]
27. Kim, J.K.; Park, S.U. Current potential health benefits of sulforaphane. *EXCLI J.* **2016**, *15*, 571–577. [CrossRef] [PubMed]
28. Kim, M.; Cho, H.J.; Kwon, G.T.; Kang, Y.H.; Kwon, S.H.; Her, S.; Park, T.; Kim, Y.; Kee, Y.; Park, J.H. Benzyl isothiocyanate suppresses high-fat diet-stimulated mammary tumor progression via the alteration of tumor microenvironments in obesity-resistant BALB/c mice. *Mol. Carcinog.* **2015**, *54*, 72–82. [CrossRef]
29. The Good Scents Company Information System. 2017. Available online: http://www.thegoodscentscompany.com (accessed on 9 December 2018).

**Sample Availability:** Samples of the compounds are not available from the authors.

 *molecules*

*Article*

# Discrimination of Cultivated Regions of Soybeans (*Glycine max*) Based on Multivariate Data Analysis of Volatile Metabolite Profiles

So-Yeon Kim [1], So Young Kim [1], Sang Mi Lee [1], Do Yup Lee [2], Byeung Kon Shin [3], Dong Jin Kang [3], Hyung-Kyoon Choi [4],* and Young-Suk Kim [1],*

[1] Department of Food Science and Engineering, Ewha Womans University, Seoul 03760, Korea; soyeon3143@daum.net (S.-Y.K.); 0303soyoung@naver.com (S.Y.K.); smlee78@ewha.ac.kr (S.M.L.)
[2] Department of Agricultural Biotechnology, Research Institute for Agricultural and Life Sciences, Seoul National University, Seoul 08779, Korea; rome73@snu.ac.kr
[3] National Agricultural Products Quality Management Service, Gimcheon 39660, Korea; sbkon1@korea.kr (B.K.S.); kdj1216@korea.kr (D.J.K.)
[4] College of Pharmacy, Chung-Ang University, Seoul 06974, Korea
* Correspondence: hykychoi@cau.ac.kr (H.-K.C.); yskim10@ewha.ac.kr (Y.-S.K.); Tel.: +82-10-7181-7922 (H.-K.C.); +82-2-3277-3091(Y.-S.K.); Fax: +82-2-812-3921(H.-K.C.); +82-2-3277-4213 (Y.-S.K.)

Academic Editor: Igor Jerković
Received: 2 January 2020; Accepted: 6 February 2020; Published: 10 February 2020

**Abstract:** Soybean (*Glycine max*) is a major crop cultivated in various regions and consumed globally. The formation of volatile compounds in soybeans is influenced by the cultivar as well as environmental factors, such as the climate and soil in the cultivation areas. This study used gas chromatography-mass spectrometry (GC-MS) combined by headspace solid-phase microextraction (HS-SPME) to analyze the volatile compounds of soybeans cultivated in Korea, China, and North America. The multivariate data analysis of partial least square-discriminant analysis (PLS-DA), and hierarchical clustering analysis (HCA) were then applied to GC-MS data sets. The soybeans could be clearly discriminated according to their geographical origins on the PLS-DA score plot. In particular, 25 volatile compounds, including terpenes (limonene, myrcene), esters (ethyl hexanoate, butyl butanoate, butyl prop-2-enoate, butyl acetate, butyl propanoate), aldehydes (nonanal, heptanal, *(E)*-hex-2-enal, *(E)*-hept-2-enal, acetaldehyde) were main contributors to the discrimination of soybeans cultivated in China from those cultivated in other regions in the PLS-DA score plot. On the other hand, 15 volatile compounds, such as 2-ethylhexan-1-ol, 2,5-dimethylhexan-2-ol, octanal, and heptanal, were related to Korean soybeans located on the negative PLS 2 axis, whereas 12 volatile compounds, such as oct-1-en-3-ol, heptan-4-ol, butyl butanoate, and butyl acetate, were responsible for North American soybeans. However, the multivariate statistical analysis (PLS-DA) was not able to clearly distinguish soybeans cultivated in Korea, except for those from the Gyeonggi and Kyeongsangbuk provinces.

**Keywords:** gas chromatography-mass spectrometry; solid-phase microextraction; soybean; origin discrimination; volatile compounds

---

## 1. Introduction

Soybean (*Glycine max*) is among the most important crops in the world and is extensively used in the production of soybean flour, soybean milk, fermented products, and oil for consumption by both humans and animals, mainly due to its high protein and fat contents [1]. It is generally accepted that soybean cultivation originated in China, but nowadays, soybeans are produced worldwide, including in North America, South America, and Asia [1]. The importing and exporting of agricultural products are increasing globally due to the expansion of the free-trade agreements. These circumstances have resulted in some foreign soybeans with unclear origins being distributed as domestic ones in Korea, which can lead to consumer distrust about the market [2]. The National Agricultural Products Quality Management Service in Korea introduced an agricultural food origin labeling system in 1991 to protect domestic agricultural producers and consumers [2]. Soybeans have been included in this system since 2017, and traders must now mark the origins of all products advertised for sale [2].

The properties and qualities of soybeans can be significantly affected by their cultivation region because each production area has different growing conditions, such as temperature, precipitation, and soil characteristics [3–5]. In particular, some previous studies have demonstrated that light and water characteristics and growth temperatures significantly affect the formation of volatile metabolites, such as alcohol and aldehydes in plants [6,7]. In addition, Grieshop et al. and Cherry et al. demonstrated that the environmental growing conditions of soybeans could change the synthesis pathways of proteins and fats, thereby affecting their chemical compositions [8,9].

Volatile components of soybeans have been found to be mainly derived from carbohydrates, proteins, and lipids via enzymatic reactions, autoxidation, and/or other chemical reactions during both storage and cultivation [10]. Lee et al. explained that major aroma constituents of soybean include hexanal, 1-octen-3-ol, γ-butyrolactone, maltol, and phenylethyl alcohol [11]. Also, Boué et al. and Dings et al. identified hexan-1-ol, octan-3-one, oct-1-en-3-ol, ethanol, octanal, 2-propanone, hexan-1-al, and 1-pentan-3-ol—most of which could be lipid oxidative degradation products—as major volatiles in soybean using Tenax trapping and solid-phase microextraction (SPME) combined by gas chromatography-mass spectrometry (GC-MS) analysis [12,13]. Since the composition of volatile compounds in soybeans can vary depending on their cultivation conditions, it might be feasible to use data sets of volatile components to distinguish where soybeans originate from.

Metabolomics is a commonly used tool for the identification and quantification of whole metabolites in biological samples [14]. Metabolite profiling has been performed using various instrumental methods, including mass spectrometry (MS) and nuclear magnetic resonance (NMR) spectroscopy [15–17]. These approaches have been successfully used to determine the geographical origins of various agricultural products [18–20]. Metabolomics approaches based on several different types of instrumental methods have been recently used to distinguish soybeans of different geographical origins [21,22]. Liquid-chromatography-orbitrap mass spectrometry (LC-Orbitrap MS) and gas-chromatography time-of-flight mass chromatography (GC-TOF MS) have been used to obtain the metabolic fingerprints of soybeans cultivated domestically in different provinces of Korea [21]. Fourier-transform infrared (FT-IR) spectroscopy has also been combined with multivariate statistical analysis to distinguish the geographical origins of Chinese and Korean soybeans [22]. However, while GC-MS has been widely used to discriminate the origins of foodstuffs and agricultural products, such as green tea, omija fruit, and honey, mainly due to its high resolution and sensitivity, in particular, in the analysis of volatile compounds, this method has not previously been applied to distinguish soybeans according to their cultivation regions based on the data sets of volatiles [14,23–25]. Therefore, we aimed to determine the feasibility of discriminating soybeans according to the cultivation regions using a GC-MS-based metabolomics approach in this study.

## 2. Results and Discussion

### 2.1. Profiling of Total Volatile Compounds in Soybeans

In total, 146 volatile compounds were identified in GC-MS data sets obtained from soybean samples of different geographical origins. Tables S1–S3 indicate that diverse lipid-derived volatile compounds and terpenes were detected in this study. Previous studies have found the major volatiles of soybeans to be ethanol, 1-octen-3-ol, maltol, phenylethyl alcohol, hexanal, octanal, 2-propanone, and γ-butyrolactone [11,12]. All of these volatile compounds were detected in the present study with the exception of maltol, which could have been due to the use of different extraction techniques [11]—the present study employed headspace extraction using SPME, which generally focuses on the detection of highly volatile compounds with low boiling points.

The 84 volatile compounds detected in the soybeans cultivated in Korea comprised of 1 acid, 23 alcohols, 9 aldehydes, 4 esters, 6 furans, 6 benzenes, 10 ketones, 3 lactones, 3 nitrogen-containing compounds, 2 sulfur-containing compounds, 10 hydrocarbons, 6 terpenes, and 1 phenol. Certain alcohols, such as 2-ethylhexan-1-ol, predominated, followed by ketones, such as propan-2-one, while terpenes were found at low levels in most samples. Unlike soybeans grown in China and North America, pyrazines were not identified in those cultivated in Korea.

The 124 volatile compounds identified in the soybeans cultivated in China comprised of 2 acids, 25 alcohols, 13 aldehydes, 13 esters, 4 furans, 8 benzenes, 17 ketones, 4 lactones, 3 nitrogen-containing compounds, 2 sulfur-containing compounds, 16 hydrocarbons, 11 terpenes, 3 phenols, and 3 pyrazines. Among them, 3-methylheptan-4-one was detected at higher levels compared to those cultivated in Korea and North America, and there was a greater diversity of terpenes in the Chinese soybeans.

The soybeans cultivated in North America contained 50 volatile compounds: 1 acid, 16 alcohols, 5 aldehydes, 4 esters, 2 furans, 4 benzenes, 4 ketones, 3 lactones, 1 nitrogen-containing compound, 1 sulfur-containing compound, 5 hydrocarbons, 2 terpenes, 1 phenol, and 1 pyrazine. The number of volatile compounds detected in North American soybeans was clearly smaller than in those from other cultivation areas, but there was a greater diversity of alcohol. Oct-1-en-3-ol was detected at higher levels, while propan-2-one and 2-methylprop-1-ene were present at lower levels in North American soybeans. The only pyrazine detected was 2-methylpyrazine. Among esters, the content of 3-hydroxy-2,4,4-trimethylpentyl 2-methylpropanoate was higher in soybeans from Indiana province (IN) than in those of other regions of North America.

Several enzymes of soybeans have been studied by various researchers, including lipoxygenase, lipase, urease, amylase, and protease [26,27]. In particular, soybeans are known to be a rich source of lipoxygenase [27], which is one of several enzymes used to produce aldehydes and alcohols via enzymatic oxidation [28]. This study found hexanal (13-linoleate hydroperoxide) and heptanal (11-linoleate hydroperoxide)—known as the major oxidative products from linoleate hydroperoxides— in most of the cultivation regions, as were octanal (11-oleate hydroperoxide) and nonanal (9-/10-oleate hydroperoxide) [29], which are known to be decomposition products of oleate hydroperoxides [30].

Benelli et al. found that the amount of hexanal was related to precipitation and light conditions in the cultivation area [7]. Table 1 [31] presents the differences in precipitation between the cultivation regions, whereas the amount of hexanal did not differ significantly between the geographical regions studied. In this study, alcohols—which are known to be secondary oxidative products of unsaturated fatty acids [29]—predominated in soybeans from Korea, China, and North America, among which pentan-1-ol and hexan-1-ol (both are derived from 13-linoleate hydroperoxide [30,32]) were observed in most samples. As mentioned above, oct-1-en-3-ol (produced from 10-linoleate hydroperoxide [32]) was the most abundant alcohol in soybeans cultivated in North America. On the other hand, furans can be produced from the oxidation of polyunsaturated fatty acids and carotenoids [33], and 2-alkylfurans are commonly derived from lipid degradation [34]. 2-Methylfuran, 2-ethylfuran, and 2-pentylfuran were detected in soybeans from Korea and China, whereas 2-methylfuran was not found in soybeans from North America.

Several ketones were also identified in soybeans from Korea, China, and North America. Other diverse ketones that are mainly formed from unsaturated fatty acids (e.g., linoleic acid) by lipoxygenase were found in soybeans from China [35,36]. Certain ketones, such as propan-2-one, butan-2-one, and 3-methylheptane-4-one, were commonly found in samples from China. Cheesbroug et al. and Gulen et al. explained that the activities of enzymes, such as peroxidase, increased with the temperature at which the plants were grown [37,38]. Also, some previous studies have reported that lipoxygenase activity is affected by the minimum mean temperature from flowering to maturity [39], which affects the formation of volatile compounds [40]. Table 1 indicates that the annual mean temperature was higher in China (excluding the northeast region) than in other cultivation areas (Korea and North

America). It could, therefore, be assumed that the formation of various ketones in soybeans from China is due to high lipoxygenase activity related to the temperatures of their cultivation areas.

**Table 1.** Climatic conditions and geographic coordinates of Korea, China, and North America.

| Province | Latitude | Longitude | Annual Mean Temperature (°C) | Annual Mean Precipitation (mm) |
|---|---|---|---|---|
| **Korea** | | | | |
| Gyeongi | 11° 7′ 15.744″ N | 105° 32′ 0.5748″ E | 11.7 | 1240 |
| Gangwon | 37° 52′ 52.7268″ N | 37° 52′ 52.7268″ N | 10.9 | 1307 |
| Chungcheongbuk | 36° 56′ 10.068″ N | 127° 41′ 44.736″ E | 10.8 | 1239 |
| Chungcheongnam | 36° 48′ 33.5196″ N | 127° 9′ 36.1512″ E | 11.8 | 1229 |
| Jeollabuk | 35° 47′ 52.8432″ N | 126° 53′ 31.9632″ E | 12.8 | 1251 |
| Jeollnam | 35° 1′ 37.308″ N | 126° 43′ 15.024″ E | 13.9 | 1264 |
| Kyeongsangbuk | 35° 59′ 18.312″ N | 128° 56′ 31.2″ E | 12.6 | 1026 |
| Kyeongsangnam | 35° 31′ 48.792″ N | 128° 30′ 28.116″ E | 13.3 | 1248 |
| **North America** | | | | |
| Illinois | 40° 37′ 59.25″ N | 89° 23′ 54.7044″ W | 11.42 | 947.93 |
| Indiana | 40° 16′ 1.8948″ N | 86° 8′ 5.6508″ W | 11.1 | 1011 |
| Minnesota | 46° 43′ 46.3908″ N | 94° 41′ 9.2328″ W | 7.3 | 807 |
| Michigan | 44° 18′ 53.4312″ N | 85° 36′ 8.5104″ W | 8.6 | 890 |
| Quebec | 52° 56′ 23.694″ N | 73° 32′ 56.8788″ W | 4.8 | 1001 |
| Ontario | 51° 15′ 13.5972″ N | 85° 19′ 23.5632″ W | 7.1 | 775.9 |
| **China** | | | | |
| Heilongjian | 45° 37′ 17.9832″ N | 126° 14′ 35.3466″ E | 3.4 | 562 |
| Jilin | 42° 31′ 40.44″ N | 125° 40′ 40.7994″ E | 4.9 | 784 |
| Liaoning | 40° 1′ 44.1114″ N | 124° 17′ 4.4484″ E | 9.0 | 1040 |
| Hebei | 38° 16′ 53.5578″ N | 114° 41′ 29.7276″ E | 13.2 | 517 |
| Shandong | 41° 1′ 59.0874″ N | 113° 6′ 25.6314″ E | 14.1 | 676 |
| Hubei | 30° 13′ 35.3634″ N | 115° 3′ 49.4634″ E | 17.0 | 1396 |
| Anhui | 33° 57′ 22.248″ N | 116° 47′ 20.5434″ E | 15.2 | 728 |
| Zhejiang | 30° 42′ 1.8″ N | 121° 0′ 37.3314″ E | 16.2 | 1118 |
| Fujian | 25° 6′ 50.796″ N | 99° 9′ 44.28″ E | 20.7 | 1677 |
| Jiangxi | 31° 21′ 54.6474″ N | 118° 23′ 22.8114″ E | 17.2 | 1475 |
| Guangdong | 24° 48′ 4.068″ N | 113° 35′ 33.7554″ E | 21.0 | 1499 |

Diverse terpenes that occur naturally as metabolites are commonly found in plants [41]. In general, terpenes are produced from isopentenyl diphosphate, which is elongated to geranyl diphosphate, farnesyl diphosphate, and geranylgeranyl diphosphate [42]. Those terpenes were identified in all of the present cultivated areas but showed the greatest abundance and variety in China. The 11 terpenes of α-pinene, α-thujene, sabinene, l-phellandrene, myrcene, α-terpinene, limonene, β-phellandrene, γ-terpinene, terpinolene, and α-cedrene were detected in soybeans from China. The formation of terpenes could depend on various factors, such as cultivar and region [43]. Marais reported that certain factors, such as increased temperature and acidic conditions, could affect the concentration and diversity of terpenes formed [43]. Also, terpene synthases could be affected by $CO_2$ levels [40]. According to Planbureau voor de Leefomgeving (PBL) Netherland Environmental Assessment Agency, China showed the largest $CO_2$ emissions in 2016 [44]. In particular, limonene derived from geranyl pyrophosphate was identified in all samples from China. A previous study suggested that a higher $CO_2$ concentration could enhance the activity of limonene synthase [40]. Therefore, the formation of limonene could be significantly affected by $CO_2$ concentration as well as other factors, such as temperature.

*2.2. Discrimination of Soybeans by Different Geographical Origins*

In order to discriminate soybeans according to their geographical origins, the relationship between soybeans from different cultivation regions and their volatile profiles was investigated. GC-MS data sets were processed using unsupervised statistical analysis (principal components analysis (PCA) and hierarchical clustering analysis (HCA)) as well as supervised statistical analysis (partial least square-discriminant analysis (PLS-DA)) [45]. PCA, HCA, and PLS-DA were performed to identify the differences in volatiles profiles obtained from GC-MS analyses of soybeans of different geographical origins.

The results of PCA were distinguished by their geographical origins (data not shown). Since both results of PCA and PLS-DA on score plots were similar, only PLS-DA results were presented to show the separation of samples according to the cultivation area (Figure 1). In addition, partial least square (PLS) components 1, 2, and 3 in the PLS PLS-DA 3D score plot for soybeans of different origins together explained 37.9% of the total variance: 24.66%, 6.84%, and 6.40%, respectively (Figure 1a). The PLS-DA score plot for PLS component 1 and PLS component 2 is presented (Figure 1b). The parameters of the cross-validation modeling were component 3, with $R^2X = 0.379$, $R^2Y = 0.788$, and $Q^2(cum) = 0.709$. After 100 times permutations, $R^2 = 0.177$ and $Q^2 = -0.219$ were obtained.

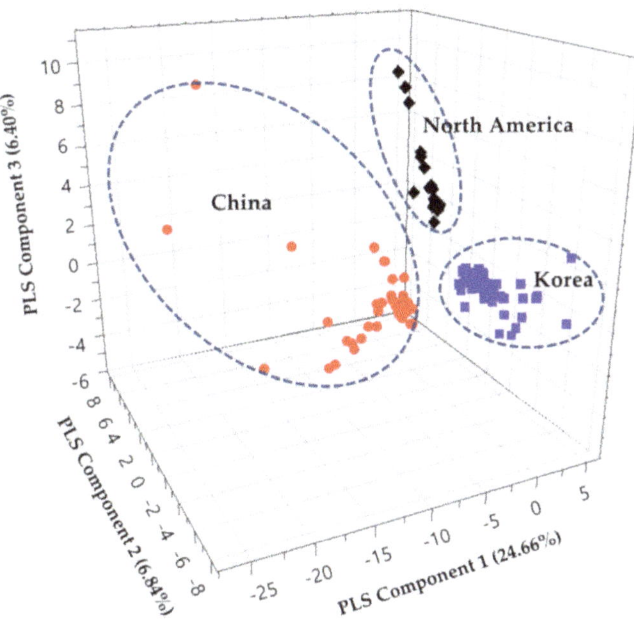


(a)

**Figure 1.** *Cont.*

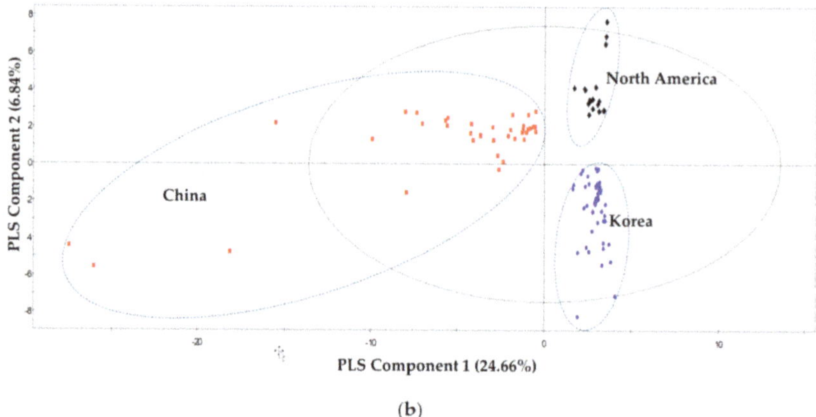



(**b**)

**Figure 1.** Partial least square-discriminant analysis (PLS-DA) score plot of soybean samples from different cultivation areas; (**a**) 3D score plot; (**b**) score plot PLS[1]-PLS[2], indicating the separation between different cultivation areas.

Some previous studies have shown that the chemical compositions of soybeans can vary significantly with differences in soils, fertilizer treatment, and climatic conditions, as well as other environmental factors [46–48]. Grieshop and Fashey showed that soybeans from China had greater crude protein content than those from North America [8]. Also, Shi et al. [47] demonstrated that soybeans from Korea contained more protein and less oil than those from North America. On the other hand, soybeans from China have been shown to have lower lipid concentration than those from North America [9]. Volatile compounds of soybeans are produced by nonvolatile precursors, such as lipids, sugars, and proteins [49]. In particular, oxidative degradation of lipids can lead to the formation of diverse volatiles. Certain lipid-derived compounds, such as oct-1-en-3-ol, differed significantly between soybeans from North America and those cultivated in other regions, which could be due to the higher lipid concentration of North American soybeans. On the other hand, the amounts of benzaldehyde, 2,6-dimethylpyrazine, and 2,5-dimethylpyrazine, which are known to be mainly produced by amino acids as major precursors [50], differed significantly between soybeans from China and those cultivated in other regions. This could be at least partially due to the differences in protein content between soybeans from different cultivation regions [46,48]. However, their exact formation mechanisms remain unclear, and they could involve both biological and chemical mechanisms during the cultivation and storage of the soybeans.

Medic et al. reported that the constituents of soybeans could be significantly altered by diverse environmental factors exerting complex combined effects [50]. This situation makes it difficult to explain how specific environmental factors influence the formation of volatile components in soybeans. As shown in Figure 1, the soybean samples could be divided into three groups for Korea, China, and North America. Soybeans from China are located in the area of negatively-related PLS component 1 in this score plot, whereas those from Korea and North America are located in the positions of both positively-related PLS component 1. Soybeans from North America are located in the positions of positively-related PLS component 2 in this score plot, whereas those from Korea are located in the positions of negatively-related PLS component 2.

Tables 2 and 3 list the main volatile metabolites identified according to the variable importance plot (VIP) values of >1.20. A VIP value >1 suggests that a compound plays a predominant role in the separation of groups [51]. The major volatile metabolites contributing to the positive PLS 1 axis were 2-ethylhexan-1-ol, while those in the negative axis of PLS component 1 were heptan-4-ol, butan-1-ol, butyl butanoate, octanal, butyl prop-2-enoate, 5-methyl-2-propan-2-ylcyclohexan-1-ol, butyl acetate, butyl propanoate, nonanal, toluene, heptanel, heptan-4-one, 5-ethyloxolan-2-one,

1,2,3-trimethylbenzene, heptan-2-one, (*E*)-hex-2-enal, ethyl hexanoate, (*E*)-hept-2-enal, limonene, 1-butoxybutane, 2-pentylfuran, acetaldehyde, myrcene, and 3-hydroxybutan-2-one, whereas those in the negative axis of PLS component 1 were found in all soybeans from China. On the other hand, the main volatile metabolites that contribute to the negative PLS 2 axis were 2-ethylhexan-1-ol, 2,5-dimethylhexan-2-ol, styrene, 2-methylfuran, 2- methylprop-2-ene, propan-2-one, 2-methylprop-2-enal, hexane, methyl acetate, 2-methylpentan-1-ol, octanal, butyl prop-2-enoate, 1-methyoxypropan-2-ol, heptanal, and toluene, whereas those in the positive PLS 2 axis were oct-1-en-3-ol, nonane, 4-methyloxolan-2-one, heptan-4-ol, butan-1-ol, octan-3-one, butyl butanoate, 3-hydroxy-2,4,4-trimethylpentyl 2-methylpropanoate, 5-methyl-2-propan-2-ylcyclohexan-1-ol, butyl acetate, butyl propanoate, and nonanal. In Figure 2, soybeans from each country are clustered according to their cultivation regions. The figure shows that soybeans from Korea were clustered more closely than the others, which is possibly due to the much smaller land area of that country (100,339 $km^2$) compared to China (9,596,951 $km^2$), Canada (9,984,670 $km^2$), and North America (9,826,676 $km^2$).

**Table 2.** The major volatile metabolites identified in soybeans from Korea, China, and North America according to variables importance plot (VIP > 1.20) list for partial least square (PLS) component 1.

| Retention Index (RI) Cal [1] | RI Ref [2] | Volatile Compounds | VIP Values | Identification (ID) [3] |
|---|---|---|---|---|
| | | Negative direction | | |
| 1289 | 1288 | Heptan-4-ol | 2.29 | B |
| 1151 | | Butan-1-ol | 2.19 | A |
| 1217 | | Butyl butanoate | 2.00 | A |
| 1285 | 1287 | Octanal | 1.97 | B |
| 1175 | | Butyl prop-2-enoate | 1.89 | A |
| 1642 | 1631 | 5-Methyl-2-propan-2-ylcyclohexan-1-ol | 1.88 | B |
| 1067 | | Butyl acetate | 1.88 | A |
| 1141 | | Butyl propanoate | 1.84 | A |
| 1391 | | Nonanal | 1.83 | A |
| 1027 | | Toluene | 1.80 | A |
| 1181 | | Heptanal | 1.75 | A |
| 1122 | | Heptan-4-one | 1.74 | A |
| 1688 | 1694 | 5-Ethyloxolan-2-one | 1.67 | B |
| 1273 | | 1,2,3-Trimethylbenzene | 1.63 | A |
| 1178 | | Heptan-2-one | 1.55 | A |
| 995 | | Acetonitrile | 1.54 | A |
| 1210 | | (*E*)-Hex-2-enal | 1.54 | A |
| 1234 | | Ethyl hexanoate | 1.49 | A |
| 1317 | | (*E*)-Hept-2-enal | 1.45 | A |
| 1190 | | Limonene | 1.40 | A |
| 959 | 968 | 1-Butoxybutane | 1.37 | B |
| 1230 | | 2-Pentylfuran | 1.30 | A |
| 605 | | Acetaldehyde | 1.27 | A |
| 1162 | | Myrcene | 1.27 | A |
| 1279 | | 3-Hydroxybutan-2-one | 1.22 | A |
| | | Positive direction | | |
| 1493 | | 2-Ethylhexan-1-ol | 1.75 | A |

[1] Retention indices were determined using *n*-alkanes $C_6$ to $C_{30}$ as an external standard; [2] Retention indices were obtained from national institute of standards and technology (NIST) database (http://webbook.nist.gov/chemistry); [3] Identification of the compounds was based as follows; A, mass spectrum and retention index agree with the authentic compounds under similar conditions (positive identification); B, mass spectrum and retention index were consistent with those from NIST database.

**Table 3.** The major volatile metabolites identified in soybeans from Korea, China, and North America according to variables importance plot (VIP > 1.20) list for partial least square PLS component 2.

| RI Cal [1] | RI Ref [2] | Volatile Compounds | VIP Values | ID [3] |
|---|---|---|---|---|
| | | Negative direction | | |
| 1493 | | 2-Ethylhexan-1-ol | 2.73 | A |
| 1194 | | 2,5-Dimethylhexan-2-ol | 2.29 | C |
| 1250 | | Styrene | 2.16 | A |
| 850 | | 2-Methylfuran | 2.02 | A |
| <600 | | 2-Methylprop-1-ene | 1.93 | C |
| 792 | | Propan-2-one | 1.88 | A |
| 861 | | 2-Methylprop-2-enal | 1.85 | A |
| 600 | | Hexane | 1.80 | A |
| 810 | | Methyl acetate | 1.48 | A |
| 1304 | 1312 | 2-Methylpentan-1-ol | 1.40 | B |
| 1285 | 1287 | Octanal | 1.37 | B |
| 1175 | | Butyl prop-2-enonate | 1.29 | A |
| 1129 | | 1-Methoxypropan-2-ol | 1.23 | A |
| 1181 | | Heptanal | 1.22 | A |
| 1027 | | Toluene | 1.21 | A |
| | | Positive direction | | |
| 1449 | | Oct-1-en-3-ol | 2.32 | A |
| 900 | 900 | Nonane | 2.11 | B |
| 1598 | | 4-Methyloxolan-2-one | 1.67 | C |
| 1289 | 1288 | Heptan-4-ol | 1.54 | B |
| 1151 | | Butan-1-ol | 1.48 | A |
| 1252 | | Octan-3-one | 1.43 | A |
| 1220 | | Butyl butanoate | 1.35 | A |
| 1850 | | 3-Hydroxy-2,4,4-trimethylpentyl 2-methylpropanoate | 1.33 | C |
| 1642 | 1631 | 5-Methyl-2-propan-2-ylcyclohexan-1-ol | 1.26 | B |
| 1067 | | Butyl acetate | 1.26 | A |
| 1140 | | Butyl propanoate | 1.24 | A |
| 1391 | | Nonanal | 1.23 | A |

[1] Retention indices were determined using *n*-alkanes $C_6$ to $C_{30}$ as an external standard; [2] Retention indices were obtained from NIST database (http://webbook.nist.gov/chemistry); [3] Identification of the compounds was based as follows; A, mass spectrum and retention index agree with the authentic compounds under similar conditions (positive identification); B, mass spectrum and retention index were consistent with those from NIST database; C, mass spectrum was consistent with that of W9N08 (Wiley and NIST) and manual interpretation (tentative identification).

Figure 2 shows the HCA dendrogram with its associated heatmap in which all of the samples are grouped in terms of their nearness or similarity [52]. The figure shows that all of the samples could be clustered into two groups except for Kyeongsangnam province Changnyeong (KNCN): group I consisted of 13 soybean samples cultivated in China, and group II comprised of 22 soybean samples from Korea and North America. The amounts of terpenes and esters were greater in group I than in group II. In group II, soybean samples from Korea—except for Kyeongsangnam province Changnyeong (KNCN)—and North America were classified into the subgroup. Among soybean samples grown in North America, those from Illinois (IL) and Indiana (IN) provinces could be distinguished from the others. Table S3 indicates that the samples from Illinois and Indiana provinces were found to contain greater amounts of alcohol than other North American soybeans (samples MI, MN, ON, and QB). The annual mean precipitations were similar across North America, but the annual mean temperatures were higher in Illinois and Indiana than in the other regions. Wills et al. reported that the concentration of esters and alcohols was positively related to temperature [53]. Therefore, it could be inferred that the formation of volatile compounds was affected by the cultivation temperature in soybeans from North America.

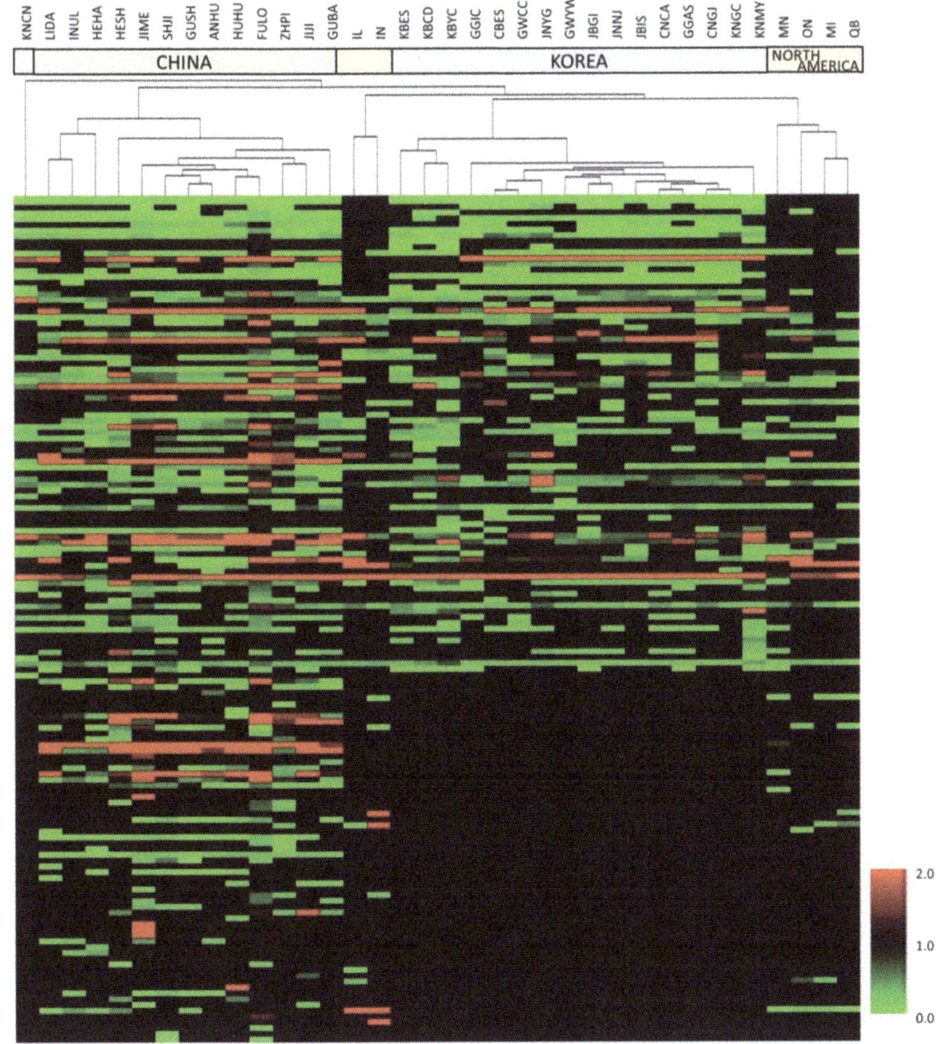

**Figure 2.** Heatmap generated by a hierarchical clustering analysis of 146 metabolites.

When the multivariate statistical analysis was performed only on domestic samples in Korea to investigate the possibility of our method to the discrimination of samples cultivated in the regions close to each other, it could not distinguish soybeans according to the region in the results of PCA (data not shown) and PLS-DA. Figure 3a shows that PLS 1, 2, and 3 together explained 43.7% of the total variance (19.09%, 15.36%, and 8.92%, respectively), while Figure 3b shows that two PLS components (PLS components 1 and 2) explained 33.86%. The parameters of the cross-validation modeling were component 3, with $R^2X = 0.437$, $R^2Y = 0.169$, and $Q^2(\text{cum}) = 0.0535$. After 100 times permutations, $R^2 = 0.0951$ and $Q^2 = -0.0676$ were obtained.

(a)

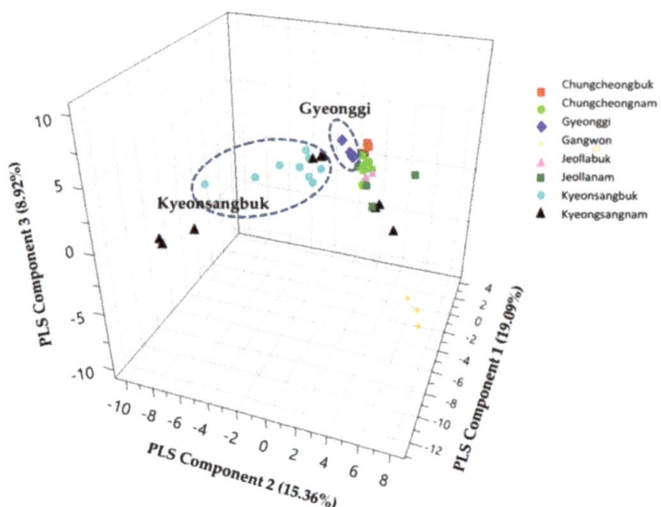

(b)

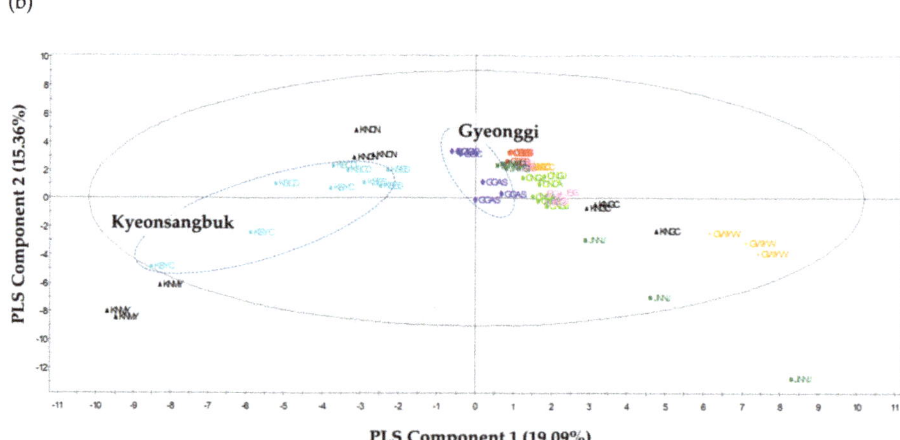

**Figure 3.** PLS-DA score plot of soybeans from Korea on the basis of volatile metabolites: (**a**) 3D score plot; (**b**) score plot PLS[1]-PLS[2]. GGIC—Gyeonggi province Anseong, GGAS—Gyeonggi province Icheon, GWCC—Gangwon province Chuncheon, GWYW—Gangwon province Yeongwol, CBES—Chungcheongbuk province Eumseong, CNCA— Chungcheongnam province Cheonan, CNGJ—Chungcheongnam province Gongju, JBGJ—Jeollabuk province Gimje, JBIS—Jeollabuk province Imsil, JNNJ—Jeollanam province Naju, JNYG—Jeollanam province Yeonggwang, KBCD—Kyeongsangbuk province Cheongdo, KBES—Kyeongsangbuk province Uiseong, KBYC—Kyeongsangbuk province Yeongcheon, KNCN—Kyeongsangnam province Changnyeong, KNMY—Kyeongsangnam province Miryang, KNGC—Kyeongsangnam province Geochang.

Soybean samples from the Gyeonggi and Kyeongsangbuk provinces were clustered according to their regions, whereas other samples were not clearly clustered in the PLS-DA score plot. As shown in Table 1, the climatic conditions varied with the cultivation area. The mean temperatures in 2016 showed similar tendencies in all of the cultivation regions studied, but with slight differences in the total precipitation and sun exposure times. Various plant volatiles can be affected by changing biotic and abiotic factors [54]. Vallat et al. explained that the concentrations of nonanal and benzaldehyde were both positively related to precipitation, and positively and negatively related to temperature, respectively [54]. This variety of climate factors could together affect the volatile metabolites formed in soybeans cultivated in different regions. However, the relationships between climate and the amounts of nonanal and benzaldehyde formed were not clear in this study. Other domestic samples except those from the Gyeonggi and Kyeongsangbuk provinces were not clearly grouped in the PLS-DA score plot.

## 3. Materials and Methods

### 3.1. Materials

Thirty-six different soybean samples (17 from Korea, 13 from China, and 6 from North America) cultivated in 2016 (Figures S1–S3, Table 4) were used. Soybeans from Korea were provided by the National Agricultural Products Quality Management Service, whereas those from China were obtained from Chinese markets (Figures S4 and S5, Table S4). Soybeans from North America were gifts from a soybean processing company in Korea (Figure S6, Table S4). All samples were stored at −70 °C in a deep freezer before they were analyzed. Solid-phase microextraction (SPME) fibers and holders were purchased from Supelco (Bellefonte, PA, USA), whereas vials and screw caps (Ultraclean 18 mm) were purchased from Agilent Technologies (Santa Clara, CA, USA). L-Borneol was purchased from Sigma-Aldrich (St. Louis, MO, USA). Authentic standard compounds for positive identification of volatile compounds were purchased as follows: 3-methylphenol and hexan-1-ol were purchased from Supelco (Bellefonte, PA, USA), 1,3-benzothiazole, acetaldehyde, α-terpinene were obtained from Fluka (St. Gallen, Switzerland), and acetonitrile was bought from J.T. Baker (Phillipsburg, NJ, USA), while all of the other authentic standards were purchased from Sigma-Aldrich (St. Louis, MO, USA).

**Table 4.** The origins of the 36 soybean samples from Korea, China, and North America.

| Nation | Province | Location | Labeling [1] |
|---|---|---|---|
| Korea | Gyeonggi | Anseong | GGIC |
| | | Icheon | GGAS |
| | Gangwon | Chuncheon | GWCC |
| | | Yeongwol | GWYW |
| | Chungcheongbuk | Eumseong | CBES |
| | Chungcheongnam | Cheonan | CNCA |
| | | Gongju | CNGJ |
| | Jeollabuk | Gimje | JBGJ |
| | | Imsil | JBIS |
| | Jeollanam | Naju | JNNJ |
| | | Yeonggwang | JNYG |
| | Kyeongsangbuk | Cheongdo | KBCD |
| | | Uiseong | KBES |
| | | Yeongcheon | KBYC |
| | Kyeongsangnam | Changnyeong | KNCN |
| | | Miryang | KNMY |
| | | Geochang | KNGC |

**Table 4.** *Cont.*

| Nation | Province | Location | Labeling [1] |
|---|---|---|---|
| China | Neimenggu | Ulanhot | INUL |
| | Heilongjiang | Harbin | HEHA |
| | Jilin | Meihekou | JIME |
| | Liaoning | Dandong | LIDA |
| | Hebei | Shijiazhuang | HESH |
| | Shandong | Jining | SHJI |
| | Anhui | Huaibei | ANHU |
| | Hubei | Huangshi | HUHU |
| | Zhejiang | Pinghu | ZHPI |
| | Jiangxi | Jiujiang | JIJI |
| | Fujian | Longyan | FULO |
| | Guangdong | Shaoguan | GUSH |
| | Guangxi | Hechi | GUBA |
| The United States (North America) | Illinois | | IL |
| | Indiana | | IN |
| | Minnesota | | MN |
| | Michigan | | MI |
| Canada (North America) | Quebec | | QB |
| | Ontario | | ON |

[1] GGIC—Gyeonggi province Anseong, GGAS—Gyeonggi province Icheon, GWCC—Gangwon province Chuncheon, GWYW—Gangwon province Yeongwol, CBES—Chungcheongbuk province Eumseong, CNCA— Chungcheongnam province Cheonan, CNGJ—Chungcheongnam province Gongju, JBGJ—Jeollabuk province Gimje, JBIS—Jeollabuk province Imsil, JNNJ—Jeollanam province Naju, JNYG—Jeollanam province Yeonggwang, KBCD—Kyeongsangbuk province Cheongdo, KBES—Kyeongsangbuk province Uiseong, KBYC—Kyeongsangbuk province Yeongcheon, KNCN—Kyeongsangnam province Changnyeong, KNMY—Kyeongsangnam province Miryang, KNGC—Kyeongsangnam province Geochang, INUL—Neimenggu province Ulanhot, HEHA—Heilongjiang province Harbin, JIME—Jilin province Meihekou, LIDA—Liaoning province Dandong, HESH—Hebei province Shijiazhuang, SHJI—Shandong province Jining, ANHU—Anhui province Huaibei, HUHU—Hubei province Huangshi, ZHPI—Zhejiang province Pinghu, JIJI—Jiangxi province Jiujiang, FULO—Fujian province Longyan, GUSH—Guangdong province Shaoguan, GUBA—Guangxi province Hechi, IL—Illinois province, IN—Indiana province, MN—Minnesota province, MI—Michigan province, QB—Quebec province, ON—Ontario province.

### 3.2. Extraction of Volatile Metabolites Using SPME

L-Borneol was prepared at 200 mg/L with *tert*-butanol. Then, distilled water was added at a final concentration of 1 mg/L before soybean (5 g) was placed in a 20 mL screw vial with a screw cap. SPME was used to obtain volatile metabolites of soybeans. The sample was maintained at 40 °C for 30 min to reach the equilibrium state. SPME fiber coated with carboxen/polydimethylsiloxane/ divinylbenzene (CAR/PDMS/DVB) was used to adsorb volatile compounds at 40 °C for 20 min, and desorption was executed at 200 °C in a GC injector for 5 min while cryo-trapping at −80 °C. For every other ten runs in GC-MS analysis, we included quality control (QC) soybean samples to confirm the relative peak areas and retention times of several main volatile compounds.

### 3.3. GC-MS Analysis

The GC-MS analysis was performed using a 7890A series gas chromatograph (Agilent Technologies, Santa Clara, CA, USA) and a 5975C mass detector (Agilent Technologies, Santa Clara, CA, USA) equipped with a DB-Wax column (30 m length × 0.25 mm i.d. × 0.25 μm film thickness, J&W Scientific,

Folsom, CA, USA). GC oven temperature was programmed as follows; initial temperature was maintained 40 °C for 10 min, raised to 42 °C at a rate of 2 °C/min and held for 3 min, and increased to 100 °C at a rate of 4 °C/min and kept for 5 min, and raised 180 °C at a rate of 4 °C/min, and the ramped to 200 °C at a rate of 10 °C/min. The flow rate of helium, carrier gas, was constant at 0.8 mL/min, whereas mass spectra were obtained with a mass scan rage of 35–350 atomic mass unites (a.m.u.) at a rate of 4.5 scans/sec, and the electron impact (EI) mode was 70 eV. All sample preparations and analyses were independently performed in triplicate. In the preliminary study, we confirmed the repeatability and precision of our method on the results of the main volatile compounds in soybean in more than six replicates.

*3.4. Identification and Quantification of Volatile Metabolites*

The identification of each volatile compound was positively confirmed by comparison of retention time and mass spectral data with those of authentic standard compounds. When standard compounds were not available, each volatile compound was identified on the basis of its mass spectral data using the NIST.08 and Wiley.9 mass spectral libraries and the retention index (RI) values in the previous literature. The RI value of volatile compounds was calculated with *n*-alkane from $C_6$ to $C_{30}$ as an external standard. The quantification of the volatile components was calculated to obtain relative peak areas by comparing their peak areas with that of the internal standard compound on the total ion chromatogram of GC-MS. Five microliters of L-borneol (1 mg/L in *tert*-butanol/distilled water solvents mixture (1:200, *v/v*)) was used as an internal standard.

*3.5. Statistical Analysis*

All the datasets obtained were processed by multivariate statistical analysis, such as principal components analysis (PCA) and partial least square-discriminant analysis (PLS-DA) using SIMCA-P (version 11.0, Umetrics, Umea, Sweden), to determine the discrimination of soybeans according to different geographic origins. Heatmap visualization and hierarchical clustering analysis were performed based on Pearson's correlation and average linkage method using Multi Experimental Viewer (MeV) software (version 4.9, The Institute for Genomic Research (TIGR)) [55].

## 4. Conclusions

This study applied GC-MS analysis combined with the multivariate statistical analysis to distinguish the geographical origins of soybeans. The profiles of volatile compounds in the soybean samples varied with their cultivation regions. In the PLS-DA results, all soybean samples were clearly discriminated by their geographical origins. However, those cultivated in Korea (except for the samples from the Gyeonggi and Kyeongsangbuk provinces) could not be clearly separated according to the region on the PLS-DA score plot. We also determined the major volatile metabolites that contributed to the discrimination of geographical origins on the basis of PLS-DA. This study has the advantage of being able to distinguish the geographical origin of soybeans without any sample pretreatment on the basis of volatile metabolite profiles, which are highly related to their quality. However, we did not have enough sample information on post-harvest practices, such as drying and storage conditions, which could affect volatiles' profiles in some way. Nevertheless, our result could be applied to the discrimination of soybeans distributed and commercially available in Korea, the main objective of this study.

In summary, the findings of this study suggested that combining GC-MS-based analysis of volatile compounds with multivariate data analysis is a useful tool for discriminating the geographical origins of soybeans, but with some limitations for domestically cultivated ones.

**Supplementary Materials:** The following are available online at http://www.mdpi.com/1420-3049/25/3/763/s1, Figures S1–S6, Tables S1–S4.

**Author Contributions:** Conceptualization, Y.-S.K. and H.-K.C.; methodology, S.Y.K., D.Y.L., H.-K.C., and Y.-S.K.; analysis, S.-Y.K. and S.Y.K.; investigation, S.Y.K.; resources, B.K.S. and D.J.K.; data curation, S.-Y.K., and S.M.L.; writing—original draft preparation, S.-Y.K., S.M.L., and Y.-S.K.; writing—review and editing, S.-Y.K. and Y.-S.K.; visualization, S.-Y.K. and D.Y.L.; supervision, Y.-S.K. All authors have read and agree to the published version of the manuscript.

**Funding:** This work was supported by Korea Institute of Planning and Evaluation for Technology in Food, Agriculture, Forestry (IPET) through Advanced Production Technology Development Program funded by Ministry of Agriculture, Food and Rural Affairs (MAFRA) (No. 316081-04), and the BK21 Plus program funded by the National Research Foundation of Korea (NRF) (No. 22A20130012233).

**Conflicts of Interest:** The authors declare no conflict of interest.

## References


1. Singh, G. *The Soybean: Botany, Production and Uses*; CABI: Wallingford, UK, 2010; Volume 1, pp. 1–7.
2. National Agricultural Products Quality Management Service. Available online: http://www.naqs.go.kr/eng/contents/contents.do?menuId=MN20581 (accessed on 15 October 2019).
3. Kaplan, L. What is the Origin of the Common Bean? *Econ. Bot.* **1981**, *35*, 240–254. [CrossRef]
4. Dong, Y.; Zhao, L.; Liu, B.; Wang, Z.; Jin, Z.; Sun, H. The Genetic Diversity of Cultivated Soybean Grown in China. *Appl. Genet.* **2004**, *108*, 931–936. [CrossRef] [PubMed]
5. Pongsuwan, W.; Fukusaki, E.; Bamba, T.; Yonetani, T.; Yamahara, T.; Kobayashi, A. Prediction of Japanese Green Tea Ranking by Gas chromatography/mass Spectrometry-Based Hydrophilic Metabolite Fingerprinting. *J. Agric. Food Chem.* **2007**, *55*, 231–236. [CrossRef] [PubMed]
6. Bertrand, B.; Boulanger, R.; Dussert, S.; Ribeyre, F.; Berthiot, L.; Descroix, F.; Joët, T. Climatic Factors Directly Impact the Volatile Organic Compound Fingerprint in Green Arabica Coffee Bean as Well as Coffee Beverage Quality. *Food Chem.* **2012**, *135*, 2575–2583. [CrossRef] [PubMed]
7. Benelli, G.; Caruso, G.; Giunti, G.; Cuzzola, A.; Saba, A.; Raffaelli, A.; Gucci, R. Changes in Olive Oil Volatile Organic Compounds Induced by Water Status and Light Environment in Canopies of *Olea Europaea*, L. Trees. *J. Sci. Food Agric.* **2015**, *95*, 2473–2481. [CrossRef]
8. Grieshop, C.M.; Fahey, G.C. Comparison of Quality Characteristics of Soybeans from Brazil, China, and the United States. *J. Agric. Food Chem.* **2001**, *49*, 2669–2673. [CrossRef]
9. Cherry, J.H.; Bishop, L.; Hasegawa, P.M.; Lefflert, H. Differences in the Fatty Acid Composition of Soybean Seed Produced in Northern and Southern Areas of the USA. *Phytochemistry* **1985**, *24*, 237–241. [CrossRef]
10. Min, S.; Yu, Y.; Yoo, S.; Martin, S.S. Effect of Soybean Varieties and Growing Locations on the Flavor of Soymilk. *J. Food Sci.* **2005**, *70*, C1–C11. [CrossRef]
11. Lee, K.; Shibamoto, T. Antioxidant Properties of Aroma Compounds Isolated from Soybeans and Mung Beans. *J. Agric. Food Chem.* **2000**, *48*, 4290–4293. [CrossRef]
12. Boué, S.M.; Shih, B.Y.; Carter-Wientjes, C.H.; Cleveland, T.E. Identification of Volatile Compounds in Soybean at various Developmental Stages using Solid Phase Microextraction. *J. Agric. Food Chem.* **2003**, *51*, 4873–4876. [CrossRef]
13. Dings, L.; Aprea, E.; Odake, S. Comparison of Volatile Flavour Profiles of Kidney Beans and Soybeans by GC-MS and PTR-MS. *Food Sci. Technol. Res.* **2005**, *11*, 63–70.
14. Oldiges, M.; Lütz, S.; Pflug, S.; Schroer, K.; Stein, N.; Wiendahl, C. Metabolomics: Current State and Evolving Methodologies and Tools. *Appl. Microbiol. Biotechnol.* **2007**, *76*, 495–511. [CrossRef] [PubMed]
15. Cubero-Leon, E.; Peñalver, R.; Maquet, A. Review on Metabolomics for Food Authentication. *Food Res. Int.* **2014**, *60*, 95–107. [CrossRef]
16. Kopka, J. Current Challenges and Developments in GC–MS Based Metabolite Profiling Technology. *J. Biotechnol.* **2006**, *124*, 312–322. [CrossRef] [PubMed]
17. Krishnan, P.; Kruger, N.; Ratcliffe, R. Metabolite Fingerprinting and Profiling in Plants using NMR. *J. Exp. Bot.* **2004**, *56*, 255–265. [CrossRef]
18. Kim, J.; Jung, Y.; Song, B.; Bong, Y.; Lee, K.; Hwang, G. Discrimination of Cabbage (*Brassica Rapa* ssp. Pekinensis) Cultivars Grown in Different Geographical Areas using 1H NMR-Based Metabolomics. *Food Chem.* **2013**, *137*, 68–75. [CrossRef]

19. Son, H.; Hwang, G.; Kim, K.M.; Ahn, H.; Park, W.; Van Den Berg, F.; Hong, Y.; Lee, C. Metabolomic Studies on Geographical Grapes and their Wines using 1H NMR Analysis Coupled with Multivariate Statistics. *J. Agric. Food Chem.* **2009**, *57*, 1481–1490. [CrossRef]

20. Jung, Y.; Lee, J.; Kwon, J.; Lee, K.; Ryu, D.H.; Hwang, G. Discrimination of the Geographical Origin of Beef by 1H NMR-Based Metabolomics. *J. Agric. Food Chem.* **2010**, *58*, 10458–10466. [CrossRef]

21. Lee, E.M.; Park, S.J.; Lee, J.; Lee, B.M.; Shin, B.K.; Kang, D.J.; Choi, H.; Kim, Y.; Lee, D.Y. Highly Geographical Specificity of Metabolomic Traits among Korean Domestic Soybeans (*Glycine Max*). *Food Res. Int.* **2019**, *120*, 12–18. [CrossRef]

22. Lee, B.; Zhou, Y.; Lee, J.S.; Shin, B.K.; Seo, J.; Lee, D.; Kim, Y.; Choi, H. Discrimination and Prediction of the Origin of Chinese and Korean Soybeans using Fourier Transform Infrared Spectrometry (FT-IR) with Multivariate Statistical Analysis. *PLoS ONE* **2018**, *13*, e0196315. [CrossRef]

23. Ye, N.; Zhang, L.; Gu, X. Discrimination of Green Teas from Different Geographical Origins by using HS-SPME/GC–MS and Pattern Recognition Methods. *Food Anal. Methods* **2012**, *5*, 856–860. [CrossRef]

24. Lee, H.J.; Cho, I.H.; Lee, K.E.; Kim, Y. The Compositions of Volatiles and Aroma-Active Compounds in Dried Omija Fruits (*Schisandra Chinensis* Baillon) According to the Cultivation Areas. *J. Agric. Food Chem.* **2011**, *59*, 8338–8346. [CrossRef] [PubMed]

25. Aliferis, K.A.; Tarantilis, P.A.; Harizanis, P.C.; Alissandrakis, E. Botanical Discrimination and Classification of Honey Samples Applying Gas chromatography/mass Spectrometry Fingerprinting of Headspace Volatile Compounds. *Food Chem.* **2010**, *121*, 856–862. [CrossRef]

26. Markley, K.S.; Goss, W.H. *Soybean Chemistry and Technology*; Chemical Publishing Company: Palm Springs, CA, USA, 1944; pp. 15–19.

27. Rice, R.; Wei, L.; Steinberg, M.; Nelson, A. Effect of Enzyme Inactivation on the Extracted Soybean Meal and Oil. *J. Am. Oil Chem. Soc.* **1981**, *58*, 578. [CrossRef]

28. Arai, S.; Noguchi, M.; Kaji, M.; Kato, H.; Fujimaki, M. N-Hexanal and some Volatile Alcohols: Their Distribution in Raw Soybean Tissues and Formation in Crude Soy Protein Concentrate by Lipoxygenase. *Agric. Biol. Chem.* **1970**, *34*, 1420–1423. [CrossRef]

29. Concepcion, J.C.T.; Ouk, S.; Riedel, A.; Calingacion, M.; Zhao, D.; Ouk, M.; Garson, M.J.; Fitzgerald, M.A. Quality Evaluation, Fatty Acid Analysis and Untargeted Profiling of Volatiles in Cambodian Rice. *Food Chem.* **2018**, *240*, 1014–1021. [CrossRef]

30. Snyder, J.; Frankel, E.; Selke, E.; Warner, K. Comparison of Gas Chromatographic Methods for Volatile Lipid Oxidation Compounds in Soybean Oil. *J. Am. Oil Chem. Soc.* **1988**, *65*, 1617–1620. [CrossRef]

31. Climate-Data. Available online: http://en.climate-data.org (accessed on 15 October 2019).

32. Zhang, Y.; Guo, S.; Liu, Z.; Chang, S.K. Off-Flavor Related Volatiles in Soymilk as Affected by Soybean Variety, Grinding, and Heat-Processing Methods. *J. Agric. Food Chem.* **2012**, *60*, 7457–7462. [CrossRef]

33. Yaylayan, V. Precursors, Formation and Determination of Furan in Food. *J. Verbrauch. Lebensm.* **2006**, *1*, 5–9. [CrossRef]

34. Min, D.; Callison, A.; Lee, H. Singlet Oxygen Oxidation for 2-pentylfuran and 2-pentenyfuran Formation in Soybean Oil. *J. Food Sci.* **2003**, *68*, 1175–1178. [CrossRef]

35. Whitfield, F.B.; Mottram, D.S. Volatiles from Interactions of Maillard Reactions and Lipids. *Crit. Rev. Food Sci. Nutr.* **1992**, *31*, 1–58. [CrossRef] [PubMed]

36. Leffingwell, J.C.; Alford, E.; Leffingwell, D. Identification of the Volatile Constituents of Raw Pumpkin (*Cucurbita Pepo*, L.) by Dynamic Headspace Analyses. *Leffingwell Rep.* **2015**, *7*, 1–14.

37. Cheesbrough, T.M. Changes in the Enzymes for Fatty Acid Synthesis and Desaturation during Acclimation of Developing Soybean Seeds to Altered Growth Temperature. *Plant Physiol.* **1989**, *90*, 760–764. [CrossRef] [PubMed]

38. Gulen, H.; Eris, A. Effect of Heat Stress on Peroxidase Activity and Total Protein Content in Strawberry Plants. *Plant Sci.* **2004**, *166*, 739–744. [CrossRef]

39. Kumar, V.; Rani, A.; Tindwani, C.; Jain, M. Lipoxygenase Isozymes and Trypsin Inhibitor Activities in Soybean as Influenced by Growing Location. *Food Chem.* **2003**, *83*, 79–83. [CrossRef]

40. Yuan, J.S.; Himanen, S.J.; Holopainen, J.K.; Chen, F.; Stewart, C.N., Jr. Smelling Global Climate Change: Mitigation of Function for Plant Volatile Organic Compounds. *Trends Ecol. Evol.* **2009**, *24*, 323–331. [CrossRef] [PubMed]

41. Croteau, R.; Davis, E.; Hartmann, T.; Hemscheidt, T.; Sanz-Cervera, J.; Shen, B.; Stocking, E.; Williams, R. *Biosynthesis: Aromatic Polyketides, Isoprenoids, Alkaloids*; Springer: Berlin, Germany, 2003; pp. 54–91.

42. Cho, I.H.; Lee, H.J.; Kim, Y. Differences in the Volatile Compositions of Ginseng Species (*Panax* sp.). *J. Agric. Food Chem.* **2012**, *60*, 7616–7622. [CrossRef]

43. Marais, J. Terpenes in the Aroma of Grapes and Wines: A Review. *S. Afr. J. Enol. Vitic.* **1983**, *4*, 49–58. [CrossRef]

44. Olivier, J.G.; Schure, K.; Peters, J. *Trends in Global CO₂ and Total Greenhouse Gas Emissions*; PBL Netherlands Environmental Assessment Agency: Hague, The Netherlands, 2017.

45. Wishart, D.S. Metabolomics: Applications to Food Science and Nutrition Research. *Trends Food Sci. Technol.* **2008**, *19*, 482–493. [CrossRef]

46. Shi, A.; Chen, P.; Zhang, B.; Hou, A. Genetic Diversity and Association Analysis of Protein and Oil Content in food-grade Soybeans from Asia and the United States. *Plant Breed* **2010**, *129*, 250–256. [CrossRef]

47. Maestri, D.M.; Labuckas, D.O.; Meriles, J.M.; Lamarque, A.L.; Zygadlo, J.A.; Guzmán, C.A. Seed Composition of Soybean Cultivars Evaluated in Different Environmental Regions. *J. Sci. Food Agric.* **1998**, *77*, 494–498. [CrossRef]

48. Dornbos, D.; Mullen, R. Soybean Seed Protein and Oil Contents and Fatty Acid Composition Adjustments by Drought and Temperature. *J. Am. Oil Chem. Soc.* **1992**, *69*, 228–231. [CrossRef]

49. MacLeod, G.; Ames, J.; Betz, N.L. Soy Flavor and Its Improvement. *Crit. Rev. Food Sci. Nutr.* **1988**, *27*, 219–400. [CrossRef] [PubMed]

50. Medic, J.; Atkinson, C.; Hurburgh, C.R. Current Knowledge in Soybean Composition. *J. Am. Oil Chem. Soc.* **2014**, *91*, 363–384. [CrossRef]

51. Oberg, J.; Spenger, C.; Wang, F.; Andersson, A.; Westman, E.; Skoglund, P.; Sunnemark, D.; Norinder, U.; Klason, T.; Wahlund, L. Age Related Changes in Brain Metabolites Observed by 1H MRS in APP/PS1 Mice. *Neurobiol. Aging* **2008**, *29*, 1423–1433. [CrossRef] [PubMed]

52. Granato, D.; Santos, J.S.; Escher, G.B.; Ferreira, B.L.; Maggio, R.M. Use of Principal Component Analysis (PCA) and Hierarchical Cluster Analysis (HCA) for Multivariate Association between Bioactive Compounds and Functional Properties in Foods: A Critical Perspective. *Trends Food Sci. Technol.* **2018**, *72*, 83–90. [CrossRef]

53. Wills, R.; McGlasson, W. Effect of Storage Temperature on Apple Volatiles Associated with Low Temperature Breakdown. *J. Hortic. Sci.* **1971**, *46*, 115–120. [CrossRef]

54. Vallat, A.; Gu, H.; Dorn, S. How Rainfall, Relative Humidity and Temperature Influence Volatile Emissions from Apple Trees In Situ. *Phytochemistry* **2005**, *66*, 1540–1550. [CrossRef]

55. The Institute for Genomic Research (TIGR). Available online: http://mev.tm4.org (accessed on 10 October 2019).

**Sample Availability:** Not available.

*Article*

# The Antibacterial Activity of Lavender Essential Oil Alone and In Combination with Octenidine Dihydrochloride against MRSA Strains

Paweł Kwiatkowski [1,*], Łukasz Łopusiewicz [2], Mateusz Kostek [1], Emilia Drozłowska [2], Agata Pruss [3], Bartosz Wojciuk [1], Monika Sienkiewicz [4], Hanna Zielińska-Bliźniewska [4] and Barbara Dołęgowska [3]

1   Department of Diagnostic Immunology, Chair of Microbiology, Immunology and Laboratory Medicine, Pomeranian Medical University in Szczecin, 72 Powstańców Wielkopolskich Avenue, 70-111 Szczecin, Poland; mkosa9406@gmail.com (M.K.); bartosz.wojciuk@pum.edu.pl (B.W.)
2   Center of Bioimmobilisation and Innovative Packaging Materials, Faculty of Food Sciences and Fisheries, West Pomeranian University of Technology Szczecin, Janickiego 35, 71-270 Szczecin, Poland; lukasz.lopusiewicz@zut.edu.pl (Ł.Ł.); emilia_drozlowska@zut.edu.pl (E.D.)
3   Department of Laboratory Medicine, Chair of Microbiology, Immunology and Laboratory Medicine, Pomeranian Medical University in Szczecin, 72 Powstańców Wielkopolskich Avenue, 70-111 Szczecin, Poland; agata.pruss@pum.edu.pl (A.P.); barbara.dolegowska@pum.edu.pl (B.D.)
4   Department of Allergology and Respiratory Rehabilitation, Medical University of Łódź, Żeligowskiego 7/9, 90-752 Łódź, Poland; monika.sienkiewicz@umed.lodz.pl (M.S.); hanna.zielinska-blizniewska@umed.lodz.pl (H.Z.-B.)
*   Correspondence: pawel.kwiatkowski@pum.edu.pl; Tel.: +48-91-466-1659

Academic Editor: Igor Jerković
Received: 23 November 2019; Accepted: 24 December 2019; Published: 26 December 2019

**Abstract:** In the post-antibiotic era the issue of bacterial resistance refers not only to antibiotics themselves but also to common antiseptics like octenidine dihydrochloride (OCT). This appears as an emerging challenge in terms of preventing staphylococcal infections, which are both potentially severe and easy to transfer horizontally. Essential oils have shown synergisms both with antibiotics and antiseptics. Therefore the aim of this study was to investigate the impact of lavender essential oil (LEO) on OCT efficiency towards methicillin-resistant *S. aureus* strains (MRSA). The LEO analyzed in this study increased the OCT's susceptibility against MRSA strains. Subsequent FTIR analysis revealed cellular wall modifications in MRSA strain cultured in media supplemented with OCT or LEO/OCT. In conclusion, LEO appears to be a promising candidate for an efficient enhancer of conventional antiseptics.

**Keywords:** MRSA; lavender essential oil; octenidine dihydrochloride; synergistic activity; FTIR

---

## 1. Introduction

*Staphylococcus aureus* is a commensal bacterium that can colonize the skin and the mucoses of humans, but is also a pathogenic microorganism responsible for many types of infections. The pathogenicity of this bacterium is primarily associated with the variety of virulence factors like enzymes and toxins [1]. Virulence factors combined with the efficient evasion mechanisms make *S. aureus* a formidable opponent. Asymptomatic carriage, mostly present in the nasal vestibule, may directly influence the development of infection under favorable circumstances. Skin and soft tissue infections (SSTIs) are the most common forms of *S. aureus* etiology [2]. These occur in both outpatients and inpatients. These infections are associated with the disruption of natural protective barriers in the skin and mucous membranes. After invasion, the bacteria multiply, the expression of their virulence genes as well as toxins production increases, and this results in the development of clinical symptoms such as SSTIs and surgical site infections (SSIs) [3].

In each of these, microbes can enter the vascular bed and cause severe systemic infection. This applies especially to inpatients in intensive care units and surgical departments but also to hemodialyzed outpatients for whom long lasting endovascular catheterization is applied. For these patients, experiencing circulatory failure, respiratory failure, severe surgical trauma, hypothermia, or hypovolemia increases the degree of tissue hypoxia. Additionally, diabetes mellitus frequently coexisting with hemodialysis is predisposed to bacterial colonization in the sites exposed to iatrogenic skin damage. As a result, the risk of systemic infection increases significantly [4].

As *S. aureus* potentially contributes to the skin microflora, the horizontal transmission of pathogenic strains appears to be critical. Most of the aforementioned infections can spread easily via skin-to-skin contact or via contaminated everyday use items. Consequently, it is crucial to deliver efficient and safe hygiene measures in order to disrupt the transmission process.

Moreover, due to increasingly common resistance to β-lactam antibiotics among *S. aureus*, the treatment of infections with these microorganisms, including the SSIs, has become more challenging. Methicillin-resistant *Staphylococcus aureus* (MRSA) strains are not only found in hospital environments or in inpatients, but can also develop in outpatients [5]. In inpatients with a high risk of colonization by MRSA due to having implanted artificial valves or vascular grafts, vancomycin is given as an alternative drug [6]. In order to avoid complications, various types of antiseptic agents are also used in wound care. One of them is *N,N'*-(1,10-decanediyldi-1[4*H*]-pyridinyl-4-ylidene)-bis-(1-octanamine) dihydrochloride, also known as octenidine dihydrochloride (OCT) [7]. OCT is a cationic active compound that exhibits a broad bactericidal spectrum, including MRSA. This antiseptic agent works by interacting with bacterial cell structures, which consequently results in lysis and cell death. OCT is light-resistant, and is chemically stable in a broad range of pH (1.6–12.2) and temperatures [8]. In addition to high antibacterial efficacy, OCT neither adversely affects epithelial cells, nor impedes the wound healing process. OCT is used only topically and is not absorbed into general circulation, so it does not cause any systemic effects. Due to its properties, OCT works well when applied to wounds, mucous membranes, and skin. OCT shows a synergistic effect with phenoxyethanol, hence phenoxyethanol as an aqueous solution in combination with OCT is applied in medical practice [7,9], e.g., to decolonize vulnerable patients with MRSA, which is an indispensable element of hospital-acquired infection prevention.

Still, the problem of bacterial resistance also applies to antiseptics such as OCT. Hardy et al. [10] observed a correlation between the use of these antiseptics and a staphylococcal sensitivity decrease. They stated that Minimum Inhibitory Concentration (MIC) and Minimum Bactericidal Concentration (MBC) values of OCT increased rapidly after OCT's introduction to widespread use. The authors pointed out the mutations in *norA* and *norB* genes encoding efflux pump proteins as a possible reason for bacterial tolerance towards antiseptics. Hence, investigating the preparation methods which support the activity of antiseptics seems to be an important and interesting research area.

This is particularly so with regard to common exposure and the severity of staphylococcal infections as described above. Essential oils (EOs) represent a major example of this [11]. Firstly, the combination of EOs and antiseptic agents can contribute to reducing the risk of infection in healing wounds caused by MRSA strains. Secondly, a synergistic effect between the active compounds can enable a dose reduction and a concomitant alleviation of side effects typically associated with these EOs and antiseptic agents. Finally, some EOs have a pleasant fragrance which can provide psychological benefits facilitating wound healing.

It seems that lavender essential oil (LEO) extracted from the flowering tops of *Lavandula angustifolia* Mill. (*Lamiaceae*) is a promising candidate for a natural product which can increase the synergistic effect of some antiseptic agents such as OCT. LEO has a wide range of applications in pharmaceutical products and as a fragrance ingredient in the cosmetics industry [12]. It has been also proven that LEO has beneficial immunomodulatory effects on wound healing [13]. In addition, this oil has various pharmacological effects described in the available literature, such as antibacterial, antifungal, antioxidant, anxiolytic, anticonvulsant, and anticholinesterase properties [14–19]. According to Malcolm and Tallian [20], LEO is classified as Generally Recognized as Safe (GRAS) by the Food and Drugs Administration (FDA) (21CFR182.20 2015).

The exact mode of action of LEO is still not fully recognized. It is hypothesized that it influences bacterial wall ultrastructure and therefore modifies whole bacterial cell susceptibility. Hence, the aim of this study was to investigate the influence of LEO on the antibacterial activity of OCT against MRSA strains. Special attention was paid to the possible effect of LEO on bacterial cell wall modification.

## 2. Results

### 2.1. Chemical Analysis of LEO

The qualitative and quantitative chemical composition of LEO analyzed using GC-FID-MS are listed in Table 1. The total number of compounds identified in LEO was 29, representing 98.5% of the total oil content. The remaining compounds (1.5%) appeared in trace amounts. The main constituents of tested LEO were linalool (34.1%) and linalyl acetate (33.3%) (Figure 1) followed by lavandulyl acetate (3.2%), (Z)-β-ocimene (3.2%), (E)-β-ocimene (2.7%), β-caryophyllene (2.7%), 1,8-cineole (2.5%), terpinene-4-ol (2.5%), and myrcene (2.4%).

### 2.2. The Antibacterial Activity of Chemicals against MRSA Strains

As determined using the microdilution method, the control strain was susceptible to both LEO and OCT. The obtained MIC values were 1.95 ± 0.00 µg/mL and 18.29 ± 7.92 mg/mL for OCT and LEO, respectively. It was also found that both OCT and LEO showed antibacterial activity against MRSA clinical strains. The MIC of OCT inhibiting growth of these strains ranged between 3.52 ± 0.00 µg/mL to 3.91 ± 0.00 µg/mL, whereas the MIC of LEO was slightly higher (13.72 ± 0.00 mg/mL). Moreover, it was also observed that the addition of Tween 80 (1%, *v/v*) or DMSO (2%, *v/v*) had no impact on the growth of any of the strains. The results of the MICs and MBCs of OCT and LEO against MRSA strains are summarized in Table 2.

### 2.3. Synergistic Effect of LEO and OCT

The study showed that LEO presented synergistic activity in combination with OCT against MRSA reference strain and clinical isolates (the FICI values ranged from 0.11 to 0.26). The detailed results of a checkerboard assay against MRSA strains are summarized in Table 2.

### 2.4. Effect of LEO Alone and In Combination With OCT against MRSA Reference Strain

#### 2.4.1. Time-Killing Curves

The time-kill kinetics profile of MRSA reference strain grown in different media (A–G) are shown in Figure 2. The MRSA strain cultured in medium G showed a reduction in the number of viable cells within the first 5 h when compared to the medium E as well as medium F.

#### 2.4.2. FTIR Analysis

The complete FTIR spectra of the samples are shown in Figures 3 and 4. No qualitative differences were observed between samples isolated from media B-E and the control sample (medium A). However, the analysis targeting in particular cellular wall components was revealed. The differences in FTIR spectra between the sample isolated from media E-G in comparison to the control sample (medium A) were observed. In the E sample, no changes at 3280 cm$^{-1}$, 2959 cm$^{-1}$, 2927 cm$^{-1}$, 1454 cm$^{-1}$, and 1394 cm$^{-1}$ were noticed. A noticeable growth of absorbance at bands 1636 cm$^{-1}$, 1532 cm$^{-1}$, and 1230 cm$^{-1}$ was observed. Moreover, an increase of absorbance at 1057 cm$^{-1}$ was also observed. Sample F showed a more multi-faceted influence on the chemical composition of *S. aureus* cells (Figure 3). Noticeable growth of absorption peaks were observed. Moreover, the new peaks at 895 cm$^{-1}$ and 837 cm$^{-1}$ were noticed (Figure 4). The FTIR spectrum of a sample exposed to both E and F samples showed that all changes observed in the cells under the influence of both compounds are also separately found when these compounds are used together.

**Table 1.** Chemical composition of volatile constituents of commercial lavender essential oil from the flowering herb of *Lavandula angustifolia* Mill. (*Lamiaceae*).

| Compound | RI | Relative Concentration (%) |
|---|---|---|
| **Monoterpenes** | | |
| α-Pinene | 936 | 0.1 |
| Camphene | 950 | 0.1 |
| Myrcene | 987 | 2.4 |
| *p*-Cymene | 1015 | 0.2 |
| 1,8-Cineole | 1024 | 2.5 |
| Limonene | 1025 | 0.6 |
| (Z)-β-Ocimene | 1029 | 3.2 |
| (E)-β-Ocimene | 1041 | 2.7 |
| γ-Terpinene | 1051 | 0.1 |
| Terpinolene | 1082 | 0.2 |
| **Monoterpene isoprenoids** | | |
| Linalool | 1086 | 34.1 |
| Camphor | 1123 | 1.2 |
| Izoborneol | 1142 | 0.2 |
| Borneol | 1150 | 1.4 |
| Lavandulol | 1151 | 1.1 |
| Terpinene-4-ol | 1164 | 2.5 |
| *cis*-Dihydrocarvone | 1172 | 0.2 |
| α-Terpineol | 1176 | 1.8 |
| Linalyl acetate | 1239 | 33.3 |
| Lavandulyl acetate | 1275 | 3.2 |
| Neryl acetate | 1342 | 0.8 |
| Geranyl acetate | 1362 | 1.3 |
| **Sesquiterpenes** | | |
| β-Caryophyllene | 1421 | 2.7 |
| Aromadendrene | 1443 | 0.1 |
| (E)-β-Farnezene | 1446 | 0.4 |
| Bicyclosesquiphellandrene | 1487 | 0.1 |
| **Sesquiterpene isoprenoids** | | |
| Caryophyllene oxide | 1578 | 0.1 |
| **Esters** | | |
| Oct-1-en-3-yl acetate | 1093 | 0.6 |
| **Ketones** | | |
| Octan-3-one | 969 | 1.3 |
| Total | | 98.5 |

RI: Retention index measured relative to *n*-alkanes (C-9 to C-26) on a non-polar Rtx-1 column.

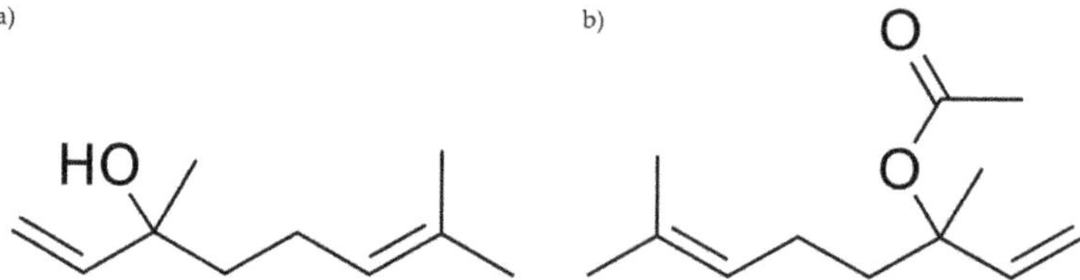



**Figure 1.** The chemical structures of the main compounds of lavender essential oil: linalool (**a**) and linalyl acetate (**b**).

**Table 2.** Fractional inhibitory concentration (FIC) and FIC indices (FICI) of octenidine dihydrochloride (OCT)—lavender essential oil (LEO) pairs against methicillin-resistant *Staphylococcus aureus* (MRSA) strains.

| | Bacteria | OCT-LEO | MICo | MBC | MICc | FIC | FICI | Type of Interaction |
|---|---|---|---|---|---|---|---|---|
| reference strain | ATCC 43300 | OCT (µg/mL) | 1.95 ± 0.00 | 5.21 ± 2.26 | 0.12 ± 0.00 | 0.06 | 0.11 | synergy |
| | | LEO (mg/mL) | 18.29 ± 7.92 | 439.00 ± 0.00 | 0.86 ± 0.00 | 0.05 | | |
| isolates | 1 | OCT (µg/mL) | 3.91 ± 0.00 | 11.72 ± 5.52 | 0.12 ± 0.00 | 0.03 | 0.16 | synergy |
| | | LEO (mg/mL) | 13.72 ± 0.00 | 27.44 ± 0.00 | 1.71 ± 0.00 | 0.13 | | |
| | 2 | OCT (µg/mL) | 3.52 ± 0.00 | 7.04 ± 0.00 | 0.24 ± 0.00 | 0.13 | 0.26 | synergy |
| | | LEO (mg/mL) | 13.72 ± 0.00 | 27.44 ± 0.00 | 1.71 ± 0.00 | 0.13 | | |
| | 3 | OCT (µg/mL) | 3.52 ± 0.00 | 7.04 ± 0.00 | 0.12 ± 0.00 | 0.06 | 0.12 | synergy |
| | | LEO (mg/mL) | 13.72 ± 0.00 | 27.44 ± 0.00 | 0.86 ± 0.00 | 0.06 | | |

Values are expressed as mean ± standard deviation. MICo, minimum inhibitory concentration of OCT or LEO; MBC, minimum bactericidal concentration; MICc, minimum inhibitory concentration of OCT/LEO combination. FIC index = FIC of OCT + FIC of LEO. FICI < 0.5, synergy; 0.5 ≤ FICI ≤ 1.0, addition; 1.1 < FICI ≤ 4.0, indifference; FICI > 4.0, antagonism. Using the known density of LEO, the final result was expressed in mg/mL.

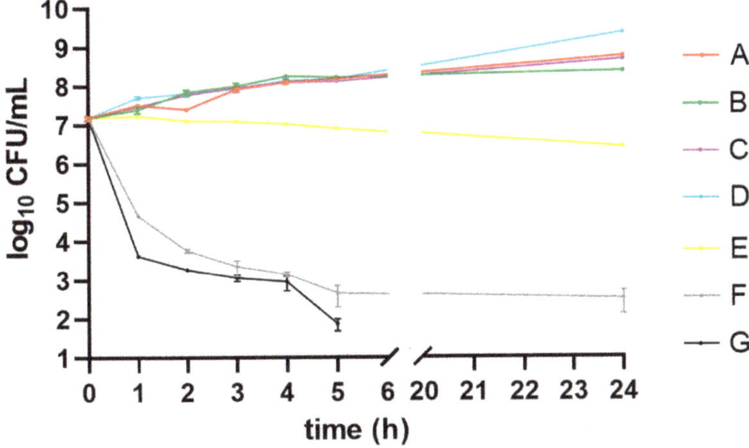


**Figure 2.** Time-kill kinetics of *Staphylococcus aureus* ATCC 43300 (MRSA) strain grown in Mueller-Hinton broth containing: no chemicals (control—medium A), Tween 80 (medium B), DMSO (medium C), Tween 80 and DMSO (medium D), lavender essential oil (LEO) at subinhibitory concentration (MIC$_{50}$) (medium E), octenidine dihydrochloride (OCT) at subinhibitory concentration (MIC$_{50}$) (medium F), LEO/OCT at subinhibitory concentrations (MICc$_{50}$) (medium G). CFU—colony forming unit.

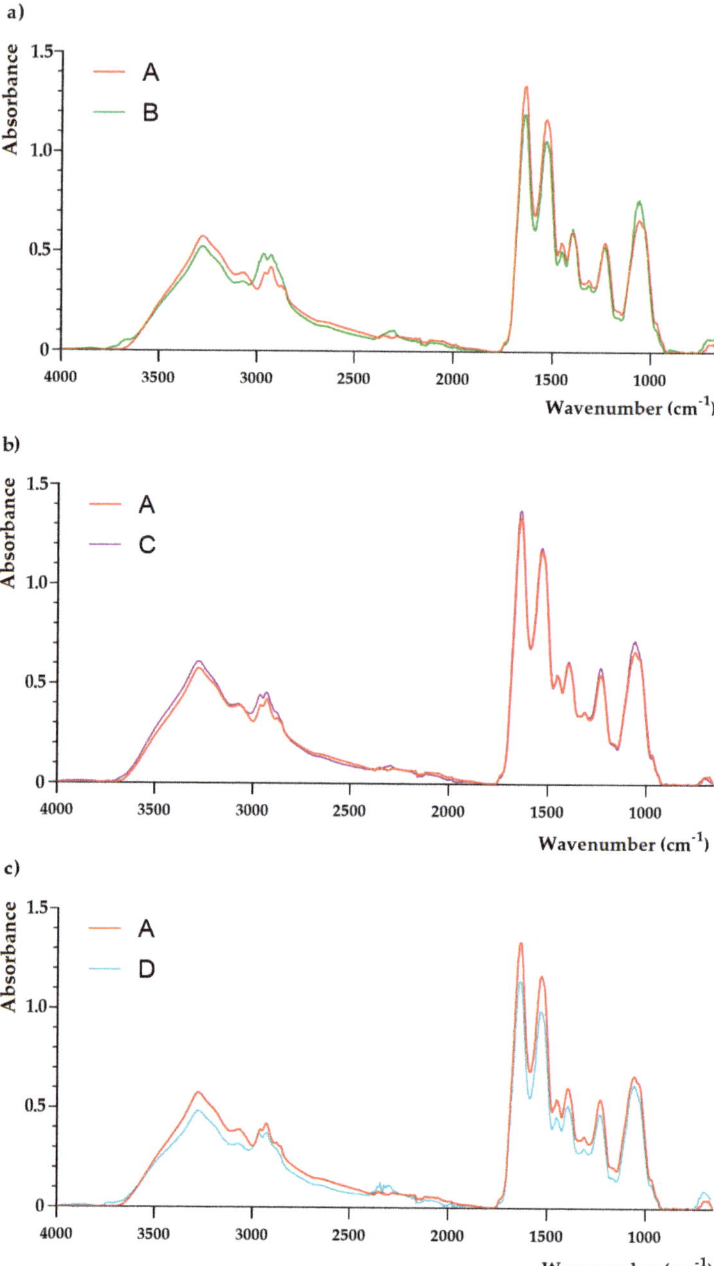

**Figure 3.** *Cont.*

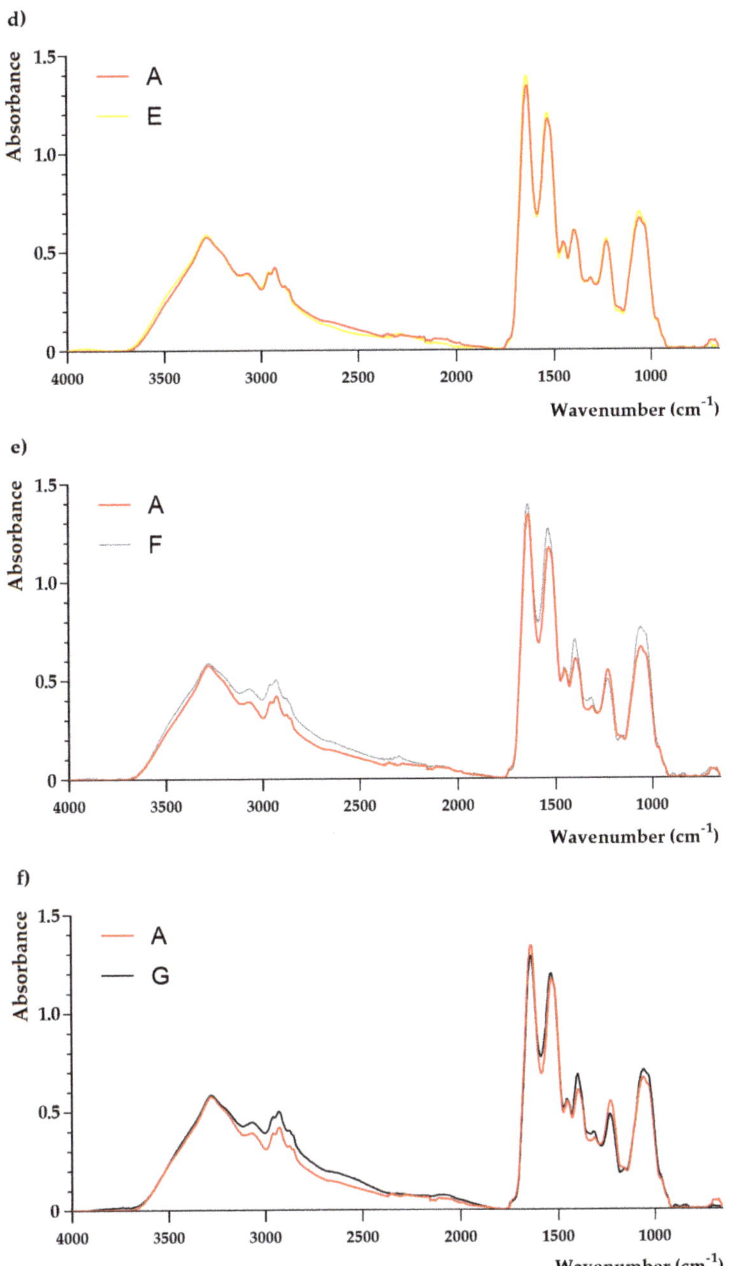

**Figure 3.** FTIR spectra of *Staphylococcus aureus* ATCC 43300 (MRSA) strain grown in Mueller-Hinton broth containing: no chemicals (control—medium A), Tween 80 (medium B), DMSO (medium C), Tween 80 and DMSO (medium D), lavender essential oil (LEO) at subinhibitory concentration ($MIC_{50}$) (medium E), octenidine dihydrochloride (OCT) at subinhibitory concentration ($MIC_{50}$) (medium F), LEO/OCT at subinhibitory concentrations ($MICc_{50}$) (medium G).

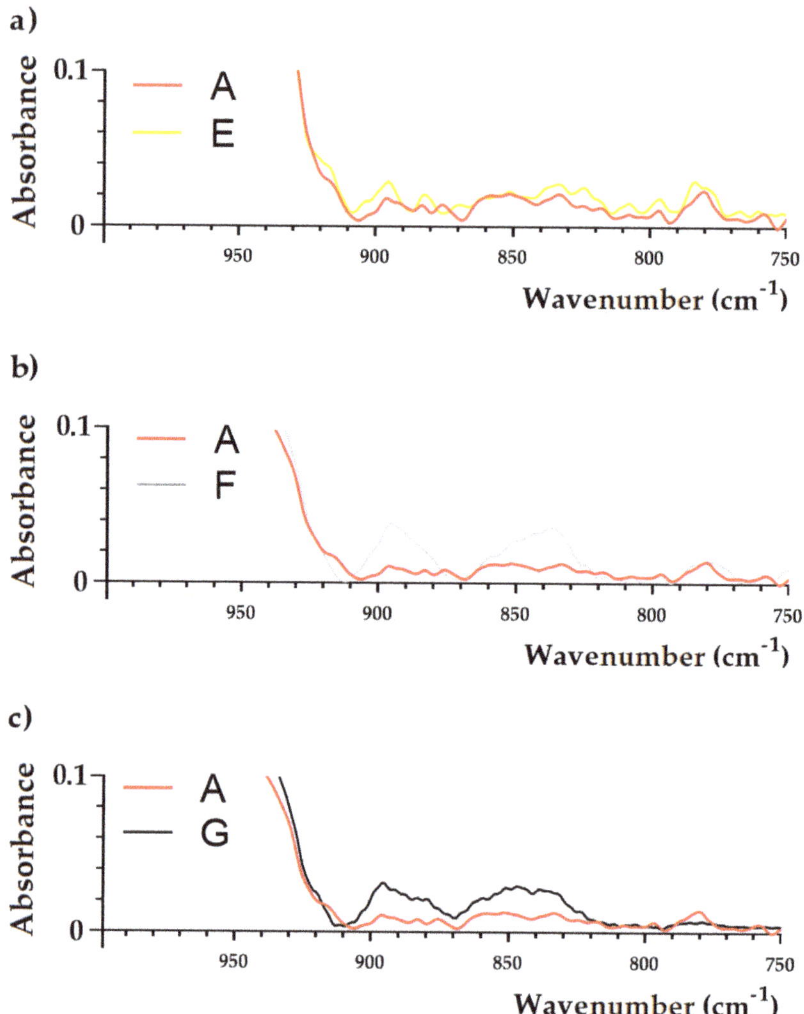

**Figure 4.** FTIR spectra (in the range: 950–700 cm$^{-1}$) of *Staphylococcus aureus* ATCC 43300 (MRSA) strain grown in Mueller-Hinton broth containing: no chemicals (control—medium A), lavender essential oil (LEO) at subinhibitory concentration (MIC$_{50}$) (medium E), octenidine dihydrochloride (OCT) at subinhibitory concentration (MIC$_{50}$) (medium F), LEO/OCT at subinhibitory concentrations (MICc$_{50}$) (medium G).

## 3. Discussion

There has been a dramatic increase in bacterial resistance to antibiotics and chemotherapeutics, which limits their therapeutic use. It was also observed that the effectiveness of new antibiotics and chemotherapeutics is rapidly decreasing. Scientific data led to the announcement in 2014 by the World Health Organization of the beginning of a post-antibiotics era. In this study, the activity of commercial LEO from the flowering herb of *L. angustifolia* Mill. (*Lamiaceae*) in combination with OCT against MRSA strains was analyzed. It has been proven that chemical analysis of LEO used in this study met the requirements outlined in the ISO Standard 11024 [21,22].

This study showed that a combination of LEO and OCT increases the effectiveness of this commonly used antiseptic agent against MRSA strains. The combination of OCT and antibiotics such as mupirocin is commonly used to eradicate the nasal carriage of *S. aureus* (especially MRSA) straight before surgical operations, especially cardiac surgery [23]. However, there are interesting data about the interactions of antiseptics with antibiotics. Hübner et al. [7] described the synergistic interaction of OCT incorporated into Mueller-Hinton agar with imipenem (against *Enterococcus faecalis, Enterococcus faecium* and *Pseudomonas aeruginosa*) and piperacillin + tazobactam (against *E. faecalis* and *E. faecium*). The same authors reported synergism between chlorhexidine digluconate (CHD) incorporated into Mueller-Hinton agar and piperacillin + tazobactam against *E. faecalis*. However, there are also reports about interaction of widely used antiseptics with essential oils and their main compounds. According to Şimşek and Duman [24], combinations of 1,8-cineole with CHD showed synergistic interactions against the following reference strains: *S. aureus* (ATCC 25923), *Escherichia coli* (ATCC 25922), *E. faecalis* (ATCC 51299), *Klebsiella pneumoniae* (ATCC 700603), and *Candida albicans* (ATCC 90028) (fractional inhibitory concentration indices—FICI values = 0.13–0.38), as well as MRSA clinical isolate (FICI = 0.05). The authors suggest that this combination may be beneficial in skin antisepsis by causing the elimination of microcolonies which are likely to exhibit increased resistance to CHD. Karpanen et al. [25] presented a similar conclusion in that essential oils, in particular eucalyptus essential oil which is more than 90% 1,8-cineole, can be used for an improved skin antisepsis when combined with CHD. They showed that in biofilm, CHD combined with eucalyptus oil demonstrated synergistic activity against the clinical isolate of *Staphylococcus epidermidis*, with an FICI value of 0.19. According to Alabdullatif et al. [26] linalool significantly enhances anti-biofilm activity of CHD with isopropyl alcohol and can potentially be used to improve skin disinfection.

According to literature data, the antimicrobial activity of essential oils depends on the content of terpenoides. Among them, phenolic compounds such as thymol or carvacrol can be distinguished by their strong action. In contrast, terpene alcohols (e.g., geraniol, citronellol, and linalool) and esters (e.g., linalyl acetate) show slightly weaker antimicrobial activity [27]. LEO owes its activity mainly to linalool and linalyl acetate, but it is known that compounds present in lower amounts are important in creating a unique mixture with a particular synergy. In this investigation, it has been proven that combination of LEO containing mostly linalool (34.1%) and linalyl acetate (33.3%) showed a synergistic effect in combination with OCT against methicillin-resistant staphylococci both the reference strain *S. aureus* ATCC 43300 and clinical isolates with FICI values between 0.11–0.26. LEO as a safe (GRAS) natural product of plants could be a good candidate to investigate its application in skin antiseptic formulation. According to literature data, LEO is well-tolerated on the surface of skin and is often administered orally or applied topically in an undiluted form [20]. However, Prashar et al. [28] showed cytotoxic activity of LEO containing mainly linalyl acetate (51%) and linalool (35%) on human skin cells (HMEC-1, HNDF, and 153BR) at a concentration of 0.25% (*v/v*). Nevertheless, in this study the most effective combination of LEO and OCT decreased the MIC of LEO from 14.86 ± 3.96 mg/mL (1.49 ± 0.4%) to 1.29 ± 0.49 mg/mL (0.13 ± 0.05%). In our previous study, it was also observed that LEO derived from the same production batch exhibited low cytotoxic activity towards HMEC-1 and glioblastoma cell (T98G) lines [29]. $IC_{50}$ values of LEO against HMEC-1 and T98G lines were 5.15 µL/mL (4.5 mg/mL) and 2.27 µL/mL (1.99 mg/mL), respectively. Moreover, a more efficient killing effect caused by synergistic LEO-OCT pairs at subinhibitory concentrations ($MICc_{50}$) was noticed. It has been shown that after five hours of incubation, there was a noticeable reduction of viable cells when compared to the control medium (without compounds). It was also observed that the addition of Tween 80 (1%, *v/v*) and DMSO (2%, *v/v*) had no impact on MRSA strains growth inhibition, and this has been also noted previously by Honório et al. [30] and Ferguson et al. [31]. Moreover, Tween 80 at the concentration of 1% is widely used as an emulsifier in cosmetics, pharmaceuticals, and food products, and has been approved by the US Food and Drug Administration for use in selected foods [32].

The present study showed the qualitative differences in FTIR spectra of samples F (Mueller-Hinton broth (MHB) containing OCT at subinhibitory concentration—$MIC_{50}$) and G (MHB containing LEO/OCT at subinhibitory concentration—MICc$_{50}$) in comparison to the control sample (MHB without chemicals). As a result of cultivation of the MRSA strain in MHB containing OCT at a subinhibitory concentration (MIC$_{50}$), two new peaks were observed at 895 cm$^{-1}$ and 837 cm$^{-1}$, which were also noticed in MRSA cells grown in medium F (MHB containing LEO/OCT at subinhibitory concentrations—MICc$_{50}$). Those changes are assignable to C-O-C glycosidic linkages and C-O-P symmetric stretching vibrations in cell wall oligosaccharides and polysaccharides, which may affect the electrostatic interactions with antibacterial molecules [33,34]. In previous studies, lower penetration of anionic antibiotic mupirocin was observed in a mupirocin-resistant MRSA strain [33]. On the other hand, OCT is a cationic, surface active antimicrobial compound (its molecular weight is approximately 624 Da), whose mode of action is based on integration with enzymatic systems. As a result, polysaccharides in the cell wall of microorganisms induce leakages in the cytoplasmic membrane and lead to cell death [7,8,35]. It has two non-interacting cation-active centers in its molecule, which are separated by a long aliphatic hydrocarbon chain [8,35]. Like other cationic antiseptics, OCT's main target appears to be glycerol phosphates in the bacterial cell membrane. It therefore binds readily onto negatively charged surfaces, such as microbial cell envelopes and eukaryotic cell membranes [7]. Thus, based on FTIR results, it can be assumed that cultivation of MRSA in medium containing subinhibitory concentration (MIC$_{50}$) of OCT resulted in changes in cell wall oligosaccharides and polysaccharides that resulted in the change of electrostatic potential of the cell surface, which therefore may affect OCT antibacterial efficacy. Similarly, the antibacterial activity of essential oils is mainly based on acting on the cytoplasmic membrane, which results in a loss of membrane stability and increased permeability [36]. In general, Gram-positive bacteria are more susceptible to essential oils in comparison to Gram-negative bacteria [36,37]. This can be linked to the fact that Gram-negative bacteria have an outer membrane which is rigid, rich in lipopolysaccharide (LPS), and more complex, thereby limiting the diffusion of hydrophobic compounds through it. This extra complex membrane is absent in Gram-positive bacteria, which instead are surrounded by a thick peptidoglycan wall that is not dense enough to resist small antimicrobial molecules, thus facilitating the access to the cell membrane [36–38]. Moreover, Gram-positive bacteria may ease the infiltration of hydrophobic compounds of EOs due to the lipophilic ends of lipoteichoic acid present in cell membrane [36,38]. Since OCT is hydrophobic compound, it requires organic solvent such a as phenoxyetanol in order to be effectively administered [8]. LEO is primarily composed of monoterpenoids and sesquiterpenoids where linalool and linalyl acetate are the most dominant, representing hydrophobic character [39,40]. Studies on the effects of the major chemical constituents of *L. angustifolia*, comprising essential oil, linalool, linalyl acetate, and terpinen-4-ol, indicate that the mechanism of action of these components damages the lipid layer of the cell membrane, which results in bacterial cell leakage [11]. Thus, based on the results of time-killing curve it may be assumed that LEO has a synergistic effect on OCT, thereby enhancing its permeation into bacterial cells.

## 4. Materials and Methods

### 4.1. Bacterial Strains and Growth Condition

The study included three methicillin-resistant *S. aureus* isolates belonging to the collection of the Chair of Microbiology, Immunology and Laboratory Medicine in Pomeranian Medical University in Szczecin, Poland. The strains were isolated from surgical wound infections. The specimens were cultivated on Columbia agar with 5% sheep blood (bioMérieux, Warsaw, Poland), incubated 18 h at 37 °C in aerobic atmosphere, and identified using the biochemical test GP Vitek 2 Compact (bioMérieux, Warsaw, Poland). A *S. aureus* ATCC 43300 (MRSA) strain was used as the control strain in this study.

### 4.2. Chemicals

#### 4.2.1. Chemical Characterization of LEO

Commercial LEO from the flowering herb of *L. angustifolia* Mill. (*Lamiaceae*) was purchased from Pollena-Aroma (Nowy Dwór Mazowiecki, Poland). The LEO was analyzed by gas chromatography-flame ionization detector-mass spectrometer (GC-FID-MS) at the Institute of General Food Chemistry, Łódź University of Technology, Poland using a Trace GC Ultra apparatus (Thermo Fisher Scientific, Waltham, MA, USA) MS DSQ II detectors, and an FID-MS splitter (SGE, Trajan Scientific Europe, Milton Keynes, UK). Identification of compounds in LEO was based on the comparison of their MS spectra with the MS spectra of computer libraries (MassFinder 3.1, Wiley Registry of Mass Spectral Data, and NIST 98.1 [41–43] along with the retention indices on a non-polar column (Rtx-1, MassFinder 3.1, Restek Corporation, Bellefonte, PA, USA) associated with a series of *n*-alkanes with linear interpolation (C-9 to C-26).

Concentrations of LEO from 500 to 0.12 µL/mL were prepared by dissolving essential oil in Tween 80 (Sigma-Aldrich, Darmstadt, Germany) (1%, *v/v*) and diluting by Mueller-Hinton broth (MHB, Sigma-Aldrich, Darmstadt, Germany).

#### 4.2.2. Octenidine Dihydrochloride (OCT)

OCT with a purity of no less than 98.0% was obtained from Schülke & Mayr GmbH (Norderstedt, Germany). Concentrations of OCT from 500 to 0.12 µg/mL were prepared by dissolving the chemical in dimethyl sulfoxide (DMSO, Loba Chemie, Mumbai, India) (2%, *v/v*) and diluting it using MHB.

### 4.3. Determination of Minimum Inhibitory Concentration (MIC) and Minimum Bactericidal Concentration (MBC) of Chemicals

The MIC of LEO or OCT was determined by the broth microdilution method according to the Clinical and Laboratory Standards Institute with a slight modification as described previously [44]. To exclude an inhibitory effect of both Tween 80 and DMSO, the control assays with MHB and MHB supplemented with Tween 80 (1%, *v/v*) or DMSO (2%, *v/v*) were performed. All tests were carried out in duplicate. At this stage, $MIC_{50}$ of each chemicals against *S. aureus* ATCC 43300 was calculated.

The MBC of chemicals was determined by transferring 20 µL of cultures in higher-than-MIC concentrations on a 96-well microplate contained MHB (100 µL) in each well, and incubating them for 18 h at 37 °C. After this period, the MBC was observed and the concentration on which transparent and verifiable medium could be found was identified. Using the known density of LEO, the final result was expressed in mg/mL.

### 4.4. Checkerboard Method

Combinations of LEO and OCT against MRSA strains were tested by using a previously described checkerboard method [44]. Using the known density of the LEO, the final result was expressed in mg/mL. All tests were performed in duplicate. Within each chemical, the lowest inhibitory concentration was considered as a minimum inhibitory concentration in combination (MICc). At this stage, $MICc_{50}$ of OCT/LEO combination against *S. aureus* ATCC 43300 was calculated. For each replicate, fractional inhibitory concentration indices (FICI) were estimated using the Equations (1) and (2):

$$FIC = \frac{MIC \text{ of LEO or OCT in combination}}{MIC \text{ of LEO or OCT alone}} \tag{1}$$

$$FICI = FIC \text{ of LEO} + FIC \text{ of OCT.} \tag{2}$$

Results were interpreted as follows: synergy (FICI < 0.5), addition (0.5 ≤ FICI ≤ 1.0), indifference (1.1 < FICI ≤ 4.0), or antagonism (FICI > 4.0).

*4.5. The Influence of LEO Alone and In Combination With OCT on the Chemical Composition of the S. aureus ATCC 43300 (MRSA) Strain*

### 4.5.1. Culture Media Preparation

One colony of the *S. aureus* ATCC 43300 (MRSA) strain was harvested from the pure culture (from Columbia agar with 5% sheep blood), inoculated into MHB, and incubated at 37 °C for 18 h with shaking (200 rpm) and the turbidity adjusted to McFarland standard number 2. Then, a 1 mL MRSA strain suspension was added to 50-mL falcon tubes and filled up with 20 mL of MHB containing: without chemicals (control—medium A), Tween 80 (1%, *v/v*) (medium B), DMSO (2%, *v/v*) (medium C), Tween 80 (1%, *v/v*) and DMSO (2%, *v/v*) (medium D), LEO at a subinhibitory concentration ($MIC_{50}$) (medium E), OCT at a subinhibitory concentration ($MIC_{50}$) (medium F), and LEO/OCT at subinhibitory concentrations ($MICc_{50}$) (medium G). The falcons were undergoing an 18 h incubation at 37 °C with shaking (200 rpm). Determination of subinhibitory concentrations ($MIC_{50}$ and $MICc_{50}$) of both LEO and OCT, as well as LEO/OCT combination were calculated in proportion to the $MIC_{100}$ and $MICc_{100}$ values obtained in Sections 4.3 and 4.4, respectively.

### 4.5.2. Time-Kill Curve Assay

A time dependent killing assay was performed to determine the killing kinetics based on the study conducted by Kang et al. [45] with a slight modification. The media A–F (25 mL) were inoculated with MRSA to obtain bacterial cells concentrations of 0.5 on the McFarland scale. After inoculation, the test tubes were incubated at 37 °C under shaking conditions (100 rpm). The viable cells were determined by counting the colonies formed from a 100-µL samples that were removed from the cultures at 0, 1, 2, 3, 4, 5, 6 12, and 24 h, which were then serially diluted, spread on Mueller-Hinton plates, and incubated for 24 h at 37 °C. Time-kill curves were constructed by plotting the mean colony counts ($Log_{10}$ CFU/mL) versus the time.

### 4.5.3. A Determination of Functional Groups in Staphylococcal Cells by the Use of Fourier Transform Infrared (FTIR) Spectroscopy

In order to confirm the presence of particular chemical moieties in MRSA reference strain incubated into different microbiological media (A–G), FTIR spectroscopy analyses was performed as described earlier [34]. FTIR is defined as a method that is sensitive to bond polarization (changes in the dipole moment), which therefore gives strong signals for polar functional groups [46,47].

The obtained spectra were normalized, baseline corrected, and analyzed using SPECTRUM software (v10, Perkin Elmer, Waltham, MA, USA).

## 5. Conclusions

LEO appears to be an efficient enhancer of the well-known antiseptic OCT against MRSA strains. It potentially influences bacterial permeation by modifying the cell wall structure. This is mirrored in phenotypic analyzes of MRSA susceptibility to OCT. Therefore LEO appears to offer therapeutic as well as preventive potential in the post-antibiotic era.

**Author Contributions:** P.K.: conceptualization, formal analysis, investigation, methodology, supervision, visualization, writing—original draft; A.P. and B.W.: formal analysis, writing—original draft; M.S.: methodology, writing—original draft; M.K. and E.D.: methodology; Ł.Ł.: formal analysis, methodology, writing—original draft; B.D. and H.Z.-B.: formal analysis, funding acquisition. All authors have read and agreed to the published version of the manuscript.

**Funding:** The project is financed from the program provided by the Polish Ministry of Science and Higher Education under the name "Regional Initiative of Excellence" in 2019–2022, project number 002/RID/2018/19, funding amount 12,000,000 PLN.

**Conflicts of Interest:** The authors declare no conflict of interest.

## References


1. Jenul, C.; Horswill, A.R. Regulation of *Staphylococcus aureus* virulence. *Microbiol. Spectr.* **2018**, *6*. [CrossRef]
2. McCaig, L.F.; McDonald, L.C.; Mandal, S.; Jernigan, D.B. *Staphylococcus aureus*-associated skin and soft tissue infections in ambulatory care. *Emerg. Infect. Dis.* **2006**, *12*, 1715–1723. [CrossRef]
3. Liu, Y.; Xu, Z.; Yang, Z.; Sun, J.; Ma, L. Characterization of community-associated *Staphylococcus aureus* from skin and soft-tissue infections: A multicenter study in China. *Emerg. Microbes Infect.* **2016**, *5*, 127. [CrossRef]
4. Poggio, J.L. Perioperative strategies to prevent surgical-site infection. *Clin. Colon Rectal Surg.* **2013**, *26*, 168–173. [CrossRef]
5. Velasco, V.; Buyukcangaz, E.; Sherwood, J.S.; Stepan, R.M.; Koslofsky, R.J.; Logue, C.M. Characterization of *Staphylococcus aureus* from humans and a comparison with isolates of animal origin, in North Dakota, United States. *PLoS ONE* **2015**, *10*, 0140497. [CrossRef]
6. Hassoun, A.; Linden, P.K.; Friedman, B. Incidence, prevalence, and management of MRSA bacteremia across patient populations—A review of recent developments in MRSA management and treatment. *Crit. Care* **2017**, *21*, 211. [CrossRef]
7. Hübner, N.O.; Siebert, J.; Kramer, A. Octenidine dihydrochloride, a modern antiseptic for skin, mucous membranes and wounds. *Skin Pharmacol. Physiol.* **2010**, *23*, 244–258. [CrossRef]
8. Szostak, K.; Czogalla, A.; Przybyło, M.; Langner, M. New lipid formulation of octenidine dihydrochloride. *J. Liposome Res.* **2018**, *28*, 106–111. [CrossRef]
9. Eisenbeiß, W.; Siemers, F.; Amtsberg, G.; Hinz, P.; Hartmann, B.; Kohlmann, T.; Ekkernkamp, A.; Albrecht, U.; Assadian, O.; Kramer, A. Prospective, double-blinded, randomised controlled trial assessing the effect of an octenidine-based hydrogel on bacterial colonisation and epithelialization of skin graft wounds in burn patients. *Int. J. Burns Trauma* **2012**, *2*, 71–79.
10. Hardy, K.; Sunnucks, K.; Gil, H.; Shabir, S.; Trampari, E.; Hawkey, P.; Webber, M. Increased usage of antiseptics is associated with reduced susceptibility in clinical isolates of *S. aureus*. *mBio* **2018**, *9*. [CrossRef]
11. De Rapper, S.; Viljoen, A.; van Vuuren, S. The in vitro antimicrobial effects of *Lavandula angustifolia* essential oil in combination with conventional antimicrobial agents. *Evid. Based Complement. Alternat. Med.* **2016**, *2016*, 2752739. [CrossRef]
12. Sarkic, A.; Stappen, I. Essential oils and their single compounds in cosmetics—A critical review. *Cosmetics* **2018**, *5*, 11. [CrossRef]
13. Mori, H.M.; Kawanami, H.; Kawahata, H.; Aoki, M. Wound healing potential of lavender oil by acceleration of granulation and wound contraction through induction of TGF-β in a rat model. *BMC Complement. Altern. Med.* **2016**, *16*, 144. [CrossRef]
14. Arzi, A.; Ahamehe, M.; Sarahroodi, S. Effect of hydroalcoholic extract of *Lavandula officinalis* on nicotine-induced convulsion in mice. *Pak. J. Biol. Sci.* **2011**, *14*, 634–640. [CrossRef]
15. Woelk, H.; Schläfke, S. A multi-center, double-blind, randomised study of the lavender oil preparation silexan in comparison to lorazepam for generalized anxiety disorder. *Phytomedicine* **2010**, *17*, 94–99. [CrossRef]
16. Costa, P.; Gonçalves, S.; Andrade, P.B.; Valentão, P.; Romano, A. Inhibitory effect of *Lavandula viridis* on $Fe^{2+}$-induced lipid peroxidation, antioxidant and anti-cholinesterase properties. *Food Chem.* **2011**, *126*, 1779–1786. [CrossRef]
17. Messaoud, C.; Chograni, H.; Boussaid, M. Chemical composition and antioxidant activities of essential oils and methanol extracts of three wild *Lavandula* L. species. *Nat. Prod. Res.* **2012**, *26*, 1976–1984. [CrossRef]
18. Varona, S.; Rojo, S.R.; Martín, Á.; Cocero, M.J.; Serra, A.T.; Crespo, T.; Duarte, C.M.M. Antimicrobial activity of lavandin essential oil formulations against three pathogenic food-borne bacteria. *Ind. Crop. Prod.* **2013**, *42*, 243–250. [CrossRef]
19. Zuzarte, M.; Vale-Silva, L.; Gonçalves, M.J.; Cavaleiro, C.; Vaz, S.; Canhoto, J.; Pinto, E.; Salgueiro, L. Antifungal activity of phenolic-rich *Lavandula multifida* L. essential oil. *Eur. J. Clin. Microbiol. Infect. Dis.* **2012**, *31*, 1359–1366. [CrossRef]
20. Malcolm, B.J.; Tallian, K. Essential oil of lavender in anxiety disorders: Ready for prime time? *Ment. Health Clin.* **2018**, *7*, 147–155. [CrossRef]
21. ISO 11024-1:1998. Essential oils—General Guidance on Chromatographic Profiles—Part 1: Preparation of Chromatographic Profiles for presentation in Standards. Available online: https://www.iso.org/obp/ui/#iso:std:iso:11024:-1:ed-1:V1:en (accessed on 25 December 2019).

22. ISO 11024-2:1998. Essential oils—General Guidance on Chromatographic Profiles—Part 2: Preparation of Chromatographic Profiles for Presentation in Standards. Available online: https://www.iso.org/obp/ui/#iso:std:iso:11024:-2:ed-1:v1:en (accessed on 25 December 2019).

23. Kohler, P.; Sommerstein, R.; Schönrath, F.; Ajdler-Schäffler, E.; Anagnostopoulos, A.; Tschirky, S.; Falk, V.; Kuster, S.P.; Sax, H. Effect of perioperative mupirocin and antiseptic body wash on infection rate and causative pathogens in patients undergoing cardiac surgery. *Am. J. Infect. Control.* **2015**, *43*, 33–38. [CrossRef]

24. Şimşek, M.; Duman, R. Investigation of effect of 1,8-cineole on antimicrobial activity of chlorhexidine gluconate. *Pharmacognosy Res.* **2017**, *9*, 234–237. [CrossRef]

25. Karpanen, T.J.; Worthington, T.; Hendry, E.R.; Conway, B.R.; Lambert, P.A. Antimicrobial efficacy of chlorhexidine digluconate alone and in combination with eucalyptus oil, tea tree oil and thymol against planktonic and biofilm cultures of *Staphylococcus epidermidis*. *J. Antimicrob. Chemother.* **2008**, *62*, 1031–1036. [CrossRef]

26. Alabdullatif, M.; Boujezza, I.; Mekni, M.; Taha, M.; Kumaran, D.; Yi, Q.L.; Landoulsi, A.; Ramirez-Arcos, S. Enhancing blood donor skin disinfection using natural oils. *Transfusion* **2017**, *57*, 2920–2927. [CrossRef]

27. Kalemba, D.; Kunicka, A. Antibacterial and antifungal properties of essential oils. *Curr. Med. Chem.* **2003**, *10*, 813–829. [CrossRef]

28. Prashar, A.; Locke, I.C.; Evans, C.S. Cytotoxicity of lavender oil its major components to human skin cells. *Cell Prolif.* **2004**, *37*, 221–229. [CrossRef]

29. Sienkiewicz, M.; Głowacka, A.; Kowalczyk, E.; Wiktorowska-Owczarek, A.; Jóźwiak-Bębenista, M.; Łysakowska, M. The biological activities of cinnamon, geranium and lavender essential oils. *Molecules* **2014**, *19*, 20929–20940. [CrossRef]

30. Honório, V.G.; Bezerra, J.; Souza, G.T.; Carvalho, R.J.; Gomes-Neto, N.J.; Figueiredo, R.C.; Melo, J.V.; Souza, E.L.; Magnani, M. Inhibition of *Staphylococcus aureus* cocktail using the synergies of oregano and rosemary essential oils or carvacrol and 1,8-cineole. *Front. Microbiol.* **2015**, *6*, 1223. [CrossRef]

31. Ferguson, S.A.; Menorca, A.; van Zuylen, E.M.; Cheung, C.Y.; McConnell, M.A.; Rennison, D.; Brimble, M.A.; Bodle, K.; McDougall, S.; Cook, G.M.; et al. Microtiter screening reveals oxygen-dependent antimicrobial activity of natural products against mastitis-causing bacteria. *Front. Microbiol.* **2019**, *10*, 1995. [CrossRef]

32. Nielsen, C.K.; Kjems, J.; Mygind, T.; Snabe, T.; Meyer, R.L. Effects of Tween 80 on growth and biofilm formation in laboratory media. *Front. Microbiol.* **2016**, *7*, 1878. [CrossRef]

33. Amiali, N.M.; Golding, G.R.; Sedman, J.; Simor, A.E.; Ismail, A.A. Rapid identification of community-associated methicillin-resistant *Staphylococcus aureus* by Fourier transform infrared spectroscopy. *Diagn. Microbiol. Infect. Dis.* **2011**, *70*, 157–166. [CrossRef]

34. Kwiatkowski, P.; Pruss, A.; Wojciuk, B.; Dołęgowska, B.; Wajs-Bonikowska, A.; Sienkiewicz, M.; Mężyńska, M.; Łopusiewicz, Ł. The influence of essential oil compounds on antibacterial activity of mupirocin-susceptible and induced low-level mupirocin-resistant MRSA strains. *Molecules* **2019**, *24*, 3105. [CrossRef]

35. Assadian, O.; Pilcher, M.; Antunes, J.N.P.; Boulton, Z.; von Hallern, B.; Hämmerle, G.; Hunt, G.; Jeffery, S.; Lahnsteiner, E.; Price, J.; et al. Facilitating wound bed preparation: Properties and clinical efficacy of octenidine and octenidine-based products in modern wound management. *J. Wound Care* **2016**, *25*, 1–28. [CrossRef]

36. Chouhan, S.; Sharma, K.; Guleria, S. Antimicrobial activity of some essential oils—Present status and future perspectives. *Medicines (Basel)* **2017**, *4*, 58. [CrossRef]

37. Predoi, D.; Iconaru, S.L.; Buton, N.; Badea, M.L.; Marutescu, L. Antimicrobial activity of new materials based on lavender and basil essential oils and hydroxyapatite. *Nanomaterials* **2018**, *8*, 291. [CrossRef]

38. Hyldgaard, M.; Mygind, T.; Meyer, R.L. Essential oils in food preservation: Mode of action, synergies, and interactions with food matrix components. *Front. Microbiol.* **2012**, *3*, 12. [CrossRef]

39. Hossain, S.; Heo, H.; De Silva, B.C.J.; Wimalasena, S.H.M.P.; Pathirana, H.N.K.S.; Heo, G.J. Antibacterial activity of essential oil from lavender (*Lavandula angustifolia*) against pet turtle-borne pathogenic bacteria. *Lab. Anim. Res.* **2017**, *33*, 195–201. [CrossRef]

40. Dhifi, W.; Bellili, S.; Jazi, S.; Bahloul, N.; Mnif, W. Essential oils' chemical characterization and investigation of some biological activities: A critical review. *Medicines* **2016**, *3*, 25. [CrossRef]

41. Hochmuth, D. *Mass Spectral Library "Terpenoids and Related Constituents of Essential oils"*; Library of MassFinder 3: Hamburg, Germany, 2006.

42. Wiley Science Solutions. *Wiley Registry of Mass Spectral Data*, 8th ed.; John Wiley & Sons Inc.: Hoboken, NJ, USA, 2008.

43. Stein, S.E. *NIST 1. NIST/EPA/NIH Mass Spectral LibraryNIST '98 ASCII Version*; National Institute of Standards and Technology: Gaithersburg, MD, USA, 2016.

44. Kwiatkowski, P.; Pruss, A.; Grygorcewicz, B.; Wojciuk, B.; Dołęgowska, B.; Giedrys-Kalemba, S.; Kochan, E.; Sienkiewicz, M. Preliminary study on the antibacterial activity of essential oils alone and in combination with gentamicin against extended-spectrum β-lactamase-producing and New Delhi metallo-β-lactamase-1-producing *Klebsiella pneumoniae* isolates. *Microb. Drug Resist.* **2018**, *24*, 1368–1375. [CrossRef]

45. Kang, J.; Liu, L.; Wu, X.; Sun, Y.; Liu, Z. Effect of thyme essential oil against *Bacillus cereus* planktonic growth and biofilm formation. *Appl. Microbiol. Biotechnol.* **2018**, *102*, 10209–10218. [CrossRef]

46. Orsini, F.; Ami, D.; Villa, A.M.; Sala, G.; Bellotti, M.G.; Doglia, S.M. FT-IR microspectroscopy for microbiological studies. *J. Microbiol. Methods* **2000**, *42*, 17–27. [CrossRef]

47. Salachna, P.; Łopusiewicz, Ł.; Meller, E.; Grzeszczuk, M.; Piechocki, R. Modulation of bioelement concentration and macromolecule conformation in leaves of *Salvia coccinea* by salicylic acid and salt stress. *Fresenius Environ. Bull.* **2019**, *28*, 9339–9347.

**Sample Availability:** Samples of the compounds (*Lavandula angustifolia* essential oil and octenidine dihydrochloride) are available from the authors.

 *molecules*

*Article*

# Antimicrobial Activity of Chamomile Essential Oil: Effect of Different Formulations

Sourav Das [1,4,†], Barbara Horváth [2,†], Silvija Šafranko [3], Stela Jokić [3], Aleksandar Széchenyi [2] and Tamás Kőszegi [1,4,*]

1   Department of Laboratory Medicine, Faculty of Medicine, University of Pécs, H-7624 Pécs, Hungary; pharma.souravdas@gmail.com
2   Institute of Pharmaceutical Technology and Biopharmacy, Faculty of Pharmacy, University of Pécs, H-7624 Pécs, Hungary; barbara.horvath@aok.pte.hu (B.H.); szechenyi.aleksandar@gytk.pte.hu (A.S.)
3   Faculty of Food Technology Osijek, University of Osijek, Franje Kuhaca 20, 31000 Osijek, Croatia; silvija.safranko@ptfos.hr (S.Š.); stela.jokic@ptfos.hr (S.J.)
4   János Szentágothai Research Center, University of Pécs, H-7624 Pécs, Hungary
*   Correspondence: koszegi.tamas@pte.hu; Tel.: +36-72-501-500 (ext. 29249)
†   These authors contributed equally to this work.

Academic Editor: Igor Jerković
Received: 31 October 2019; Accepted: 21 November 2019; Published: 26 November 2019

**Abstract:** Essential oils (EOs) are highly lipophilic, which makes the measurement of their biological action difficult in an aqueous environment. We formulated a Pickering nanoemulsion of chamomile EO ($C_{Pe}$). Surface-modified Stöber silica nanoparticles (20 nm) were prepared and used as a stabilizing agent of $C_{Pe}$. The antimicrobial activity of $C_{Pe}$ was compared with that of emulsion stabilized with Tween 80 ($C_{T80}$) and ethanolic solution ($C_{Et}$). The antimicrobial effects were assessed by their minimum inhibitory concentration ($MIC_{90}$) and minimum effective ($MEC_{10}$) concentrations. Besides growth inhibition (CFU/mL), the metabolic activity and viability of Gram-positive and Gram-negative bacteria as well as *Candida* species, in addition to the generation of oxygen free radical species (ROS), were studied. We followed the killing activity of $C_{Pe}$ and analyzed the efficiency of the EO delivery for examined formulations by using unilamellar liposomes as a cellular model. $C_{Pe}$ showed significantly higher antibacterial and antifungal activities than $C_{T80}$ and $C_{Et}$. Chamomile EOs generated superoxide anion and peroxide related oxidative stress which might be the major mode of action of Ch essential oil. We could also demonstrate that $C_{Pe}$ was the most effective in donation of the active EO components when compared with $C_{T80}$ and $C_{Et}$. Our data suggest that $C_{Pe}$ formulation is useful in the fight against microbial infections.

**Keywords:** chamomile essential oil; Pickering emulsion; antimicrobial activity; free radical generation

## 1. Introduction

Essential oils (EOs) have been widely used in folk medicine throughout the history of humankind. The application of EOs covers a wide range from therapeutic, hygienic, and spiritual to ritualistic purposes. EOs are aromatic, volatile, lipophilic liquids extracted from different parts of plant materials such as barks, buds, flowers, fruits, seeds, and roots [1]. EOs are mixtures of complex compounds with variable individual chemical composition and concentrations that includes primarily terpenoids, like monoterpenes (C10), sesquiterpenes (C15), diterpenes (C20), acids, alcohols, aldehydes, aliphatic hydrocarbons, acyclic esters or lactones, rare nitrogen- and sulfur-containing compounds, coumarin, and homologues of phenylpropanoids [1,2]. The biological effects of EOs cover a wide range of effects, including antioxidant, antimicrobial, antitumor, anti-inflammatory, and antiviral activity [3].

The increase in demand for the use of aromatherapy as complementary and alternative medicine has led people to believe in the myth that EOs are harmless because they are natural and have been used for a long time [4]. However, there might be several side effects of EOs even if topical administration is applied and, among these, allergic reactions are the most frequent (many EOs can cause, e.g., rashes on the skin). Some of them can be poisonous if absorbed through the skin, breathed, or swallowed. Previous studies also report the interaction of EOs with other drugs [5]. The continuous production of new aroma chemicals and their widespread and uncontrolled usage as alternative therapies together with many carrier diluents have brought serious problems, especially among children. In this regard, it is of utmost importance to study the mode of action of essential oils and to find a proper, unharmful formulation. Another serious problem is the highly lipophilic nature of EOs, which makes it impossible to measure their biological effects in an aqueous environment [1,6,7].

One major characteristic and application of EOs are their strong antimicrobial activity, including antibacterial and antifungal effects without the development of microbial resistance. Numerous studies are found in the literature describing the antimicrobial activities of a large variety of EOs [8–13]. Most of these assays include conventional broth dilution method, disk diffusion method, and bioautography assay to measure the antimicrobial activity of EOs. Efforts have been made to overcome the lipophilic nature of the oils usually by application of EOs diluted in seemingly suitable solvents/detergents. In the case of natural lipophilic volatile compounds like EOs, solvents of varying polarity, e.g., DMSO, ethanol, and methanol, are most commonly used. However, previous studies have reported the antimicrobial effects of the solvents themselves (DMSO, ethanol, and other solvents in various microbial assays) or their influence on the true antimicrobial effects of EOs [14]. The usage of solubilizing agents limits the precise determination of the antimicrobial activities of EOs. Also, a major problem might arise in the classical assays due to the evaporation of EOs during the assay or the inability of the test microbes to reach the lipophilic range of the tested EOs (in bioautography, as an example) [15,16].

Therefore, new formulations have been determined to increase the solubility or to emulsify the EOs in an aqueous environment. These efforts help to stabilize the oils, to produce an even release of the active components into the required environment, and to maintain their antioxidant and antimicrobial activities [6,11,17,18]. Detergents and organic solvents are not welcome in this regard. Attempts have been made to entrap EOs by modified cyclodextrins for the exact determination of their antimicrobial characteristics [19,20].

The application of Pickering nanoemulsion is a quite novel approach to stabilize oil-in-water (O/W) and water-in-oil (W/O) emulsions by solid particles instead of surfactants. The mechanism involves the adsorption of solid particles on the oil–water interface, causing a significant decrease in the interfacial surface tension that results in high emulsion stability [18]. Previous studies have reported decreased evaporation of EOs from O/W emulsion of nanoparticle-stabilized formulations versus EO–surfactant systems to be a beneficial factor [21,22].

Despite the numerous existing studies on EO–Pickering emulsion, we could not find any literature data on chamomile volatile oil–nanoparticle formulation [7,23]. The main aim of the present work is to use Pickering emulsion of chamomile EO stabilized with modified Stöber silica nanoparticles and characterize its antimicrobial effect using Gram-positive and Gram-negative bacteria as well as *Candida* fungal species. We could demonstrate the strong antimicrobial effects of the chamomile EO–Pickering emulsion and suggest a plausible mode of action of this formulation. Experimental efforts were made to support the suggested mode of action.

## 2. Results

### 2.1. Characteristics of Stöber Silica Nanoparticles

The mean diameter, PDI value (polydispersity index), and zeta potential of modified Stöber silica nanoparticles (SNPs) were determined by dynamic light scattering (DLS), and these values were 20 nm, 0.01, and −21.3 mV, respectively. The size and morphology of SNPs were examined by TEM (see

Figure 1). The size distribution obtained by DLS was confirmed by TEM, which showed that the mean diameter of silica samples was 20 nm; they are highly monodisperse and have a spherical shape.

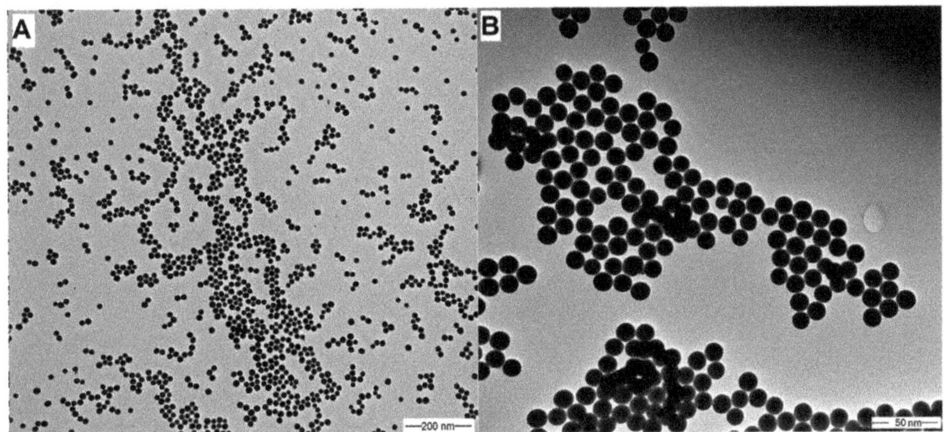


**Figure 1.** TEM images of silica nanoparticles (SNPs) with different resolutions: 100,000× magnification (**A**) and 500,000× magnification (**B**), accelerating voltage: 80 kV; $d_{TEM}$ = 20 nm. PDI = 0.015.

## 2.2. Nanoemulsion Stability

We have prepared a Pickering nanoemulsion with surface-modified silica nanoparticles as a stabilizing agent; the particle concentration was 1 mg/mL in every case. The chamomile EO concentration was 100 µg/mL. To compare properties of chamomile EO–Pickering nanoemulsion ($C_{Pe}$) with the conventional, surfactant-stabilized nanoemulsions, and an emulsion with the Tween 80 stabilizing agent was also prepared. The concentration of surfactant was the same as nanoparticles, 1 mg/mL. The emulsions were stored at room temperature (25 °C).

We considered the emulsion to be stable when its droplet size does not change and sedimentation, aggregation of particles, or phase separation cannot be observed. The results show that the prepared Pickering emulsion is more stable than conventional emulsion (see Table 1). When the volume fraction of chamomile EO was very low, we assumed that all emulsions were of O/W type and this was confirmed by filter paper tests with $CoCl_2$ and dye test with Sudan Red G.

**Table 1.** Parameters of Pickering nanoemulsion and conventional emulsion.

| Stabilizing Agent | $D_{droplet}$ (nm) | Stability |
| --- | --- | --- |
| SNP | 290 ± 4.5 | 3 months |
| Tw80 | 210 ± 10.5 | 1 month |

## 2.3. Antibacterial and Antifungal Activities (MIC$_{90}$) of Prepared Emulsions

The effect of chamomile Pickering nanoemulsion, conventional emulsion, and essential oil in ethanol on the growth of some foodborne microbes and opportunistic fungi have been evaluated. The $C_{Pe}$ has been shown to have good antibacterial and antifungal activities (MIC$_{90}$) on *Escherichia coli* (*E. coli*) (2.19 µg/mL), *Pseudomonas aeruginosa* (*P. aeruginosa*) (1.02 µg/mL), *Bacillus subtilis* (*B. subtilis*) (1.13 µg/mL), *Staphylococcus aureus* (*S. aureus*) (1.06 µg/mL), *Streptococcus pyogenes* (*S. pyogenes*) (2.45 µg/mL), *Schizosaccharomyces pombe* (*S. pombe*) (1.28 µg/mL), *Candida albicans* (*C. albicans*) (2.65 µg/mL), and *Candida tropicalis* (*C. tropicalis*) (1.69 µg/mL), respectively when compared to $C_{T80}$ counterpart (*P < 0.01*). $C_{Pe}$ showed antimicrobial activity on the selected microbes at an average of fourteen-fold less concentration compared with free essential oil in ethanol ($C_{Et}$). Simultaneously, $C_{Pe}$ showed a similar antifungal effect as caspofungin (Cas) on *Candida tropicalis.* The comparative dose–response curves are shown in Figures 2 and 3 for bacteria and fungi, respectively.

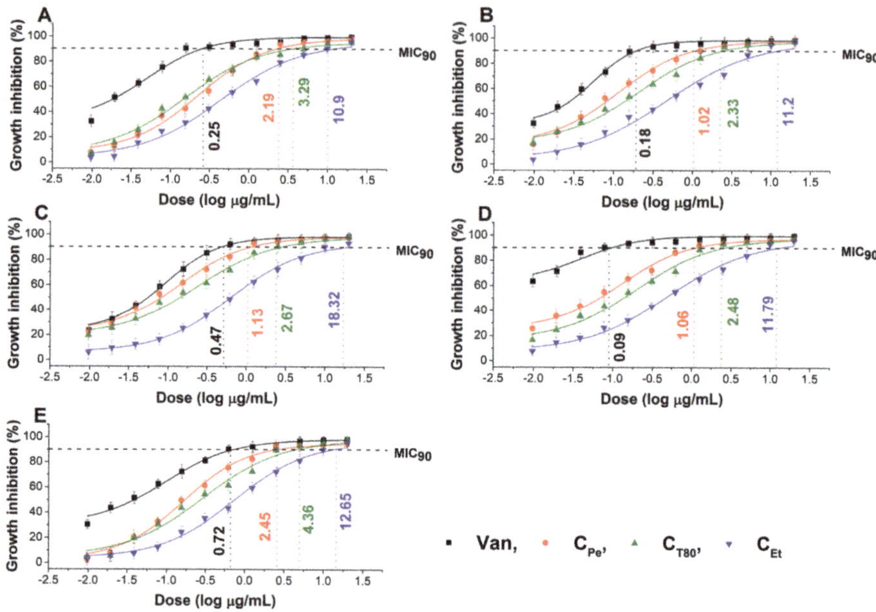

**Figure 2.** Minimum inhibitory concentration (MIC$_{90}$) of C$_{Pe}$, C$_{T80}$, C$_{Et}$, and vancomycin (Van, μg/mL) on *E. coli* (**A**), *S. aureus* (**B**), *B. subtilis* (**C**), *P. aeruginosa* (**D**), and *S. pyogenes* (**E**).

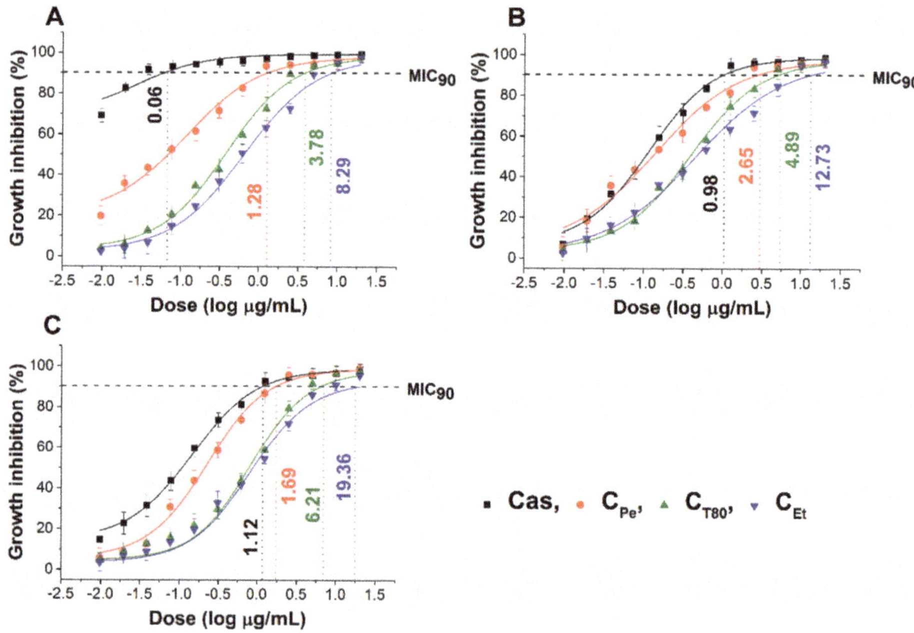

**Figure 3.** Minimum inhibitory concentration (MIC$_{90}$) of C$_{Pe}$, CT$_{80}$, C$_{Et}$, and caspofungin (Cas, μg/mL) on *S. pombe* (**A**), *C. albicans* (**B**), and *C. tropicalis* (**C**).

### 2.4. Minimum Effective Concentrations (MEC$_{10}$) for Tested Bacteria and Fungi

The minimum effective concentration (MEC$_{10}$) of C$_{Pe}$, C$_{T80}$, and C$_{Et}$ on foodborne Gram-positive and Gram-negative bacteria as well as fungi have been determined. The dose–response curve shows a slow killing effect (≤10% of the population) of C$_{Pe}$ after 1 h of treatment at a two-fold higher concentration compared with MIC$_{90}$ data. The MEC$_{10}$ also highlights the effective killing effect of C$_{Pe}$ when compared to C$_{T80}$ and C$_{Et}$ (Figures 4 and 5) ($P < 0.01$).

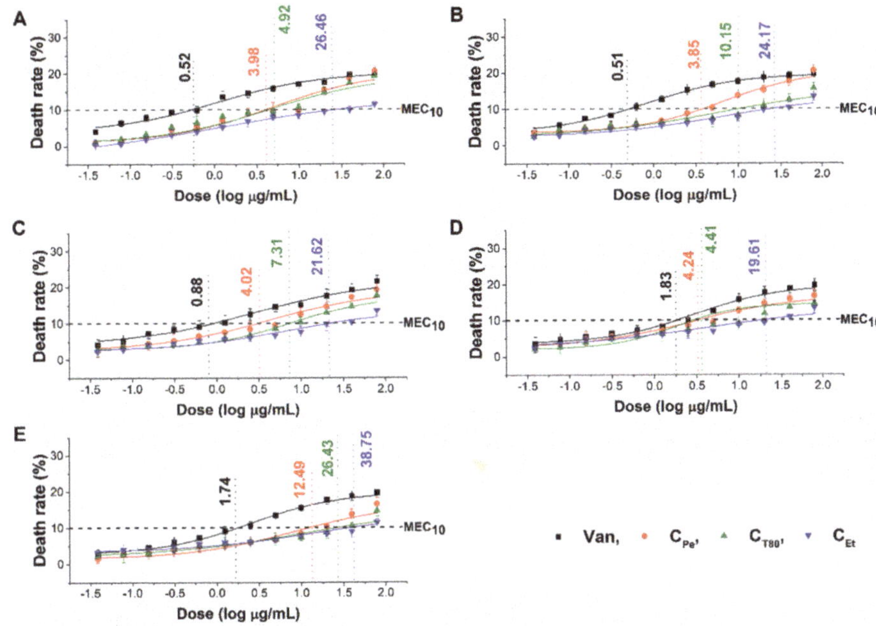

**Figure 4.** Minimum effective concentration (MEC$_{10}$) of C$_{Pe}$, C$_{T80}$, C$_{Et}$, and Van (µg/mL) on *E. coli* (**A**), *S. aureus* (**B**), *B. subtilis* (**C**), *P. aeruginosa* (**D**), and *S. pyogenes* (**E**).

### 2.5. Effect on Microbial Oxidative Balance

Reactive oxygen species (ROS) production and accumulation in the cells initiates oxidative stress, leading to cellular structural damage followed by induced apoptosis [6]. We have investigated the relationship between oxidative stress generation after 1 h of treatment and microbial killing activity. The results are demonstrated in Figures 6 and 7 for bacteria and fungi, respectively. Data expressed as % of the control are as follows: the ROS (1085.86 ± 126.36), peroxide (1229.86 ± 164.52) and superoxide (1276.86 ± 165.42) generation were the highest in case of *S. aureus*. The C$_{Pe}$ showed an effective increment of ROS, peroxide, and superoxide generation in both Gram-positive and -negative bacteria when compared to C$_{T80}$ and C$_{Et}$ ($P < 0.01$). C$_{Pe}$ showed increased oxidative stress in both bacteria and fungi at least seven-fold higher than the negative control whereas the positive control (menadione) produced an eight- to nine-fold increase in 1 h. The C$_{Et}$ has generated a two to four-fold increment in oxidative stress which is the lowest among all tested compounds.

### 2.6. Time–Kill Kinetics Study

The time–kill kinetics curve was performed to quantify living populations after a definite time interval under different sample MEC$_{10}$ concentrations. A significant reduction (four log-fold) in the cell survivability has been observed in case of C$_{Pe}$ when compared to Gc ($P < 0.01$) (Figures 8 and 9). Fifty percent of cell death occurred by C$_{Pe}$ at 16 and 36 h in the case of bacteria and fungi, and was most effective in reducing living colonies in case of *C. albicans* (1.73 ± 0.15 CFU/mL) after 48 h of treatment. At an average of a two-fold higher concentration, $C_{Et}$ was able to show a killing effect compared to $C_{Pe}$ ($P < 0.01$).

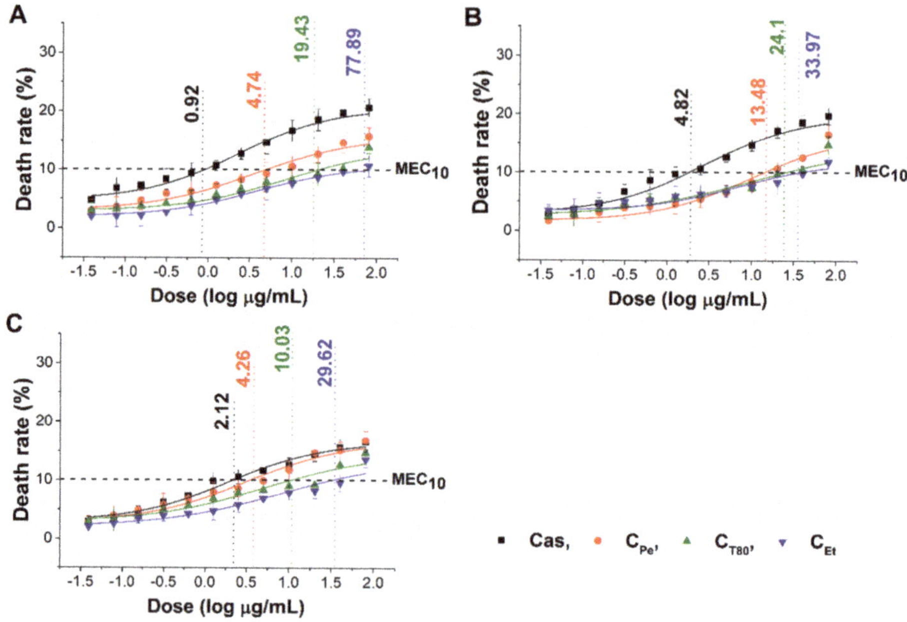

**Figure 5.** Minimum effective concentration ($MEC_{10}$) of $C_{Pe}$, $C_{T80}$, $C_{Et}$, and Cas (μg/mL) on *S. pombe* (**A**), *C. albicans* (**B**), and *C. tropicalis* (**C**).

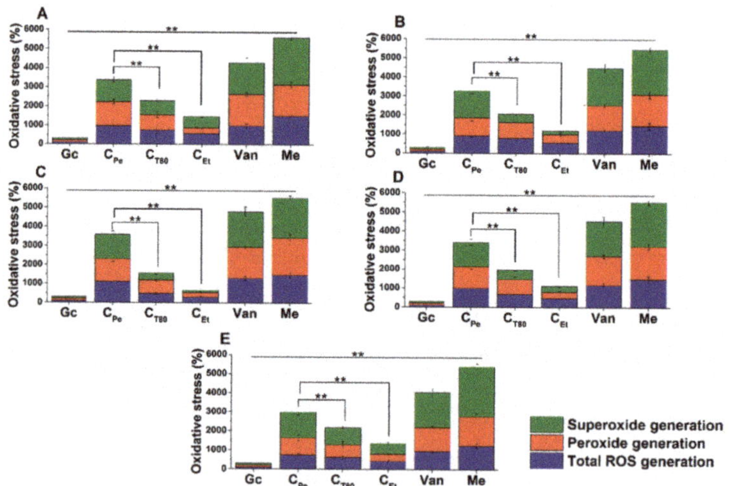

**Figure 6.** Percentage oxidative stress generation by $C_{Pe}$, $C_{T80}$, $C_{Et}$, and Van on *E. coli* (**A**), *S. aureus* (**B**), *B. subtilis* (**C**), *P. aeruginosa* (**D**), and *S. pyogenes* (**E**). Six independent experiments, each with 3 replicates, compared with menadione (Me) and growth control (Gc) as controls after 1 h of treatment (*******P* < 0.01).

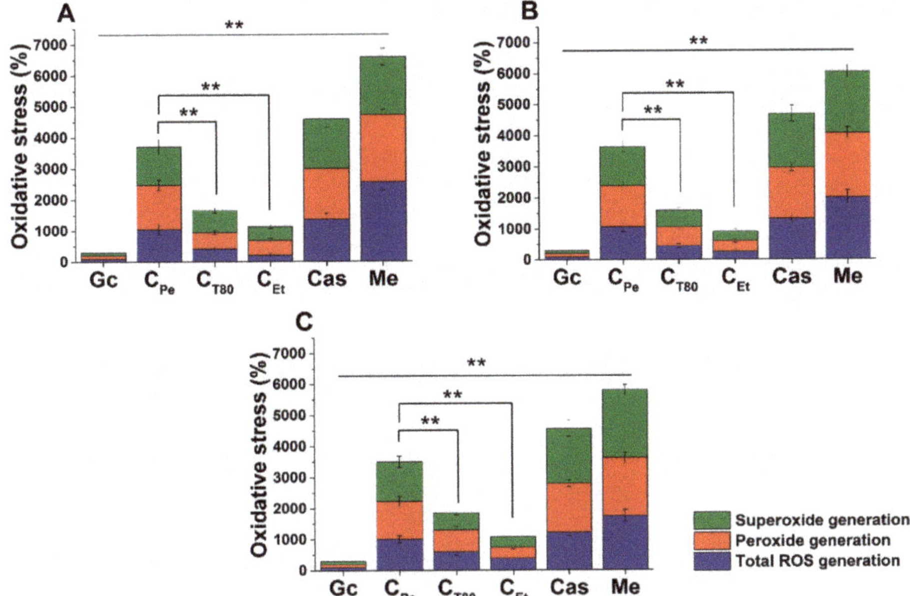



**Figure 7.** Percentage oxidative stress generation by $C_{Pe}$, $C_{T80}$, $C_{Et}$, and Cas on *S. pombe* (**A**), *C. albicans* (**B**), and *C. tropicalis* (**C**). Six independent experiments, each with 3 replicates, compared with Me and Gc as positive and growth controls after 1 h of treatment (**$P < 0.01$).

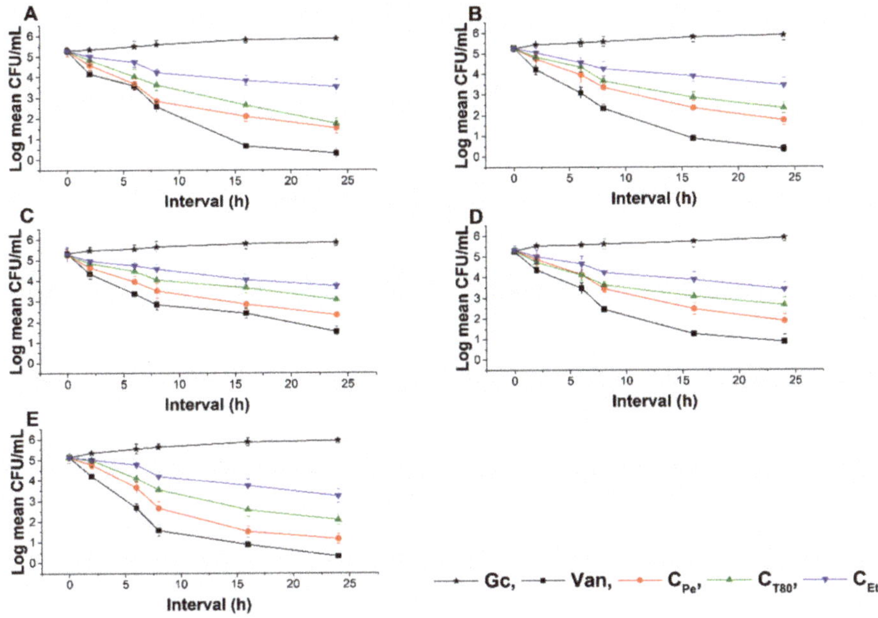



**Figure 8.** Colony-forming unit (CFU/mL) of $C_{Pe}$, $C_{T80}$, and $C_{Et}$ on *E. coli* (**A**), *S. aureus* (**B**), *B. subtilis* (**C**), *P. aeruginosa* (**D**), and *S. pyogenes* (**E**). Six independent experiments, each with 3 replicates, compared with Van and Gc as positive and growth controls.

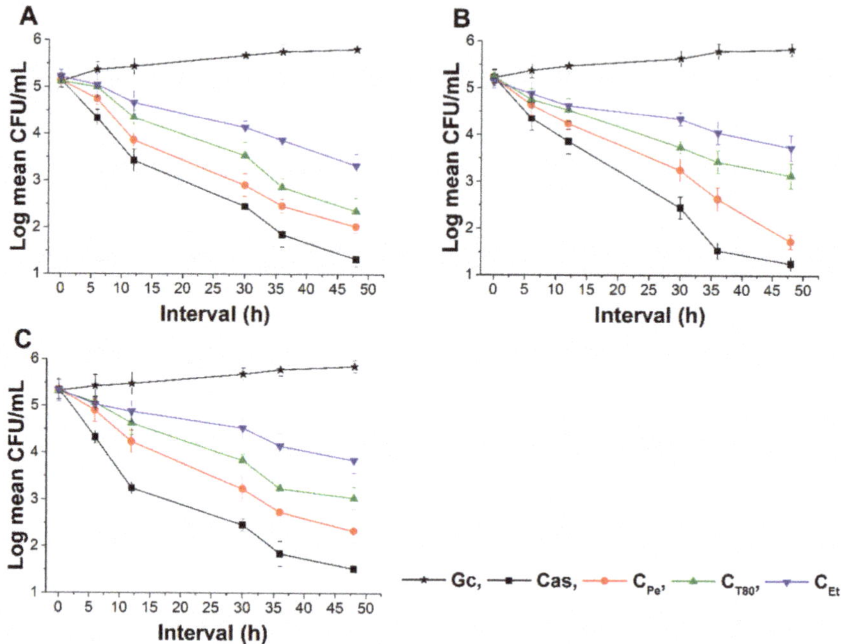

**Figure 9.** Colony-forming unit (CFU/mL) of $C_{Pe}$, $C_{T80}$, and $C_{Et}$ on *S. pombe* (**A**), *C. albicans* (**B**), and *C. tropicalis* (**C**). Six independent experiments, each with 3 replicates, compared with Cas and Gc as positive and growth controls.

### 2.7. Live/dead Cell Viability Discrimination

The effect of $C_{Pe}$, $C_{T80}$, and $C_{Et}$ on the viability of selected bacteria and fungi were tested (Figures 10 and 11). $C_{Pe}$ decreases the viability of the tested bacteria and fungi with an average viability reduction to 42.36% ± 3.74% and 49.62% ± 5.25% of mean percentage viability compared to Gc after 16 and 36 h of treatments in bacteria and fungi respectively ($P < 0.01$), whereas $C_{T80}$ and $C_{Et}$ were less effective than $C_{Pe}$ with mean percentage viabilities of ≥60% and 70%, respectively.

### 2.8. Interaction Study between Cell Model and Different Formulations of Chamomile EO

The unilamellar liposomes (ULs), consisting of a single phospholipid, can be used as artificial cells or biological membrane model for studying the interactions between cells or cell membranes and drugs or biologically active components [24]. This study was conducted to determine the intracellular delivery ability of active components from chamomile EO for different formulations.

We have studied the interaction of ULs and different forms of chamomile EO for 24 h at 35 °C. After 1 h of interaction, 27.2% of EO have penetrated the liposomes from Pickering nanoemulsion, while conventional emulsion and the ethanolic solution did not provide a measurable amount. The next sampling was after 2 h, where Pickering emulsion has delivered 48.3% of EO, conventional emulsion 0.5%, while the amount of EO delivered by the ethanolic solution was not measurable. Final sampling was after 24 h, with 82.2% of chamomile EO found in the ULs when it was introduced in Pickering nanoemulsion form, and this value was 66.8% for conventional emulsion and 32.5% for the ethanolic solution (see Table 2).

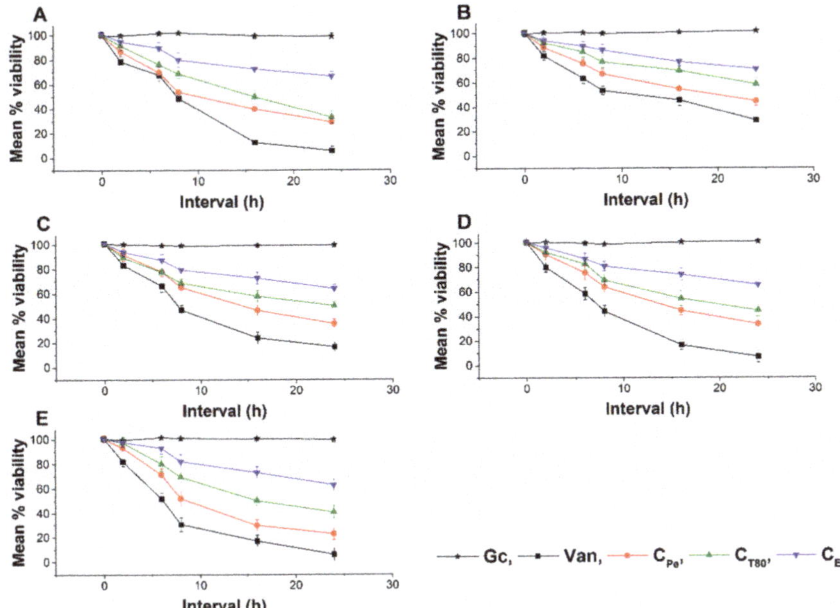



**Figure 10.** Mean percentage viability of C$_{Pe}$, C$_{T80}$, and C$_{Et}$ on *E. coli* (**A**), *S. aureus* (**B**), *B. subtilis* (**C**), *P. aeruginosa* (**D**), and *S. pyogenes* (**E**). Six independent experiments, each with 3 replicates, compared with Van and Gc as positive and growth controls.

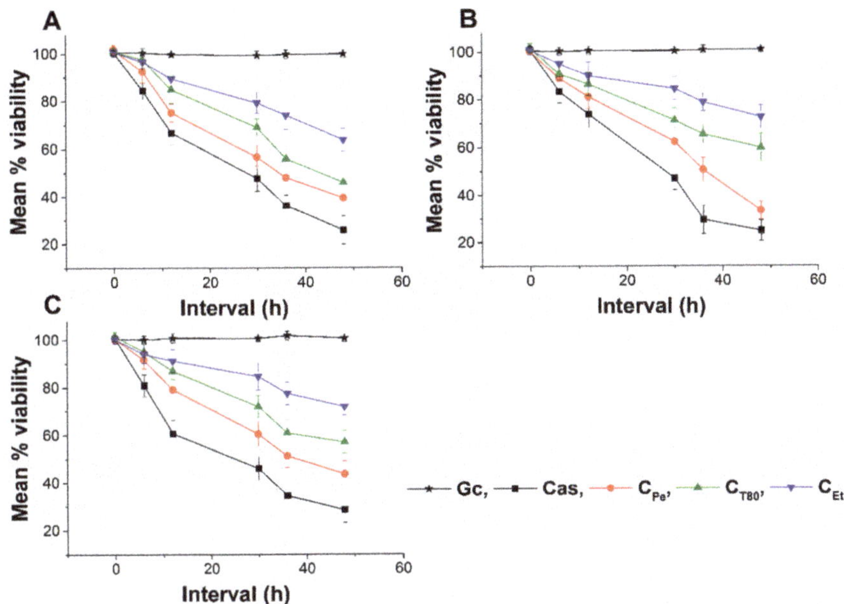



**Figure 11.** Mean percentage viability of C$_{Pe}$, C$_{T80}$, and C$_{Et}$ on *S. pombe* (**A**), *C. albicans* (**B**) and *C. tropicalis* (**C**). Six independent experiments, each with 3 replicates, compared with Cas and Gc as positive and growth controls.

**Table 2.** Results of the interaction study between unilamellar liposomes (Uls) and different formulations of chamomile EO. PA means the amount in the percentage of penetrated chamomile EO. Standard deviations were calculated from 3 independent experiments.

| Formulation | $PA_{1h}$ | $PA_{2h}$ | $PA_{24h}$ |
|:---:|:---:|:---:|:---:|
| $C_{Pe}$ | $27.2 \pm 3.7$ | $48.3 \pm 5.1$ | $82.2 \pm 4.9$ |
| $C_{T80}$ | - | $0.5 \pm 0.1$ | $66.8 \pm 3.6$ |
| $C_{Et}$ | - | - | $35.5 \pm 1.0$ |

## 3. Discussion

The effect of three different formulations on antimicrobial activity of chamomile essential oil has been examined. We have successfully prepared stable Pickering nanoemulsion using silica nanoparticles with appropriate lipophilicity. The emulsion was stable for three months. The effectiveness of Pickering emulsion was compared with conventional emulsion and ethanolic solution.

Based on our antimicrobial activity analyses, $C_{Pe}$ shows higher growth inhibitory action and consequently lower MICs compared to $C_{T80}$ and $C_{Et}$. Many researchers have studied the antimicrobial activity of chamomile oil [3,7,13,23,25], however, the mechanism of action at subinhibitory concentrations has not previously been studied. Our data suggest an effective killing activity of $C_{Pe}$ on selected bacteria and fungi. It is believed that EOs act against cell cytoplasmic membrane and induce stress in microorganisms [26–29]. To visualize the effects of $C_{Pe}$, $C_{T80}$, and $C_{Et}$, we introduced different staining methods to understand their mechanism. $C_{Pe}$ was able to generate higher oxidative stress compared to the conventional emulsion and ethanolic solution followed by metabolic interference and cell wall disruption and finally caused cell death at subinhibitory concentration [21,30–33].

The results obtained in the model experiment show that $C_{Pe}$ is the most effective form for the intracellular delivery of chamomile EO. Based on these results it can be established that the different antibacterial and antifungal effects may be caused by the difference of adsorption properties of EO forms to the microbial cells. The mechanism of delivery has not been revealed in this study, but evidence for the adsorption of Pickering emulsion droplets on the cell membrane has been previously reported [18]. Assuming the adsorption of $C_{Pe}$ droplets on the cell membrane of investigated microbes, intracellular delivery of active components from EO is feasible in two ways. Passive diffusion is caused by the higher local concentration gradient of EO on the cell membrane, or fusion of $C_{Pe}$ droplets with microbial cells. Overall, our observations demonstrate that $C_{Pe}$ facilitates chamomile oil to permeate cells, inducing oxidative stress and disrupting the membrane integrity because of higher adsorption efficacy of chamomile EO. SNP acts as a stabilizer, inhibiting the easy escape of EOs from the emulsion system compared to the conventional emulsion and free oils.

## 4. Materials and Methods

### 4.1. Synthesis, Surface Modification, and Characterization of Stöber Silica Nanoparticles

We have performed the synthesis of 20 nm hydrophilic silica nanoparticles with the previously reported modified Stöber method [9]. Briefly, a solution of tetraethoxysilane (TEOS) and ultrapure water in ethanol was prepared by using tetraethoxysilane, (Thermo Fisher GmbH, Kandel, Germany, pur. 98%); absolute ethanol AnalaR Normapur ≥99.8% purity (VWR Chemicals, Debrecen, Hungary) and water (membraPure Astacus Analytical with UV, VWR Chemicals, Debrecen, Hungary). The solution was stirred for 20 min and sonicated for another 20 min (Bandelin Sonorex RK 52H, BANDELIN electronic GmbH & Co. KG, Berlin, Germany). An appropriate amount of $NH_3$ solution (28% (*w/w*) ammonium solution, VWR Chemicals, Debrecen, Hungary) was added to the reaction mixture and was stirred at 1000 rpm for 24 h at room temperature (25 °C). The molar ratio of components was water/ethanol/TEOS/$NH_3$ = 100:300:5.2:1. The surface of hydrophilic silica nanoparticles was modified with propyltriethoxysilane (PTES Alfa Aesar, Haverhill, MA, USA, pur. 99%) in a post-synthesis modification reaction [32]. The ethanolic solution of the modifying agent was added to the freshly prepared hydrophilic silica nanoparticle suspensions; the mixtures were stirred for 6 h with 1000 rpm at room temperature. Before further use of the SNPs, the ammonium hydroxide and ethanol were always removed from the reaction mixture by distillation (Heidolph Laborota 4000, Heidolph Instruments GmbH & CO. KG, Germany). The water content was supplemented three times. The concentration of silica nanoparticle water-based suspension was finally adjusted to 1 mg/cm$^3$.

The size distribution and zeta potential of silica nanoparticles were determined by dynamic light scattering (DLS) using Malvern Zetasizer NanoS and NanoZ instruments (Malvern Instruments-Malvern Panalytical, Worcester, UK). The morphology and size distribution were also examined with transmission electron microscopy (TEM), (JEOL JEM-1200 EX II and JEM-1400, JEOL Ltd., Tokyo, Japan). The samples were dropped onto 200 mesh copper grids coated with carbon film (EMR Carbon support grids, Micro to Nano Ltd, Haarlem, The Netherlands) from diluted suspensions.

### 4.2. Preparation and Characterization of Pickering Nanoemulsion

As stabilizing agents, surface-modified silica nanoparticles or Tween 80 surfactant (Polysorbate80, Acros Organics, New Jersey, NJ, USA) were used. The concentration of stabilizing agents and chamomile essential oil (bluish *Matricaria chamomilla* oil, Aromax Ltd., Budapest, Hungary) was kept constant for all experiments, the values were 1 mg/mL and 100 µg/mL, respectively. The first step of the emulsification process was sonication for 2 minutes (Bandelin Sonorex RK 52H, BANDELINelectronic GmbH & Co. KG, Germany), then emulsification using UltraTurrax (IKA Werke T-25 basic, IKA®-Werke GmbH & Co. KG, Germany) for 5 min at 21,000 rpm. To compare the different formulations, an ethanolic solution was also prepared; chamomile essential oil was added to absolute ethanol at 100 µg/mL concentration, and the solution was sonicated for 5 min.

The stability of Pickering emulsion was studied from periodical droplet size determination using DLS measurements (Malvern Zetasizer Nano S, Malvern Panalytical Ltd, Worcester, UK).

### 4.3. Materials for Biological Experiments

In these experiments, the sterile 96-well microtiter plates were from Greiner Bio-One (Kremsmunster, Austria), potassium phosphate monobasic, glucose, adenine, 96% ethanol (Et), peptone, yeast extract, agar-agar, and Mueller Hinton agar were from Reanal Labor (Budapest, Hungary), modified RPMI 1640 (contains 3.4% MOPS, 1.8% glucose, and 0.002% adenine), SYBR green I 10,000×, propidium iodide, dihydrorhodamine 123 (DHR 123), 2′,7′-dichlorofluorescin diacetate (DCFDA), dihydroethidine (DHE) and menadione (Me) were from Sigma-Aldrich Chemie GmbH (Steinheim, Germany), disodium phosphate and dimethyl sulfoxide (DMSO) were from Chemolab Ltd. (Budapest, Hungary), sodium chloride from VWR Chemicals (Debrecen, Hungary), potassium chloride was from Scharlau Chemie S.A (Barcelona, Spain), 3-(*N*-morpholino) propanesulfonic acid (MOPS) was from Serva Electrophoresis GmbH (Heidelberg, Germany), caspofungin (Cas) from Merck Sharp & Dohme Ltd (Hertfordshire, UK), vancomycin (Van) from Fresenius Kabi Ltd. (Budapest, Hungary), 0.22 µm vacuum filters from Millipore (Molsheim, France) and the cell spreader was from Sarstedt AG & Co. KG (Numbrecht, Germany). All other chemicals used in the study were of analytical or spectroscopic grade. For fungi, we used an in-house nutrient agar medium [34] while phosphate-buffered saline (PBS, pH 7.4) was from Life Technologies Ltd. (Budapest, Hungary). Highly purified water (<1.0 µS) was applied throughout the studies.

### 4.4. Determination of Minimum Inhibitory Concentration (MIC$_{90}$)

#### 4.4.1. Microorganisms

*Escherichia coli* (*E. coli*) PMC 201, *Pseudomonas aeruginosa* (*P. aeruginosa*) PMC 103, *Bacillus subtilis* (*B. subtilis*) SZMC 0209, *Staphylococcus aureus* (*S. aureus*) ATCC 29213, *Streptococcus pyogenes* (*S. pyogenes*) SZMC 0119, *Schizosaccharomyces pombe* (*S. pombe*) ATCC 38366, *Candida albicans* (*C. albicans*) ATCC 1001, and *Candida tropicalis* (*C. tropicalis*) SZMC 1368 were obtained from Szeged Microbial Collection,

Department of Microbiology, University of Szeged, Hungary (SZMC) and Department of General and Environmental Microbiology, Institute of Biology, University of Pecs, Hungary (PMC).

### 4.4.2. Antimicrobial Activity Tests

The antibacterial activity of the tested drugs was separately evaluated on *E. coli*, *P. aeruginosa*, *B. subtilis*, *S. aureus*, and *S. pyogenes* according to our previously published protocol [35]. In brief, bacterial populations of ~$10^5$ CFU/mL were inoculated into RPMI media and incubated for 16 h at $35 \pm 2$ °C with test compounds ($C_{Pe}$, $C_{T80}$, $C_{Et}$, and Van) over a wide concentration range (0.3–0.01 µg/mL). The absorbance was measured by a Thermo Scientific Multiskan EX 355 plate reader (InterLabsystems, Budapest, Hungary) at 600 nm.

The antifungal activity against *S. pombe*, *C. albicans*, and *C. tropicalis* species were also carried out according to our previously published method [35]. Briefly, ~$10^3$ cells/mL were incubated for 48 h at $30 \pm 2$ °C with test compounds ($C_{Pe}$, $C_{T80}$, $C_{Et}$ and Cas) at wide concentration range (20–0.01 µg/mL) in modified RPMI media. The absorbance was measured by a Thermo Scientific Multiskan EX 355 plate reader (InterLabsystems, Budapest, Hungary) at 595 nm. Absorbance values were converted to percentages compared to growth control (~100%) and data were fitted by nonlinear dose–response curve method to calculate the dose producing ≥90% growth inhibition ($MIC_{90}$). All the measurements were performed by applying three technical replicates in six independent experiments. Van and Cas were used as a standard antibacterial and antifungal drug, respectively, throughout the experiments.

### 4.5. Determination of Minimum Effective Concentration (MEC$_{10}$)

The assay has been designed to determine the killing effects of the test compounds for a certain period of time. In brief, a wide concentration range (0.03–80 µg/mL) of the test samples were used to treat ~$10^6$ cells/mL for one hour. One milliliter of treated and untreated samples were taken and were diluted $10^5$ times followed by spreading 50 µL samples onto 20 mL nutrient agar media and incubated for 24 h at 35 °C (bacteria) and 30 °C (fungi) for colony-forming unit (CFU/mL) quantification. The data were compared with growth control and positive control (Van for bacteria and Cas for fungi) for percentage mortality (~10% death) determination. It is noteworthy that in the $MEC_{10}$ experiments, the inoculated microbial cell number was $10^3$ times higher than was used in the $MIC_{90}$ determinations ($10^6$ vs. $10^3$). However, in both cases, the same formula was used (see Section 4.7). All the measurements were performed by applying three technical replicates in six independent experiments.

### 4.6. Determination of Microbial Oxidative Generation and Killing Activity

### 4.6.1. Quantification of Total ROS Generation

Total ROS generation was assayed according to previously published protocols [27,28,34,36]. Briefly, ~$10^6$ cells/mL were collected and centrifuged at 1500 *g* for 5 min and were suspended in PBS. The cells were stained with a 20 mM stock solution of DCFDA in PBS (pH 7.4) to achieve an end concentration of 25 µM, and were incubated at 35 °C (for bacteria) and 30 °C (for fungi) for 30 min in the dark with mild shaking. The cells were centrifuged (Hettich Rotina 420R, Auro-Science Consulting Ltd., Budapest, Hungary) and suspended in RPMI media. The cells were treated with $C_{Pe}$, $C_{T80}$, $C_{Et}$, Van (bacteria), and Cas (fungi) at their respective $MEC_{10}$ concentrations for one hour. The fluorescence signals were recorded at Ex/Em = 485/535 nm wavelengths by a Hitachi F-7000 fluorescence spectrophotometer/plate reader (Auro-Science Consulting Ltd., Budapest, Hungary). The percentage increase in oxidative balance was measured by comparing the signals to those of the growth controls (Gc). Six independent experiments were done with three technical replicates for each treatment.

### 4.6.2. Detection of Peroxide ($O_2^{2-}$) and Superoxide Anion ($O_2^{\bullet-}$) Generation

The previously described protocol was adapted for peroxide [36,37] and superoxide anion radicals [34,38] with modifications. A positive control, Me (0.5 mmol/L as end concentration), $C_{Pe}$, $C_{T80}$, $C_{Et}$, Van (bacteria), and Cas (fungi) at their respective $MEC_{10}$ concentrations were used to treat the cell suspensions (~$10^6$ cells/mL) for an hour at 35 °C (bacteria) and 30 °C (fungi) in RPMI media. Thereafter, the cells were centrifuged at 1500 *g* for 5 min at room temperature followed by resuspension of pellets in PBS of the same volume. DHR 123 (10 µmol/L, end concentration) and DHE (15 µmol/L, end concentration) were added separately to the cell samples for peroxide and superoxide determination. The stained cells were further incubated at 35 °C (bacteria) and 30 °C (fungi) in the dark with mild shaking. The samples were centrifuged and resuspended in PBS followed by the distribution of the samples into the wells of 96-well microplates. The fluorescence was measured at excitation/emission wavelengths of 500/536 nm for peroxides and 473/521 nm for superoxide detection by a Hitachi F-7000 fluorescence spectrophotometer/plate reader (Auro-Science Consulting Ltd., Budapest, Hungary). The percentage increase in oxidative stress was measured by comparing the signals to those of the growth controls (Gc). Six independent experiments were done with three technical replicates for each treatment.

### 4.6.3. Time–Kill Kinetics Assay

We followed a protocol previously published by T. Appiah et. al., with modifications [39]. In brief, $C_{Pe}$, $C_{T80}$, $C_{Et}$, Van (bacteria), and Cas (fungi) at their respective $MEC_{10}$ concentrations were used to treat the microbial population of ~$10^6$ CFU/mL and were incubated at 35 °C (bacteria) and 30 °C (fungi). One milliliter of the treated and untreated samples was pipetted at time intervals of 0, 2, 6, 8, 16, and 24 h for bacteria, and 0, 6, 12, 30, 36, and 48 h for fungi, and were diluted $10^5$ times followed by spreading 50 µL onto 20 mL nutrient agar media using a cell spreader and incubated at 35 °C (bacteria) and 30 °C (fungi) for 24 h. Van and Cas were used as reference controls for bacteria and fungi. Control without treatment was considered as growth control (Gc). The colony-forming unit (CFU/mL) of the microorganisms were determined, performed in triplicate and was plotted against time (h). Six independent experiments were done with three technical replicates for each treatment.

### 4.6.4. Live/dead Discrimination of Microbial Cells

For live/dead cell discrimination, we followed the protocol published previously [35]. In brief, the cell population of ~$10^6$ cells/mL were treated with $C_{Pe}$, $C_{T80}$, $C_{Et}$, Van (bacteria), and Cas (fungi) at their respective $MEC_{10}$ concentrations and were incubated at 35 °C (bacteria) and 30 °C (fungi). Treated and untreated samples were pipetted at time intervals of 0, 2, 6, 8, 16, and 24 h for bacteria, and 0, 6, 12, 30, 36 and 48 h for fungi followed by centrifugation at 1000 *g* for 5 min, washed, and resuspended in PBS (100 µL/well). One hundred microliters of freshly prepared working dye solution in PBS (using 20 µL SYBR green I and 4 µL propidium iodide diluted solutions as described earlier) were added to the samples. The plate was incubated at room temperature for 15 min in the dark with mild shaking. A Hitachi F-7000 fluorescence spectrophotometer/plate reader (Auro-Science, Consulting Ltd., Budapest, Hungary) was used to measure the fluorescence intensities of SYBR green I (excitation/emission wavelengths: 490/525 nm) and propidium iodide (excitation/emission wavelengths: 530/620 nm), respectively. A green to red fluorescence ratio for each sample and for each dose was achieved and the % of dead cells with the response to the applied dose was plotted against the applied test compound doses using a previously published formula [35]. All treatments were done in triplicates and six independent experiments were performed.

### 4.7. Statistical Analysis of Microbiological Experiments

All data were given as mean ± SD. Graphs and statistical analyses were conducted using OriginPro 2016 (OriginLab Corp., Northampton, MA, USA). All experiments were performed independently six times and data were analyzed by one-way ANOVA test. $P < 0.01$ was considered statistically significant. The growth inhibition concentration ($MIC_{90}$) and minimum effective concentration ($MEC_{10}$) were calculated using a nonlinear dose–response sigmoidal curve function as follows:

$$y = A_1 + \frac{A_2 - A_1}{1 + 10^{(\log_x 0 - x)p}}$$

where A1, A2, $LOG_x0$ and p as the bottom asymptote, top asymptote, center, and hill slope of the curve have been considered.

### 4.8. Interaction Study between the Cell Model (Unilamellar Liposomes) and Different Formulations of Chamomile EO

Unilamellar liposomes (ULs) have been prepared from phosphatidylcholine (Phospolipon 90G, Phospholipid GmbH, Berlin, Germany) by the modified method described before by Alexander Moscho et al. [40]. Phosphatidylcholine was dissolved in chloroform ($\geq$98% stabilized, VWR Chemicals, Debrecen, Hungary) in 0.1 M concentration, and 150 μL of this solution was diluted in a mixture of 6 mL chloroform and 1 mL of methanol. This solution was added dropwise to 40 mL of PBS buffer while stirring on a magnetic stirrer at 600 rpm (VELP Scientifica Microstirrer, Magnetic Stirrer, Usmate Velate MB, Italy). The solvents were removed on a rotational evaporator at 40 °C (Heidolph Laborota 4000, Heidolph Instruments GmbH & CO. KG, Germany). The resulting suspension volume was set to 25 mL with PBS buffer and stored in the refrigerator at 8 °C until further use. A 5 mL suspension of ULs was mixed with 3 mL Pickering nanoemulsion, conventional emulsion, or ethanolic solution, and the chamomile EO concentration was 100 μg/mL for the different formulations. The mixture was stirred at 600 rpm for 24 h at 35 °C, and 1 mL aliquots were taken after 1, 2, and 24 h. The samples were centrifuged at 3000 rpm and 20 °C for 5 min, and the ULs were collected and dissolved in absolute ethanol. The chamomile–EO content of samples was determined with UV/VIS Spectroscopy at 290 nm (Jasco V-550 UV/VIS Spectrophotometer; Jasco Inc., Easton, MD, USA). For UV/VIS measurements we have prepared samples without chamomile–EOs, i.e., ULs with SNP suspension, Tween 80 solution, or ethanol were also mixed and centrifuged and were used as blanks.

### 4.9. GC-MS Analysis of Chamomile EO

Gas chromatography and mass spectrometry (GC–MS) analyses were carried out on an Agilent Technologies (Palo Alto, CA, USA) gas chromatograph model 7890A with 5975C mass detector. Operating conditions were as follows: column HP-5MS (5% phenylmethyl polysiloxane), 30 m $\times$ 0.25 mm i.d., 0.25 μm coating thickness. Helium was used as the carrier gas at 1 mL/min, injector temperature was 250 °C. HP-5MS column temperature was programmed at 70 °C isothermal for 2 min, and then increased to 200 °C at a rate of 3 °C/min and held isothermal for 18 min. The split ratio was 1:50, ionization voltage 70 eV; ion source temperature 230 °C; mass scan range: 45–450 mass units. The percentage composition was calculated from the GC peak areas using the normalization method (without correction factors). The component percentages were calculated as mean values from duplicate GC–MS analyses of the oil sample. The results of GC–MS analysis can be seen in Supplementary Materials.

**Supplementary Materials:** The followings are available online: Composition of chamomile essential oil analyzed by GC-MS method. Results can be seen in Supplementary Table S1 and Figure S1.

**Author Contributions:** Conceptualization T.K., and A.S.; methodology S.D., B.H., T.K. and A.S.; software, S.D.; formal analysis, S.D. and B.H.; investigation, S.D. and B.H.; GC analysis S.Š. and S.J. resources, A.S.; data curation, S.D., B.H.; writing—original draft preparation, S.D. and B.H.; writing—review and editing, T.K. and A.S.; visualization, S.D. and B.H.; supervision, T.K. and A.S.; funding acquisition A.S. and T.K.

**Funding:** This work was supported by EFOP 3.6.1-16-2016-00004 project (Comprehensive Development for Implementing Smart Specialization Strategies at the University of Pécs), University of Pécs, Medical School, KA-2018-17 and NKFI-EPR K/115394/2015 grants.

**Conflicts of Interest:** The authors declare no conflict of interest.

## References


1. Manion, C.R.; Widder, R.M. Essentials of essential oils. *Am. J. Health-Syst. Pharm.* **2017**, *74*, e153–e162. [CrossRef]
2. Omonijo, F.A.; Ni, L.; Gong, J.; Wang, Q.; Lahaye, L.; Yang, C. Essential oils as alternatives to antibiotics in swine production. *Anim. Nutr.* **2018**, *4*, 126–136. [CrossRef]
3. McKay, D.L.; Blumberg, J.B. A Review of the bioactivity and potential health benefits of chamomile tea (*Matricaria recutita* L.). *Phytother. Res.* **2006**, *20*, 519–530. [CrossRef]
4. Lis-Balchin, M. *Aromatherapy Science: A Guide for Healthcare Professionals*; Pharmaceutical Press: London, UK, 2006; ISBN 978-0-85369-578-3.
5. Pfaller, M.A.; Sheehan, D.J.; Rex, J.H. Determination of Fungicidal Activities against Yeasts and Molds: Lessons Learned from Bactericidal Testing and the Need for Standardization. *Clin. Microbiol. Rev.* **2004**, *17*, 268–280. [CrossRef]
6. Donsì, F.; Ferrari, G. Essential oil nanoemulsions as antimicrobial agents in food. *J. Biotechnol.* **2016**, *233*, 106–120. [CrossRef] [PubMed]
7. Singh, O.; Khanam, Z.; Misra, N.; Srivastava, M. Chamomile (*Matricaria chamomilla* L.): An overview. *Pharmacogn. Rev.* **2011**, *5*, 82. [CrossRef] [PubMed]
8. Alizadeh Behbahani, B.; Tabatabaei Yazdi, F.; Vasiee, A.; Mortazavi, S.A. Oliveria decumbens essential oil: Chemical compositions and antimicrobial activity against the growth of some clinical and standard strains causing infection. *Microb. Pathog.* **2018**, *114*, 449–452. [CrossRef]
9. Balázs, V.L.; Horváth, B.; Kerekes, E.; Ács, K.; Kocsis, B.; Varga, A.; Böszörményi, A.; Nagy, D.U.; Krisch, J.; Széchenyi, A.; et al. Anti-Haemophilus Activity of Selected Essential Oils Detected by TLC-Direct Bioautography and Biofilm Inhibition. *Molecules* **2019**, *24*, 3301. [CrossRef] [PubMed]
10. Ćavar, S.; Maksimović, M.; Vidic, D.; Parić, A. Chemical composition and antioxidant and antimicrobial activity of essential oil of *Artemisia annua* L. from Bosnia. *Ind. Crop. Prod.* **2012**, *37*, 479–485.
11. Fasihi, H.; Noshirvani, N.; Hashemi, M.; Fazilati, M.; Salavati, H.; Coma, V. Antioxidant and antimicrobial properties of carbohydrate-based films enriched with cinnamon essential oil by Pickering emulsion method. *Food Packag. Shelf Life* **2019**, *19*, 147–154. [CrossRef]
12. Tang, X.; Shao, Y.-L.; Tang, Y.-J.; Zhou, W.-W. Antifungal Activity of Essential Oil Compounds (Geraniol and Citral) and Inhibitory Mechanisms on Grain Pathogens (Aspergillus flavus and Aspergillus ochraceus). *Molecules* **2018**, *23*, 2108. [CrossRef] [PubMed]
13. Stanojevic, L.P.; Marjanovic-Balaban, Z.R.; Kalaba, V.D.; Stanojevic, J.S.; Cvetkovic, D.J. Chemical Composition, Antioxidant and Antimicrobial Activity of Chamomile Flowers Essential Oil (*Matricaria chamomilla* L.). *J. Essent. Oil Bear. Plants* **2016**, *19*, 2017–2028. [CrossRef]
14. Alastruey-Izquierdo, A.; Gomez-Lopez, A.; Arendrup, M.C.; Lass-Florl, C.; Hope, W.W.; Perlin, D.S.; Rodriguez-Tudela, J.L.; Cuenca-Estrella, M. Comparison of Dimethyl Sulfoxide and Water as Solvents for Echinocandin Susceptibility Testing by the EUCAST Methodology. *J. Clin. Microbiol.* **2012**, *50*, 2509–2512. [CrossRef] [PubMed]
15. Inouye, S.; Tsuruoka, T.; Uchida, K.; Yamaguchi, H. Effect of Sealing and Tween 80 on the Antifungal Susceptibility Testing of Essential Oils. *Microbiol. Immunol.* **2001**, *45*, 201–208. [CrossRef] [PubMed]
16. Thielmann, J.; Muranyi, P.; Kazman, P. Screening essential oils for their antimicrobial activities against the foodborne pathogenic bacteria *Escherichia coli* and *Staphylococcus aureus*. *Heliyon* **2019**, *5*, e01860. [CrossRef]
17. Pina-Barrera, A.M.; Alvarez-Roman, R.; Baez-Gonzalez, J.G.; Amaya-Guerra, C.A.; Rivas-Morales, C.; Gallardo-Rivera, C.T.; Galindo-Rodriguez, S.A. Application of a multisystem coating based on polymeric nanocapsules containing essential oil of *Thymus vulgaris* L. to increase the shelf life of table grapes (*Vitis vinifera* L.). *IEEE Trans. NanoBiosci.* **2019**, 549–557. [CrossRef]
18. Zhou, Y.; Sun, S.; Bei, W.; Zahi, M.R.; Yuan, Q.; Liang, H. Preparation and antimicrobial activity of oregano essential oil Pickering emulsion stabilized by cellulose nanocrystals. *Int. J. Biol. Macromol.* **2018**, *112*, 7–13. [CrossRef]
19. Abarca, R.L.; Rodríguez, F.J.; Guarda, A.; Galotto, M.J.; Bruna, J.E. Characterization of beta-cyclodextrin inclusion complexes containing an essential oil component. *Food Chem.* **2016**, *196*, 968–975. [CrossRef]

20. Ciobanu, A.; Landy, D.; Fourmentin, S. Complexation efficiency of cyclodextrins for volatile flavor compounds. *Food Res. Int.* **2013**, *53*, 110–114. [CrossRef]

21. Cossu, A.; Wang, M.S.; Chaudhari, A.; Nitin, N. Antifungal activity against *Candida albicans* of starch Pickering emulsion with thymol or amphotericin B in suspension and calcium alginate films. *Int. J. Pharm.* **2015**, *493*, 233–242. [CrossRef]

22. Wang, J.; Li, Y.; Gao, Y.; Xie, Z.; Zhou, M.; He, Y.; Wu, H.; Zhou, W.; Dong, X.; Yang, Z.; et al. Cinnamon oil-loaded composite emulsion hydrogels with antibacterial activity prepared using concentrated emulsion templates. *Ind. Crop. Prod.* **2018**, *112*, 281–289. [CrossRef]

23. Srivastava, J.K.; Shankar, E.; Gupta, S. Chamomile: A herbal medicine of the past with bright future. *Mol. Med. Rep.* **2010**, *3*, 895–901. [PubMed]

24. Rideau, E.; Dimova, R.; Schwille, P.; Wurm, F.R.; Landfester, K. Liposomes and polymersomes: A comparative review towards cell mimicking. *Chem. Soc. Rev.* **2018**, *47*, 8572–8610. [CrossRef] [PubMed]

25. Göger, G.; Demirci, B.; Ilgın, S.; Demirci, F. Antimicrobial and toxicity profiles evaluation of the Chamomile (*Matricaria recutita* L.) essential oil combination with standard antimicrobial agents. *Ind. Crop. Prod.* **2018**, *120*, 279–285.

26. Memar, M.Y.; Ghotaslou, R.; Samiei, M.; Adibkia, K. Antimicrobial use of reactive oxygen therapy: Current insights. *Infect. Drug Resist.* **2018**, *11*, 567–576. [CrossRef] [PubMed]

27. Eruslanov, E.; Kusmartsev, S. Identification of ROS Using Oxidized DCFDA and Flow-Cytometry. In *Advanced Protocols in Oxidative Stress II*; Armstrong, D., Ed.; Humana Press: Totowa, NJ, USA, 2010; Volume 594, pp. 57–72. ISBN 978-1-60761-410-4.

28. Dong, T.G.; Dong, S.; Catalano, C.; Moore, R.; Liang, X.; Mekalanos, J.J. Generation of reactive oxygen species by lethal attacks from competing microbes. *Proc. Natl. Acad. Sci. USA* **2015**, *112*, 2181–2186. [CrossRef] [PubMed]

29. Máté, G.; Gazdag, Z.; Mike, N.; Papp, G.; Pócsi, I.; Pesti, M. Regulation of oxidative stress-induced cytotoxic processes of citrinin in the fission yeast *Schizosaccharomyces pombe*. *Toxicon* **2014**, *90*, 155–166. [CrossRef]

30. Dohare, S.; Dubey, S.D.; Kalia, M.; Verma, P.; Pandey, H.; Singh, K.; Agarwal, V. Anti-biofilm activity of *Eucalyptus globulus* oil encapsulated silica nanoparticles against *E. coli* biofilm. *Int. J. Pharm. Sci. Res.* **2014**, *5*, 5013–5018.

31. Saad, N.Y.; Muller, C.D.; Lobstein, A. Major bioactivities and mechanism of action of essential oils and their components: Essential oils and their bioactive components. *Flavour Fragr. J.* **2013**, *28*, 269–279. [CrossRef]

32. Horváth, B.; Balázs, V.L.; Horváth, G.; Széchenyi, A. Preparation and in vitro diffusion study of essential oil Pickering emulsions stabilized by silica nanoparticles for *Streptococcus mutans* biofilm inhibition. *Flavour Fragr. J.* **2018**, *33*, 385–396. [CrossRef]

33. Bravo Cadena, M.; Preston, G.M.; Van der Hoorn, R.A.L.; Townley, H.E.; Thompson, I.P. Species-specific antimicrobial activity of essential oils and enhancement by encapsulation in mesoporous silica nanoparticles. *Ind. Crop. Prod.* **2018**, *122*, 582–590. [CrossRef]

34. Fujs, S.; Gazdag, Z.; Poljsak, B.; Stibilj, V.; Milacic, R.; Pesti, M.; Raspor, P.; Batic, M. The oxidative stress response of the yeast *Candida intermedia* to copper, zinc, and selenium exposure. *J. Basic Microbiol.* **2005**, *45*, 125–135. [CrossRef] [PubMed]

35. Das, S.; Gazdag, Z.; Szente, L.; Meggyes, M.; Horváth, G.; Lemli, B.; Kunsági-Máté, S.; Kuzma, M.; Kőszegi, T. Antioxidant and antimicrobial properties of randomly methylated β cyclodextrin—Captured essential oils. *Food Chem.* **2019**, *278*, 305–313. [CrossRef] [PubMed]

36. Stromájer-Rácz, T.; Gazdag, Z.; Belágyi, J.; Vágvölgyi, C.; Zhao, R.Y.; Pesti, M. Oxidative stress induced by HIV-1 F34IVpr in *Schizosaccharomyces pombe* is one of its multiple functions. *Exp. Mol. Pathol.* **2010**, *88*, 38–44. [CrossRef]

37. Gazdag, Z.; Máté, G.; Čertik, M.; Türmer, K.; Virág, E.; Pócsi, I.; Pesti, M. *tert*-Butyl hydroperoxide-induced differing plasma membrane and oxidative stress processes in yeast strains BY4741 and *erg5Δ*: *t*-BuOOH-induced cytotoxic processes in yeast. *J. Basic Microbiol.* **2014**, *54*, S50–S62. [CrossRef]

38. Henderson, L.M.; Chappell, J.B. Dihydrorhodamine 123: A fluorescent probe for superoxide generation? *Eur. J. Biochem.* **1993**, *217*, 973–980. [CrossRef]

39. Appiah, T.; Boakye, Y.D.; Agyare, C. Antimicrobial Activities and Time-Kill Kinetics of Extracts of Selected Ghanaian Mushrooms. *Evid. Based Complement. Altern. Med.* **2017**, *2017*, 1–15. [CrossRef]

40. Moscho, A.; Orwar, O.; Chiu, D.T.; Modi, B.P.; Zare, R.N. Rapid preparation of giant unilamellar vesicles. *Proc. Natl. Acad. Sci.* **1996**, *93*, 11443–11447. [CrossRef]

**Sample Availability:** Samples of the compounds chamomile EO and SNPs are available from the authors.

*Article*

# Fingerprinting, Antimicrobial, Antioxidant, Anticancer, Cyclooxygenase and Metabolic Enzymes Inhibitory Characteristic Evaluations of *Stachys viticina* Boiss. Essential Oil


**Nidal Jaradat [1],* and Nawaf Al-Maharik [2],***

[1] Department of Pharmacy, Faculty of Medicine and Health Sciences, An-Najah National University, Nablus, 00970, Palestine

[2] Department of Chemistry, Faculty of Science, An-Najah National University, Nablus, 00970, Palestine

* Correspondence: nidaljaradat@najah.edu (N.J.); na10@st-andrews.ac.uk (N.A.-M.)


Received: 20 August 2019; Accepted: 19 October 2019; Published: 28 October 2019


**Abstract:** The present study aimed to identify the chemical constituents and to assess the in-vitro, antimicrobial, anticancer, antioxidant, metabolic enzymes and cyclooxygenase (COX) inhibitory properties of essential oil (EO) of *Stachys viticina* Boiss. leaves. The *S. viticina* EO was isolated and identified using microwave-ultrasonic and GC-MS techniques, respectively. Fifty-two compounds were identified, of which *endo*-borneol was the major component, followed by eucalyptol and epizonarene. The EO was evaluated against a panel of in-vitro bioassays. The EO displayed antimicrobial activity against methicillin-resistant *Staphylococcus aureus* (MRSA), *Escherichia coli* and *Epidermophyton floccosum*, with MIC values of 0.039, 0.078 and 0.78 mg/mL, respectively. The EO exhibited cytotoxicity against HeLa (cervical adenocarcinoma) and Colo-205 (colon) cancer cell lines with percentages of inhibition of 95% and 90%, for EO concentrations of 1.25 and 0.5 mg/mL, respectively. Furthermore, it showed metabolic enzyme ($\alpha$-amylase, $\alpha$-glucosidase, and lipase) inhibitory ($IC_{50}$ = 45.22 ± 1.1, 63.09 ± 0.26, 501.18 ± 0.38 µg/mL, respectively) and antioxidant activity, with an $IC_{50}$ value of 19.95 ± 2.08 µg/mL. Moreover, the *S. viticina* EO showed high cyclooxygenase inhibitory activity against COX-1 and COX-2 with $IC_{50}$ values of 0.25 and 0.5 µg/mL, respectively, similar to those of the positive control (the NSAID etodolac). Outcomes amassed from this investigation illustrate that *S. viticina* EO represents a rich source of pharmacologically active molecules which can be further validated and explored clinically for its therapeutic potential and for the development and design of new natural therapeutic preparations.

**Keywords:** *Stachys viticina*; essential oil; cytotoxicity; antioxidant; antimicrobial; metabolic enzymes; cyclooxygenase


---

## 1. Introduction

People since ancient times have been using plant secondary metabolites in their daily life for the treatment of a wide range of diseases, as food preservatives and flavoring, and to annihilate insects. The World Health Organization has estimated that around 80% of the world population rely on plant extracts for their medication [1]. With time, synthetic chemicals have replaced plant secondary metabolites as the former provide early results. However, most synthetic drugs have various side effects and may cause many serious human health problems, therefore strategies have been implemented to replace synthetic drugs with plant secondary metabolites due to their ability to protect human body against oxidative damage and their great positive impact on human health [1–3]. Nowadays, due to the growing public awareness about the probable harmful effects of synthetic additives and the tendency of consumers to use natural foods, the use of natural bioactive compounds is witnessing significant growth [4]. Due to the wide prevalence of cancer worldwide, there is an essential and urgent need for a search for new anticancer medications [5]. In fact, majorities of used antitumor drugs are derived from natural sources [6].

The essential oils extracted from plants have been used in traditional medicine for the treatment and prevention of various diseases, in cosmetics, foods, and food supplements in addition to dental practice and hygiene products [7,8].

*Stachys* (Lamiaceae family) is a large genus of herbs and shrubs containing around 300 species, widely spread in the temperate regions, especially in the Irano-Turanian and Mediterranean areas. Recently, several studies have documented anti-inflammatory, cytotoxic, antioxidant, immune system booster and antimicrobial properties of the extracts from a number of *Stachys* species [9].

*Stachys viticina* Boiss. is a perennial herbaceous shrubby plant that can reach a height of 100 cm, with multiple branches, grey-colored and erect stems. The upper leaves are crenate, while the lower ones are large, oblong-ovate and verticillaster. In traditional medicine, various *Stachys* species have been utilized as anti-inflammatory, antimicrobial, anti-diarrheal, wound healing and astringent remedies. In addition, they were used to treat ulcers, cough, sclerosis of the spleen and genital tumors [10,11].

It is worth mentioning that the chemical constituents of *S. viticina* EO, as well as its biological and pharmacological properties, haven't been reported. Thus, the current study was designed to investigate the chemical composition of the EO extracted from *S. viticina* leaves. Additionally, the antioxidant, anticancer, antimicrobial, metabolic enzyme ($\alpha$-amylase, $\alpha$-glucosidase, and lipase) and cyclooxygenase (COX) inhibitory properties of *S. viticina* EO was assessed.

## 2. Results

### 2.1. Phytochemical Composition of S. viticina Essential Oil

In accord with our ability and experience in the isolation and identification of natural products from plants, as well as in assessing their biological activities, we decided to study the chemical components of *S. viticina* EO, as well as their biological effects, in the hope to finding new drug leads from natural sources. The isolation procedure was carried out using a microwave-ultrasonic apparatus. The phytochemical composition was identified and estimated using GC-MS. The obtained yield of the EO was 1.72 ± 0.97%. GC-MS analyses indicated the presence of 52 phytochemical compounds in *S. viticina* EO, of which *endo*-borneol was the major component, followed by eucalyptol and epizonarene, as revealed in Table 1 and Figure S1.

### 2.2. Antioxidant and Metabolic Enzymes Inhibitory Activity

The inhibitory activity of *S. viticina* EO against oxidation and metabolic enzymes ($\alpha$-amylase and $\alpha$-glucosidase and lipase) was assessed using standard biomedical assays. The results showed that *S. viticina* essential oil has $\alpha$-amylase and $\alpha$-glucosidase inhibitory activities, with $IC_{50}$ values of 45.22 ± 1.1 and 63.09 ± 0.26 µg/mL, respectively.

As indicated in Table 2 and Figures S2–S5, the EO displayed antioxidant and antilipase activities with $IC_{50}$ values of 19.95 ± 2.08 and 501.18 ± 0.38 µg/mL, respectively.

**Table 1.** *S. viticina* leaves EO phytochemical composition.

| | Essential Oil Components | R.T. | R.I. | %, ±SD |
|---|---|---|---|---|
| 1. | 2,4-Dimethyltetrahydro-2H-thiopyran 1,1-dioxide | 5.664 | 770 | 0.11 ± 0.023 |
| 2. | α-Phellandrene | 8.335 | 917 | 0.09 ± 0.001 |
| 3. | δ-3-Carene | 10.271 | 919 | 0.25 ± 0.01 |
| 4. | α-Pinene | 10.436 | 891 | 0.72 ± 0.013 |
| 5. | Benzylin | 11.126 | 909 | 5.29 ± 0.31 |
| 6. | Santolina triene | 11.361 | 837 | 0.4 ± 0.001 |
| 7. | 4-Carene | 12.111 | 893 | 0.69 ± 0.023 |
| 8. | Eucalyptol | 12.827 | 841 | 21.26 ± 1.97 |
| 9. | γ-Terpinene | 13.837 | 928 | 1.52 ± 0.014 |
| 10. | 1,5-Dimethyl-1,5-cyclooctadiene | 14.227 | 883 | 0.18 ± 0.001 |
| 11. | 5,5-Dimethyl-2-ethyl-1,3-cyclopentadiene | 14.352 | 881 | 0.18 ± 0.001 |
| 12. | Ocimene | 14.778 | 817 | 0.07 ± 0.004 |
| 13. | α-Terpinolene | 14.923 | 917 | 0.29 ± 0.001 |
| 14. | Linalool oxide | 15.018 | 803 | 0.04 ± 0.001 |
| 15. | Arachidonic acid methyl ester | 15.683 | 840 | 0.17 ± 0.02 |
| 16. | Cosmene | 15.99 | 803 | 0.01 ± 0.001 |
| 17. | Norbornane | 16.11 | 847 | 0.02 ± 0.001 |
| 18. | 4,7-Dimethyl-4,4a,5,6-tetrahydrocyclopenta[c]pyran-1,3-dione | 16.508 | 910 | 0.21 ± 0.01 |
| 19. | *cis*-Carveol | 16.843 | 829 | 0.01 ± 0.001 |
| 20. | Endo-Borneol | 18.584 | 909 | 29.06 ± 1.24 |
| 21. | Bornyl chloride | 18.769 | 829 | 5.79 ± 0.09 |
| 22. | α-Terpineol | 19.324 | 907 | 4.48 ± 0.07 |
| 23. | Menthen-9-ol | 22.41 | 808 | 4.2 ± 0.023 |
| 24. | Geranyl Phenylacetate | 24.966 | 803 | 0.48 ± 0.001 |
| 25. | Copaene | 25.547 | 900 | 0.93 ± 0.015 |
| 26. | β-Bourbonene | 25.807 | 921 | 0.78 ± 0.011 |
| 27. | Jasmone | 26.137 | 899 | 0.07 ± 0.001 |
| 28. | Caryophyllene | 26.967 | 924 | 0.95 ± 0.009 |
| 29. | β-Cubebene | 27.302 | 902 | 0.11 ± 0.004 |
| 30. | D-Germacrene | 27.753 | 871 | 0.05 ± 0.001 |
| 31. | 1,1,4,8-Tetramethyl-4,7,10-cycloundecatriene | 28.113 | 889 | 0.24 ± 0.006 |
| 32. | *epi*-Bicyclosesquiphellandrene | 29.288 | 896 | 0.78 ± 0.008 |
| 33. | 5,7-Diethyl-5,6-decadien-3-yne | 29.618 | 929 | 0.06 ± 0.002 |
| 34. | γ-Cadinene | 29.923 | 919 | 0.23 ± 0.008 |
| 35. | α-Cadinene | 30.148 | 916 | 6.09 ± 0.1 |
| 36. | α-Ferulene | 30.644 | 800 | 0.02 ± 0.001 |
| 37. | 3-Ethyl-3-hydroxyandrostan-17-one | 30.644 | 824 | 0.02 ± 0.002 |
| 38. | Cadala-1(10),3,8-Triene | 30.779 | 830 | 0.02 ± 0.001 |
| 39. | *cis*-α-Copaene-8-ol | 31.029 | 741 | 0.05 ± 0.001 |
| 40. | Bicyclo[5.2.0]nonane | 31.319 | 818 | 0.03 ± 0.001 |
| 41. | Acetic Acid,4a-Methyl-2,3,4,4a,5,6,7,8 Octahydronaphthalen-2-yl | 31.489 | 732 | 0.05 ± 0.001 |
| 42. | Benzoic acid, 3-hexenyl ester | 31.649 | 843 | 0.07 ± 0.001 |
| 43. | Retinal | 31.729 | 829 | 0.07 ± 0.001 |
| 44. | *p*-Menthane | 31.864 | 784 | 0.07 ± 0.001 |
| 45. | Aromadendrane | 32.649 | 839 | 0.06 ± 0.001 |
| 46. | [2,2-dimethyl-4-(3-methylbut-2-enyl)-6-methylidenecyclohexyl] methanol | 32.829 | 875 | 0.02 ± 0.001 |
| 47. | α-Cubebene | 33.38 | 844 | 1.48 ± 0.031 |
| 48. | Valencene | 33.515 | 827 | 0.31 ± 0.002 |
| 49. | Epizonarene | 33.845 | 855 | 7.89 ± 0.91 |
| 50. | γ-Gurjunene | 34.45 | 864 | 2.17 ± 0.09 |
| 51. | Isophorone | 34.78 | 880 | 0.53 ± 0.003 |
| 52. | Timnodonic acid | 41.238 | 738 | 0.37 ± 0.001 |
| | Total | | | 100 |

**Table 2.** The metabolic enzymes inhibitory and the antioxidant activities $IC_{50}$ values of *S. viticina* essential oil compared with the required positive controls.

| α-Amylase Inhibitory Activity $IC_{50}$ (μg/mL), ±SD | | α-Glucosidase Inhibitory Activity $IC_{50}$ (μg/mL), ±SD | | Antioxidant Activity $IC_{50}$ (μg/mL), ±SD | | Antilipase Activity $IC_{50}$ (μg/mL), ±SD | |
|---|---|---|---|---|---|---|---|
| EO | Acarbose (Positive control) | EO | Acarbose (Positive control) | EO | Trolox (Positive control) | EO | Orlistat (Positive control) |
| 45.22 ± 1.1 | 28.89 ± 1.22 | 63.09 ± 0.26 | 37.15 ± 0.33 | 19.95 ± 2.08 | 2.23 ± 1.57 | 501.18 ± 0.38 | 12.3 ± 0.33 |

### 2.3. Antimicrobial Capacity

The microdilution assay was used to determine the antimicrobial potential of *S. viticina* essential oil against Methicillin-resistant *Staphylococcus aureus* (MRSA), *S. aureus*, *E. faecium*, *S. sonnie*, *P. aeruginosa*, *E. coli*, *E. floccosum*, and *C. albicans*. The EO inhibited the growth of all of the studied bacterial and fungal strains to different degrees. The highest antibacterial activity was recorded against MRSA, followed by *E. coli* with MIC values of 0.039 and 0.078 mg/mL, respectively, while the highest antifungal activity observed was against *E. floccosum* with MIC value of 0.78 mg/mL as presented in Table 3.

**Table 3.** Antimicrobial MIC values of *S. viticina* EO.

| Microorganisms | Sources | MIC, (mg/mL) |
|---|---|---|
| *Staphylococcus aureus* | ATCC 25923 | 1.5625 |
| *Escherichia coli* | ATCC 25922 | 0.078 |
| *Pseudomonas aeruginosa* | ATCC 27853 | 3.125 |
| *Shigella sonnie* | ATCC 25931 | 3.125 |
| Methicillin-Resistant *Staphylococcus aureus* (MRSA) | Clinical isolate | 0.039 |
| *Enterococcus faecium* | ATCC 10231 | 1.5625 |
| *Candida albicans* | ATCC 90028 | 3.125 |
| *Epidermophyton floccosum* | ATCC 700221 | 0.78 |

### 2.4. Cyclooxygenase Inhibitory Effect

The cyclooxygenase enzymatic inhibitory activity assay of *S. viticina* EO was conducted utilizing Cayman Chemical ELISA kit No. 560131. The calculated percentage inhibitions of COX-2 as well as for COX-1 were 98% and 99% at EO concentrations of 0.25 and 0.5 μg/mL, respectively. The results showed that the *S. viticina* EO displayed a strong, but not selective inhibition activity towards both COX-2 and COX-1 enzymes.

### 2.5. Cytotoxic Effect

As shown in Figure 1, treatment of Colo-205 cells with 1, 0.5 and 0.25 mg/mL of EO induced cytotoxicity significantly ($p \leq 0.0001$) by approximately 90%, 90%, and 55%, respectively. Treatment of HeLa cells with 2.5, 1.25 and 0.625 mg/mL of EO induced cytotoxicity significantly ($p \leq 0.0001$) by approximately 95%, 95%, and 80%, respectively, while at a concentration of 0.1875 mg/mL did not exhibit a significant effect as presented in Figure 2.

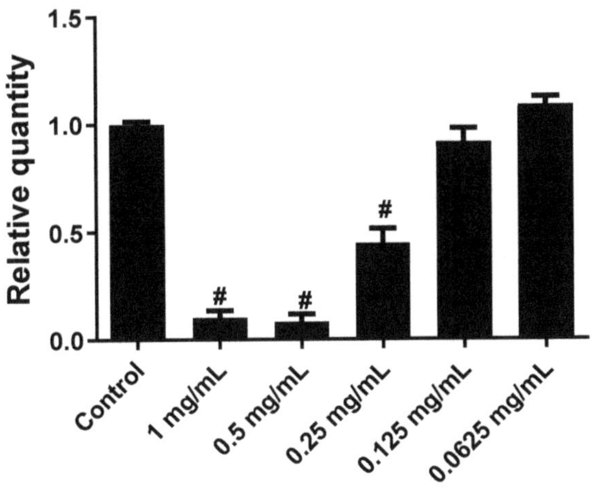



**Figure 1.** The effect of *S. viticina* EO on the cytotoxicity of Colo-205 cells. Results were depicted as relative quantities (RQs) compared to the control (only media). # where $P < 0.0001$. Error bars represent SD.

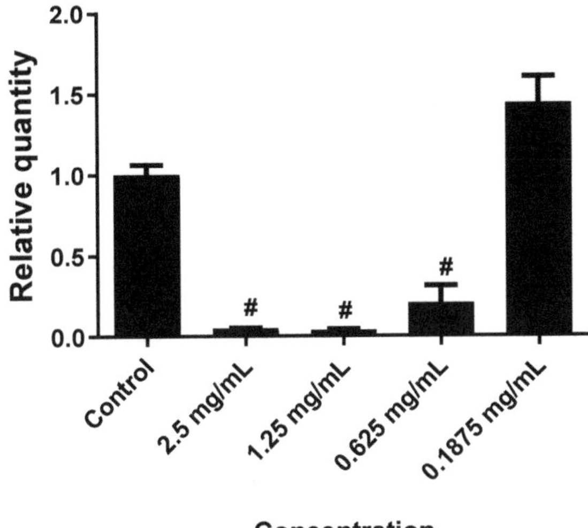



**Figure 2.** The effect of *S. viticina* EO on the cytotoxicity of HeLa cells. Results were depicted as relative quantities (RQs) compared to the control (only media). # Where $P < 0.0001$. Error bars represent SD.

## 3. Discussion

The study of plants secondary metabolites represents a worldwide strategy for the search of new therapeutic agents since plants contain a wide range of biologically active compounds. Medicinal plants are extensively used in traditional medicine due to their healing properties. In order to set up an inventory of their biological activities, it is important to link scientific studies with clinical observations and traditional knowledge. This investigation is part of this scenario and was aimed to screen the biological activity of *Stachys viticina* essential oil. To the best of the authors' knowledge, the full chemical constituents, as well as the antimicrobial, anticancer, antioxidant, metabolic enzymes and cyclooxygenase inhibitory properties of *S. viticina* EO have not been reported to date.

### 3.1. Phytochemical Components of S. viticina Essential Oil

The *S. viticina* EO was isolated using an eco-friendly microwave-ultrasonic technique and the components of the EO were identified using GC-MS technology. The primary components were determined to be *endo*-borneol, eucalyptol, epizonarene, $\alpha$-cadinene, bornyl chloride, and benzylin with percentage areas of $29.06 \pm 1.24$, $21.26 \pm 1.97$, $7.89 \pm 0.91$, $6.09 \pm 0.1$, $5.79 \pm 0.09$, and $5.29 \pm 0.31$, respectively. The chromatogram of *S. viticina* EO is shown in Figure S1. The retention times and mass fragmentation data of the 52 compounds identified are listed in Table 1.

Recently, Gören et al. reported the identification of only 20 compounds from EO chemical composition of *S. viticina* aerial parts (flowering tops, stems, and leaves) growing in Turkey, of which $\beta$-caryophyllene (62.3%), farnesyl acetate (8.9%) and $\alpha$-bisabolol (4.4%) were the major components [12]. The difference in the results could be attributed to the difference in the area of plant growth, extraction conditions and/or to the fact that only the leaves were used in this investigation, while in Goren et al. study the aerial parts (flowers, leaves and stems) were used.

### 3.2. Metabolic Enzymes Inhibitory Activity

The metabolic pancreatic $\alpha$-amylase and the intestinal $\alpha$-glucosidase enzymes play a vital role in the digestion of carbohydrates such as starch, dextrin, maltotriose, maltose, glucose and table sugar. In fact, diabetes, overweight and obesity can be controlled by the inhibition of these enzymes due to their ability to retard the carbohydrate's digestion process and to improve glucose tolerance in patients suffering from type II of diabetes [13].

As shown in Figures S3 and S4, *S. viticina* EO displayed $\alpha$-amylase pancreatic enzyme and $\alpha$-glucosidase inhibitory activity in vitro, with $IC_{50}$ values of $45.22 \pm 1.1$ µg/mL and $63.09 \pm 0.26$ µg/mL, respectively, compared to that of the antidiabetic drug acarbose which has $\alpha$-amylase and $\alpha$-glucosidase inhibitory effects with $IC_{50}$ values of $28.89 \pm 1.22$ µg/mL and $37.15 \pm 0.33$ µg/mL, respectively. This means that *S. viticina* EO inhibited 56.52% and 69.82% of $\alpha$-amylase and $\alpha$-glucosidase activity, respectively.

There are many reports on the inhibitory effect of EO extracted from different plant species on the $\alpha$-amylase and $\alpha$-glycosidase. Loizzo et al. reported that the *Cedrus libani* wood EO exhibited an $\alpha$-amylase inhibitory effect with $IC_{50}$ of 140 µg/mL [14]. Ma et al. found that the EOs extracted from *Pyrola calliantha* entire plant, *Senecio scandens* flowers and of *Schisandra chinensis* roots have $\alpha$-glycosidase inhibitory activity with $IC_{50}$ values of 22.11, 130.4 and 19.25 µg/mL, respectively [15]. Another study conducted by Dang et al. revealed that the *Citrus medica* var. *sarcodactylis* (Buddha's hand) fruits EO inhibited an $\alpha$-glucosidase enzyme with an $IC_{50}$ value of 412.2 µg/mL [16].

However, the $IC_{50}$ values showed that *S. viticina* EO displayed a stronger inhibitory activity on $\alpha$-amylase than on $\alpha$-glucosidase activity, and therefore this finding is therapeutically important especially in preventing some of the side effects associated with the use of chemical $\alpha$-glucosidase and $\alpha$-amylase inhibitory agents [17]. The observed $\alpha$-amylase and $\alpha$-glucosidase inhibitory activity of the *S. viticina* EO gives credence to the fact that *S. viticina* EO could be promising antidiabetic medicine after in vivo and clinical trials.

Therapeutic agents that can inhibit the pancreatic lipase enzyme have gained much importance in recent years. Obviously, the inhibition of pancreatic lipase enzyme retards the lipids digestion and consequently lower the rate of fats absorption and decrease the levels of triglyceride in blood serum which may also cause a reduction of body weight. The EO of *S. viticina* revealed weak porcine pancreatic lipase inhibitory activity with an $IC_{50}$ value of $501.18 \pm 0.38$ µg/mL compared to that of antiobesity available drug Orlistat ($IC_{50} = 12.3 \pm 0.33$ µg/mL). Salameh et al. reported that EO of *Micromeria fruticosa serpyllifolia* leaves (Lamiaceae) exhibited potent antilipase activity with an $IC_{50}$ value of 39.81 µg/mL, which is better than that of *S. viticina* EO ($IC_{50} = 501.18 \pm 0.38$ µg/mL) [18].

### 3.3. Antioxidant Activity

Due to its simplicity and high sensitivity, the widely used DPPH assay for radical scavenging was employed to assess the antioxidant activity of *S. viticina* EO [19]. The results of the present work showed that the EO exhibited antioxidant activity with an $IC_{50}$ value of 19.95 ± 2.08 µg/mL, less than that of the positive control (Trolox) which has an $IC_{50}$ value of 2.23 ± 1.57 µg/mL. Previous studies demonstrated that various *Stachys* species exhibited good antioxidant activity. A study conducted by Kukić et al. revealed that *S. anisochila*, *S. beckeana*, *S. plumose*, and *S. alpina* ssp. *dinarica* have antioxidant potential with $IC_{50}$ values of 17.9, 20.9, 101.61, and 26.14 µg/mL, respectively [20]. The EO of *S. inflate*, growing Iran, exhibited antioxidant activity with an $IC_{50}$ value of 89.50 ± 0.65 µg/mL [21]. Furthermore, A study performed by Conforti et al. on the antioxidant activity of the EOs extracted from various *Stachys* species from different regions of the Mediterranean area revealed that *S. palustris*, *S. cretica*, and *S. hydrophila* displayed the highest antiradical effect, with $IC_{50}$ values of 48.2, 65.2 and 66.4 µg/mL, respectively [22].

### 3.4. Cyclooxygenase Inhibitory Assay

Non-steroidal anti-inflammatory molecules have the potentials to reduce inflammation and relieve pains in the human body which is associated with the increase of prostaglandin levels. However, several traditional herbal products have been used to reduce pain sensation, inflammations and fever [23]. The current cyclooxygenase inhibitory assay revealed that *S. viticina* EO caused a 98% and 99% inhibition of COX-1 and COX-2 enzymes activity at concentrations of 0.25 and 0.5 µg/mL, respectively. *S. viticina* EO exhibited cyclooxygenase inhibitory activity against COX-1 and Cox-2 with $IC_{50}$ values of 0.25 and 0.5 µg/mL, respectively, similar to those $IC_{50}$ values of the positive control (the NSAID etodolac).

### 3.5. Cytotoxic Activity

Cytotoxicity assays have gained increasing interest over recent years due to their great value in the biological screening of insecticide, herbicides and anticancer agents [24]. These assays are proved to be reliable, quick, cheap and reproducible. Various types of cytotoxic tests, such as luminometric, fluorometric, dye exclusion and colorimetric assays, are used in the fields of pharmacology and toxicology [25]. The significant antimicrobial activity of *S. viticina* EO encouraged us to evaluate its cytotoxicity. The potential cytotoxicity of the *S. viticina* EO towards HeLa (cervical adenocarcinoma) and Colo-205 (colon) cancer cells was evaluated using the MTS assay. As shown in Figure 1; Figure 2, treatment of Colo-205 cells with 0.5 mg/mL of *S. viticina* EO induced the best cytotoxic effect (90%), while treatment of HeLa cells with 2.5 mg/mL of EO induced the highest cytotoxicity (95%). The *S. viticina* EO displayed very strong cytotoxicity against HeLa and Colo-205 cancer cells.

In a previous study conducted in our laboratory, Jaradat et al. reported that *Teucrium pruinosum* EO at a dose of 7.67 mg/mL induced cytotoxic activity on HeLa cancer cells by 90–95%. The cytotoxicity of *S. viticina* EO against HeLa cancer cells is triple that induced by *Teucrium pruinosum* EO (Lamiaceae) [26].

### 3.6. Antimicrobial Capacity

The microdilution assay was used to assess the antimicrobial activity of *S. viticina* EO against MRSA, *S. aureus*, *E. faecium*, *S. sonnie*, *P. aeruginosa*, *E. coli*, *E. floccosum*, and *C. albicans*. The *S. viticina* EO inhibited the growth of all the screened pathogens. In particular, the *S. viticina* EO exhibited potent antibacterial activity against MRSA with MIC value of 0.039 mg/mL, which is two folds more potential than the inhibition activity against *E. coli* (MIC = 0.078 mg/mL). Moreover, the EO of *S. viticina* plant inhibited the growth of all the studied fungal strains, especially inhibited potentially the growth of *E. floccosum* with MIC value of 0.78 mg/mL.

## 4. Materials and Methods

### 4.1. Chemical Reagents

All the experiments in the current study were carried out utilizing commercially available chemicals and reagents unless otherwise stated.

### 4.2. Equipment

A gas chromatograph (Clarus 500-Perkin Elmer, Singapore), Mass Spectrometer (Clarus 560D-Perkin Elmer), microscope (IX-73-inverted, Olympus (China) CO.,LTD, Beijing, China), UV-Visible Spectrophotometer (Jenway-7315, Staffordshire, UK), microwaves-ultrasonic reactor-extractor (CW-2000, Gloria Wang Zhejiang Nade Scientific Instrument Co., Ltd., Zhejiang, China), Balance (Rad-weight, International Weighing Review, Toruńska, Poland), sonicator (MRC, 2014-207, MRC Lab., Haifa, Israel), water bath (LabTech, 2011051806, Thermo Fisher Scientific, Seoul, South Korea), incubator (Nüve, 06-3376, Ankara, Turkey), vortexer (090626691, Heidolph Company, Schwabach, Germany), autoclave (MRC, A13182, Mrc lab., Haifa, Israel), grinder (Molineux I, Jiangmen, China) and microplate reader (6000, Unilab, Fort Lauderdale, FL USA) were utilized in the current investigation.

### 4.3. Herbal Material

*Stachys viticina* leaves were collected in March 2018 from the Nablus area, Palestine. The taxonomical characterization of the plant was established at the Pharmacy Department at An-Najah National University and deposited under the voucher specimen number (Pharm-PCT-2341). Later on, *S. viticina* leaves were cleaned, rinsed at least three times with distilled water and then dried in the shade at 25 ± 2 °C and 55 ± 5 RH of humidity for three weeks. The dried leaves were cut into small pieces which were kept in special paper bags for farther experimental work [27].

### 4.4. Isolation of S. viticina Essential Oil

The EO of *S. viticina* plant was extracted using the microwave and ultrasonic apparatus as reported by Jaradat et al. [28]. Dried *S. viticina* powder (100 g) and distilled water (0.5 L) were placed in a round bottomed flask and exposed to micro- and ultrasonic- waves at a fixed power of 1000 W at 100 °C for 10 min. The extracted EO was collected into an amber glass bottle and kept in the refrigerator at 2–8 °C for further experiments.

### 4.5. GC-MS Characterization of S. viticina Essential Oil

The separation was achieved on a Perkin Elmer Elite-5-MS fused-silica capillary column (30 m × 0.25 mm, film thickness 0.25 μm), using helium as carrier gas at a standard flow rate of 1.1 mL/min. The temperature of the injector was adjusted at 250 °C with an initial temperature of 50 °C, initial hold 5 min, and ramp 4.0 °C/min to 280 °C. The total running time was 62.50 min and the solvent delay was from 0 to 4.0 min. MS scan time was from 4 to 62.5 min, covering mass range 50.00 to 300.00 *m/z*. The chemical ingredients of the EO were characterized by comparing their mass spectra with the reference spectra in the MS Data Centre of the National Institute of Standards and Technology, and by matching their Kovats and retention indices with values reported in the literature [29,30]. In addition, the EOs Kovats and retention indices with values were compared with 20% of HPLC grade of reference EOs including α-pinene, eucalyptol, caryophyllene, γ-terpinene, ocimene, *endo*-borneol, α-terpineol, jasmone, α-cadinene, *p*-menthane that were purchased from Sigma-Aldrich, Hamburg, Germany) [31].

### 4.6. Porcine Pancreatic Lipase Inhibitory Assay

A porcine pancreatic lipase inhibition assay was conducted in order to assess the anti-obesity activity of *S. viticina* EO. The drug Orlistat, an anti-obesity, and anti-lipase agent, was used as a positive control. The porcine pancreatic lipase inhibitory method was performed according to the protocol described by Zheng et al. with slight modifications [32]. A 500 µg/mL stock solution from the EO was dissolved in dimethyl sulfoxide (DMSO): methanol (1:9), and five different dilutions (50, 100, 200, 300 and 400 µg/mL) were prepared. Then, a 1 mg/mL stock solution of porcine pancreatic lipase was freshly prepared before use and was dispersed in Tris-HCl buffer. The substrate used was *p*-nitrophenyl butyrate (PNPB) (Sigma-Aldrich, Hamburg, Germany), prepared by dissolving 20.9 mg of it in 2 mL of acetonitrile. 0.1 mL of porcine pancreatic lipase (1 mg/mL) and 0.2 mL of the EO from each of the concentration series were placed in 5 different working test tubes and mixed. The resulting mixture was adjusted to 1 mL by adding Tri-HCl solution and incubated at 37 °C for 15 min. thereafter, 0.1 mL of *p*-nitrophenyl butyrate solution was added to each test tube and the mixture was incubated for 30 min. at 37 °C. The pancreatic lipase activity was determined by measuring the hydrolysis of PNPB into *p*-nitrophenolate ions at 410 nm using a UV-Vis spectrophotometer. The same procedure was repeated for the positive control sample (Orlistat, Sigma-Aldrich). The inhibitory percentage of the anti-lipase activity was calculated using the following equation:

$$\text{Lipase inhibition\%} = (AB - Ats)/AB \times 100\% \tag{1}$$

where AB is the recorded absorbance of the blank solution and Ats is the recorded absorbance of the tested sample solution.

### 4.7. α-Amylase Inhibition Assay

The α-amylase inhibitory activity of *S. viticina* EO was assessed according to the standard method reported by Nyambe-Silavwe et al. with minor modifications [33]. The EO was dissolved in DMSO (Riedel-de-Haen, Hamburg, Germany) and then diluted with a buffer ((Na$_2$HPO$_4$/NaH$_2$PO$_4$ (0.02 M), NaCl (0.006 M) at pH 6.9) to a concentration of 1000 µg/mL. A concentration series of 10, 50, 70, 100 and 500 µg/mL were prepared. 0.2 mL of porcine pancreatic α-amylase enzyme solution (Sigma-Aldrich, St. Louis, MO, USA) with a concentration of 2 units/mL was mixed with 0.2 mL of the EO and incubated at 30 °C for 10 min. Thereafter, 0.2 mL of freshly prepared starch solution (1%) was added and the mixture was incubated for at least 3 min. The reaction was stopped by addition of 0.2 mL dinitrosalicylic acid (DNSA) (Alf-aAesar, Lancashire, UK), then the mixture was diluted with 5 mL of distilled water and heated in a water bath at 90 °C for 10 min. The mixture was left to cool down to room temperature and the absorbance was measured at 540 nm. A blank was prepared following the same procedure by replacing the *S. viticina* EO with 0.2 mL of the buffer.

Acarbose (Sigma-Aldrich, USA) was used as a positive control and prepared to adopt the same procedure described above. The α-amylase inhibitory activity was calculated using the following equation:

$$\% \text{ of } \alpha\text{-amylase inhibition} = (A_b - A_S)/A_b\ 100\% \tag{2}$$

where $A_b$ is the absorbance of the blank and $A_S$ is the absorbance of the tested sample or control.

### 4.8. α-Glucosidase Inhibitory Activity Assay

The α-glucosidase inhibitory activity of *S. viticina* EO was performed according to the standard protocol with a slight modification [34]. A mixture of 50 µL of phosphate buffer (100 mM, pH 6.8), 10 µL α-glucosidase (1 U/mL) (Sigma-Aldrich, USA) and 20 µL of varying concentrations of *S. viticina* EO (100, 200, 300, 400 and 500 µg/mL) were placed in 5 different test tubes. After 15 min. incubation at 37 °C, 20 µL of pre-incubated 5 mM PNPG (Sigma-Aldrich, USA) was added as the substrate to each test tube and the reaction mixtures were incubated at 37 °C for additional 20 min. The reaction was terminated by adding 50 µL of aqueous Na$_2$CO$_3$ (0.1 M). The absorbance of the released *p*-nitrophenol was measured by a UV/Vis spectrophotometer at 405 nm. Acarbose at similar concentrations as the plant EO was used as the positive control. The inhibition percentage was calculated using the following equation:

$$\% \text{ of } \alpha\text{-amylase inhibition} = (A_b - A_S)/A_b\ 100\% \tag{3}$$

where $A_b$ is the absorbance of the blank and $A_S$ is the absorbance of the tested sample or control [35].

*4.9. Antioxidant Assay*

For estimation of *S. viticina* EO antioxidant potential, a solution of EO (1 mg/mL) in methanol was serially diluted with methanol to obtain concentration of 1, 2, 3, 5, 7, 10, 20, 30, 40, 50, 80 and 100 µg/mL. Then, DPPH (2,2-diphenyl-1-picrylhydrazyl) reagent (Sigma, New York, NY, USA) was dissolved in 0.002% *w/v* methanol and mixed with the previously prepared working concentrations in a 1:1 ratio. The same procedures were repeated for Trolox (Sigma-Aldrich, Darmstadt, Denmark) which was used as a positive control. All of the solutions were kept in a dark chamber for 30 min at ordinary temperature.

Then, their absorbance values were measured at a wavelength of 517 nm utilizing a UV-Visible spectrophotometer. The DPPH inhibition potential by *S. viticina* EO and Trolox were determined employing the following equation:

$$\text{DPPH inhibition (\%)} = (\text{abs}_{\text{blank}} - \text{abs}_{\text{sample}})/\text{abs}_{\text{blank}}\ 100\% \tag{4}$$

where $\text{abs}_{\text{blank}}$ is the blank absorbance and $\text{abs}_{\text{sample}}$ is the absorbance of the samples. The antioxidant half-maximal inhibitory concentration ($IC_{50}$) of *S. viticina* EO and Trolox were assessed using BioDataFit-E1051 program [36].

*4.10. Antimicrobial Activities of S. viticina Essential Oil*

The fungal and bacterial isolates used in this study were from the American Type Culture Collection (ATCC, Manassas, VA, USA) and from selected MRSA species (), that were obtained from the Palestinian area at clinical settings and exhibited multi-antibiotic resistance. The screened microorganisms included three Gram-positive bacteria (MRSA, *Enterococcus faecium* (ATCC 700221) and *S. aureus* (ATCC 25923)), three of Gram-negative bacteria (*Shigella sonnie* (ATCC 25931), *Pseudomonas aeruginosa* (ATCC 27853) and *Escherichia coli* (ATCC 25922) and two fungal strains (*Epidermophyton floccosum* (ATCC 10231) and *Candida albicans* (ATCC 90028)).

The microdilution assay was utilized to assess the antimicrobial activity of *S. viticina* EO against bacterial and yeast strains. A total of 8.8 g of Mueller-Hinton broth were dissolved in 400 mL of distilled water under heat and boiled for 1 min. The obtained solution was autoclaved and kept at 4 °C until use. 100 µL from the broth was placed in each well of the microdilution tray. Then, 100 µL of the EO was added to the first well and the mixture was shacked. Thereafter, serial dilutions up to well #11 were performed. Well #12 did not contain any EO and was considered as a positive control for microbial growth. A fresh bacterial colony was picked from an overnight agar culture and was prepared to match the turbidity of the 0.5 McFarland standards to provide a bacterial suspension of $1.5 \times 10^8$ CFU (colony forming unit)/mL. The suspension was diluted with broth by a ratio of 1:3 to a final concentration of $5 \times 10^7$ CFU. Then, 1 µL of the bacterial suspension was added to each well except well #11, which was a negative control for microbial growth. Finally, the plate was incubated at 35 °C for 18 h.

The microdilution method for the yeast *C. albicans* was performed as described above, except that after matching the yeast suspension with the McFarland standard, it was diluted with NaCl by a ratio of 1:50, followed by 1:20, and 100 µL was placed in the wells. The plate was incubated for 48 h instead of 18 h.

To investigate anti-*Epidermatophyton floccosum* mold activity of the *S. viticina* EO, an agar dilution method was performed. Sabouraud dextrose agar (SDA) was prepared, of which 1 mL was placed in each tube and kept in a 40 °C water bath. 1 mL of the EO was mixed with 1 mL of SDA in the first tube and serial dilutions were performed in six tubes, except tube #6, which did not contain any plant material and considered as a positive control for microbial growth.

After the SDA was solidified, the spores of the mold culture were dissolved using distilled water containing 0.05% Tween 80 and scratched from the plate for comparison with McFarland turbidity. From that suspension, 20 µL was pipetted into the six tubes, except tube #5, which was considered as a negative control. The tubes were incubated at 25 °C for 14 days. The MIC is the lowest concentration of an antimicrobial agent that inhibited the visible growth of a microorganism [37,38].

### 4.11. Determination of COX Inhibition

The *S. viticina* EO inhibitory activity of human recombinant COX-2 and ovine COX-1 enzymes was assessed using a COX inhibitor screening method (Cayman Chemical, kit No.560131, Ann Arbor, MI, USA). The yellow color of the enzymatic reaction was evaluated using UV spectrophotometer in a Microplate Reader at a wavelength of 415 nm. Cyclooxygenase inhibitory assay was performed for two concentrations of EO (0.25 and 0.5 µg/mL) with celecoxib (a commercial NSAID marker). The anti-inflammatory activity of the tested EO was determined by calculating the percentage production inhibition of prostaglandin E2 (PGE2). The tested compounds concentration causing 50% inhibition ($IC_{50}$) of the formation of prostaglandin PGE2 by COX1 and COX2 enzymes were determined from the curve of concentration inhibition response by regression analysis [39].

### 4.12. Cell Culture and Cytotoxicity Assay

The HeLa (cervical adenocarcinoma) and Colo-205 (colon) cancer cells were cultured in media (RPMI-1640) and supplemented with 10% fetal bovine serum, 1% Penicillin/Streptomycin antibiotics and 1% l-glutamine. Cells were grown in a humidified atmosphere with 5% $CO_2$ at 37 °C. Cells were seeded in 96-well plates ($2.6 \times 10^4$ cells). After 48 hrs, the cells were incubated with EO at various concentrations for 24 h. Cell viability was assessed by Cell-Tilter 96® Aqueous One Solution Cell Proliferation (MTS) Assay following the instructions of the manufacturer (Promega Corporation, Madison, WI, USA). At the end of the treatment, 20 µL of MTS solution per 100 µL of media was added to each well and incubated at 37 °C for 2 h and finally, the absorbance was spectrophotometrically estimated at 490 nm [40].

### 4.13. Statistical Examination

All the data on $\alpha$-amylase, $\alpha$-glucosidase and porcine pancreatic enzymes inhibitory activity as well as on the antioxidant, COX inhibitory and cytotoxicity activities were the average of triplicate analyses. The outcomes were presented as means ± standard deviation (SD). Statistical analysis was established employing GraphPad Prism software version 6.01 (GraphPad, San Diego, CA).

## 5. Conclusions

The GC-MS analysis revealed the presence of fifty-two compounds in the *S. viticina* EO, of which *endo*-borneol was the major component, followed by eucalyptol and epizonarene. The EO of *S. viticina* showed an ability to scavenge the free radical DPPH with an $IC_{50}$ value of 19.95 µg/mL and displayed cytotoxic activity against colon and HeLa cancer cells at 0.5 mg/mL and 1.25 mg/mL, respectively. It also showed very strong antimicrobial (against three Gram-positive, three Gram-negative bacteria and two fungi) effects, of which the strongest was against MRSA, with a MIC value of 0.039 mg/mL. Moreover, the *S. viticina* EO showed high cyclooxygenase inhibitory activity against COX-1 and COX-2, with $IC_{50}$ values of 0.25 and 0.5 µg/mL, respectively, similar to those of the positive control (the NSAID etodolac). In addition, the EO showed potential in inhibiting the activity of $\alpha$-amylase (56.52% at 45.22 µg/mL) and of $\alpha$-glucosidase (69.82% at 63.09 µg/mL). Briefly, *S. viticina* EO contains pharmacologically active molecules which could be further validated and explored clinically for their therapeutic potential and for the development and design of new natural therapeutic preparations.

**Supplementary Materials:** The following are available online, Figure S1. GC-MS chromatogram of *S. viticina* EO, Figure S2. DPPH Inhibition % by Trolox (positive control) and *S. viticina* EO, Figure S3. The $\alpha$-amylase inhibitory potential by Acarbose (positive control) and *S. viticina* EO, Figure S4. The α-glucosidase inhibitory potential by Acarbose (positive control) and *S. viticina* EO, Figure S5. Lipase enzyme inhibitory activity by *S. viticina* EO and Orlistat (positive control).

**Author Contributions:** Conceptualization, N.J. and N.A.-M.; methodology, N.J. and N.A.-M.; software, N.J. and N.A.-M.; validation, N.J. and N.A.-M.; formal analysis, N.J. and N.A.-M.; investigation, N.J. and N.A.-M.; resources, N.J. and N.A.-M.; data curation, N.J. and N.A.-M.; writing—original draft preparation, N.J.; writing—review and editing, N.A.-M.; visualization, N.J. and N.A.-M; supervision, N.J. and N.A.-M; project administration, N.J. and N.A.-M.

**Funding:** This research received no external funding.

**Acknowledgments:** The authors wish to thank An-Najah National University for its support to carry out this work

**Conflicts of Interest:** The authors declare that there are no conflicts of interest.

## References

1. Dimitrios, B. Sources of natural phenolic antioxidants. *Trends Food Sci. Technol.* **2006**, *17*, 505–512. [CrossRef]
2. Locatelli, M. Anthraquinones: Analytical techniques as a novel tool to investigate on the triggering of biological targets. *Curr. Drug Targets* **2011**, *12*, 366–380. [CrossRef]
3. Locatelli, M.; Epifano, F.; Genovese, S.; Carlucci, G.; Končić, M.Z.; Kosalec, I.; Kremer, D. Anthraquinone profile, antioxidant and antimicrobial properties of bark extracts of rhamnus cathartics and r. Orbiculatus. *Nat. Prod. Commun.* **2011**, *6*, 1934578X1100600917. [CrossRef]
4. Rahimi Khoigani, S.; Rajaei, A.; Goli, S.A.H. Evaluation of antioxidant activity, total phenolics, total flavonoids and lc–ms/ms characterisation of phenolic constituents in *Stachys lavandulifolia*. *Nat Prod Res* **2017**, *31*, 355–358. [CrossRef] [PubMed]
5. Heymach, J.; Krilov, L.; Alberg, A.; Baxter, N.; Chang, S.M.; Corcoran, R.B.; Dale, W.; DeMichele, A.; Magid Diefenbach, C.S.; Dreicer, R. Clinical cancer advances 2018: Annual report on progress against cancer from the american society of clinical oncology. *J. Clin. Oncol.* **2018**, *36*, 1020–1044. [CrossRef] [PubMed]
6. Seca, A.; Pinto, D. Plant secondary metabolites as anticancer agents: Successes in clinical trials and therapeutic application. *Int. J. Mol. Sci.* **2018**, *19*, 263. [CrossRef] [PubMed]
7. Chouhan, S.; Sharma, K.; Guleria, S. Antimicrobial activity of some essential oils—present status and future perspectives. *Medicines* **2017**, *4*, 58. [CrossRef]
8. Locatelli, M.; Genovese, S.; Carlucci, G.; Kremer, D.; Randic, M.; Epifano, F. Development and application of high-performance liquid chromatography for the study of two new oxyprenylated anthraquinones produced by *Rhamnus* species. *J. Chromatogr. A* **2012**, *1225*, 113–120. [CrossRef]
9. Skaltsa, H.D.; Demetzos, C.; Lazari, D.; Sokovic, M. Essential oil analysis and antimicrobial activity of eight *Stachys* species from greece. *Phytochemistry* **2003**, *64*, 743–752. [CrossRef]
10. Goren, A.C. Use of *Stachys* species (mountain tea) as herbal tea and food. *Rec. Nat. Prod.* **2014**, *8*, 71–82.
11. Tundis, R.; Peruzzi, L.; Menichini, F. Phytochemical and biological studies of *Stachys* species in relation to chemotaxonomy: A review. *Phytochemistry* **2014**, *102*, 7–39. [CrossRef] [PubMed]
12. Goren, A.C.; Piozzi, F.; Akcicek, E.; Kılıç, T.; Çarıkçı, S.; Mozioğlu, E.; Setzer, W.N. Essential oil composition of twenty-two *Stachys* species (mountain tea) and their biological activities. *Phytochem. Lett.* **2011**, *4*, 448–453. [CrossRef]
13. Wei, J.; Zhang, X.-Y.; Deng, S.; Cao, L.; Xue, Q.-H.; Gao, J.-M. A-glucosidase inhibitors and phytotoxins from *Streptomyces xanthophaeus*. *Nat. Prod. Res.* **2017**, *31*, 2062–2066. [CrossRef] [PubMed]
14. Loizzo, M.; Saab, A.; Statti, G.; Menichini, F. Composition and α-amylase inhibitory effect of essential oils from *Cedrus libani*. *Fitoterapia* **2007**, *78*, 323–326. [CrossRef] [PubMed]
15. Ma, L.; Lin, Q.; Lei, D.; Liu, S.; Wang, X.; Zhao, Y. Alpha-glucosidase inhibitory activities of essential oils extracted from three chinese herbal medicines. *Chem. Eng. Trans.* **2018**, *64*, 61–66.
16. Dang, N.H.; Nhung, P.H.; Anh, M.; Thi, B.; Thuy, T.; Thi, D.; Minh, C.V.; Dat, N.T. Chemical composition and α-glucosidase inhibitory activity of vietnamese citrus peels essential oils. *J. Chem.* **2016**, *2016*. [CrossRef]
17. Oboh, G.; Ademosun, A.O.; Odubanjo, O.V.; Akinbola, I.A. Antioxidative properties and inhibition of key enzymes relevant to type-2 diabetes and hypertension by essential oils from black pepper. *Adv. Pharmacol. Sci.* **2013**, *2013*, 215–219. [CrossRef]

18. Salameh, N.; Shraim, N.; Jaradat, N. Chemical composition and enzymatic screening of *Micromeria fruticosa serpyllifolia* volatile oils collected from three different regions of west bank, palestine. *BioMed Res. Int.* **2018**, *2018*. [CrossRef]

19. Zheng, C.-D.; Li, G.; Li, H.-Q.; Xu, X.-J.; Gao, J.-M.; Zhang, A.-L. DPPH-scavenging activities and structure-activity relationships of phenolic compounds. *Nat. Prod. Commun.* **2010**, *5*, 1934578X1000501112. [CrossRef]

20. Kukić, J.; Petrović, S.; Niketić, M. Antioxidant activity of four endemic *Stachys* taxa. *Biol. Pharm. Bull.* **2006**, *29*, 725–729. [CrossRef]

21. Ebrahimabadi, A.H.; Ebrahimabadi, E.H.; Djafari-Bidgoli, Z.; Kashi, F.J.; Mazoochi, A.; Batooli, H. Composition and antioxidant and antimicrobial activity of the essential oil and extracts of *Stachys inflata* benth from iran. *Food Chem.* **2010**, *119*, 452–458. [CrossRef]

22. Conforti, F.; Menichini, F.; Formisano, C.; Rigano, D.; Senatore, F.; Arnold, N.A.; Piozzi, F. Comparative chemical composition, free radical-scavenging and cytotoxic properties of essential oils of six *Stachys* species from different regions of the mediterranean area. *Food Chem.* **2009**, *116*, 898–905. [CrossRef]

23. Taylor, J.; Van Staden, J. Cox-1 inhibitory activity in extracts from *Eucomis* l'herit. Species. *J. Ethnopharmacol.* **2001**, *75*, 257–265. [CrossRef]

24. Xiao, J.; Zhang, Q.; Gao, Y.-Q.; Tang, J.-J.; Zhang, A.-L.; Gao, J.-M. Secondary metabolites from the Endophytic botryosphaeria dothidea of melia azedarach and their antifungal, antibacterial, antioxidant, and cytotoxic activities. *J. Agric. Food Chem.* **2014**, *62*, 3584–3590. [CrossRef]

25. Gerets, H.; Hanon, E.; Cornet, M.; Dhalluin, S.; Depelchin, O.; Canning, M.; Atienzar, F. Selection of cytotoxicity markers for the screening of new chemical entities in a pharmaceutical context: A preliminary study using a multiplexing approach. *Toxicol. Vitr.* **2009**, *23*, 319–332. [CrossRef]

26. Jaradat, N.; Al-lahham, S.; Abualhasan, M.N.; Bakri, A.; Zaide, H.; Hammad, J.; Hussein, F.; Issa, L.; Mousa, A.; Speih, R. Chemical constituents, antioxidant, cyclooxygenase inhibitor, and cytotoxic activities of *Teucrium pruinosum* boiss. Essential oil. *BioMed Res. Int.* **2018**, *2018*. [CrossRef]

27. Shehadeh, M.; Jaradat, N.; Al-Masri, M. Rapid, cost-effective and organic solvent-free production of biologically active essential oil from mediterranean wild origanum syriacum. *Saudi Pharm. J.* **2019**, *27*, 612–618. [CrossRef]

28. Jaradat, N.A.; Zaid, A.N.; Abuzant, A.; Shawahna, R. Investigation the efficiency of various methods of volatile oil extraction from *Trichodesma africanum* and their impact on the antioxidant and antimicrobial activities. *J. Intercult. Ethnopharmacol.* **2016**, *5*, 250–256. [CrossRef]

29. Vinaixa, M.; Schymanski, E.L.; Neumann, S.; Navarro, M.; Salek, R.M.; Yanes, O. Mass spectral databases for LC/MS-and GC/MS-based metabolomics: State of the field and future prospects. *TrAC Trends Anal. Chem.* **2016**, *78*, 23–35. [CrossRef]

30. Wei, X.; Koo, I.; Kim, S.; Zhang, X. Compound identification in GC-MS by simultaneously evaluating the mass spectrum and retention index. *Analyst* **2014**, *139*, 2507–2514. [CrossRef]

31. Serfling, A.; Wohlrab, J.; Deising, H.B. Treatment of a clinically relevant plant-pathogenic fungus with an agricultural azole causes cross-resistance to medical azoles and potentiates caspofungin efficacy. *Antimicrob. Agents Chemother.* **2007**, *51*, 3672–3676. [CrossRef] [PubMed]

32. Zheng, C.-D.; Duan, Y.-Q.; Gao, J.-M.; Ruan, Z.-G. Screening for anti-lipase properties of 37 traditional chinese medicinal herbs. *J. Chin. Med Assoc.* **2010**, *73*, 319–324. [CrossRef]

33. Nyambe-Silavwe, H.; Villa-Rodriguez, J.A.; Ifie, I.; Holmes, M.; Aydin, E.; Jensen, J.M.; Williamson, G. Inhibition of human α-amylase by dietary polyphenols. *J. Funct. Foods* **2015**, *19*, 723–732. [CrossRef]

34. Ademiluyi, A.O.; Oboh, G. Soybean phenolic-rich extracts inhibit key-enzymes linked to type 2 diabetes (α-amylase and α-glucosidase) and hypertension (angiotensin i converting enzyme) *in vitro. Exp. Toxicol. Pathol.* **2013**, *65*, 305–309. [CrossRef] [PubMed]

35. Jaradat, N.A.; Al-lahham, S.; Zaid, A.N.; Hussein, F.; Issa, L.; Abualhasan, M.N.; Hawash, M.; Yahya, A.; Shehadi, O.; Omair, R. Carlina curetum plant phytoconstituents, enzymes inhibitory and cytotoxic activity on cervical epithelial carcinoma and colon cancer cell lines. *Eur. J. Integr. Med.* **2019**, *30*, 93–100. [CrossRef]

36. Jaradat, N.A.; Shawahna, R.; Hussein, F.; Al-Lahham, S. Analysis of the antioxidant potential in aerial parts of *Trigonella arabica* and *Trigonella berythea* grown widely in palestine: A comparative study. *Eur. J. Integr. Med.* **2016**, *8*, 623–630. [CrossRef]

37. Tille, P. *Bailey & Scott's Diagnostic Microbiology e-Book*; Elsevier Health Sciences: Berlin, Germany.

38. Wikler, M.A. *Performance Standards for Antimicrobial Susceptibility Testing: Seventeenth Informational Supplement*; Clinical and Laboratory Standards Institute: Wayne, PA, USA, 2007.

39. Murias, M.; Handler, N.; Erker, T.; Pleban, K.; Ecker, G.; Saiko, P.; Szekeres, T.; Jäger, W. Resveratrol analogues as selective cyclooxygenase-2 inhibitors: Synthesis and structure–activity relationship. *Bioorganic Med. Chem.* **2004**, *12*, 5571–5578. [CrossRef]

40. Mosmann, T. Rapid colorimetric assay for cellular growth and survival: Application to proliferation and cytotoxicity assays. *J. Immunol. Methods* **1983**, *65*, 55–63. [CrossRef]

*Article*

# β-Caryophyllene in the Essential Oil from *Chrysanthemum Boreale* Induces $G_1$ Phase Cell Cycle Arrest in Human Lung Cancer Cells

Kyung-Sook Chung [1,†], Joo Young Hong [1,2,†], Jeong-Hun Lee [1,2], Hae-Jun Lee [1,2], Ji Yeon Park [1], Jung-Hye Choi [2,3], Hee-Juhn Park [4], Jongki Hong [1,*] and Kyung-Tae Lee [1,2,*]

[1]  College of Pharmacy, Kyung Hee University, Seoul 02447, Korea; adella76@hanmail.net (K.-S.C.); hjisuk1206@naver.com (J.Y.H.); ztztzt08@hanmail.net (J.-H.L.); inoseant@naver.com (H.-J.L.); yoolii@daum.net (J.Y.P.)
[2]  Life and Nanopharmaceutical Science, and Kyung Hee University, Seoul 02447, Korea; jchoi@khu.ac.kr
[3]  College of Oriental Pharmaceutical Science, Kyung Hee University, Seoul 02447, Korea
[4]  Division of Applied Plant Sciences, Sang-Ji University, Wonju 220-702, Korea; hjpark@sangji.ac.kr
*  Correspondence: jhong@khu.ac.kr (J.H.); ktlee@khu.ac.kr (K.-T.L.); Tel.: +82-2-961-9255 (J.H.); +82-2-961-0860 (K.-T.L.)
†  These authors contributed equally to this study.

Academic Editor: Igor Jerković
Received: 5 October 2019; Accepted: 17 October 2019; Published: 18 October 2019

**Abstract:** *Chrysanthemum boreale* is a plant widespread in East Asia, used in folk medicine to treat various disorders, such as pneumonia, colitis, stomatitis, and carbuncle. Whether the essential oil from *C. boreale* (ECB) and its active constituents have anti-proliferative activities in lung cancer is unknown. Therefore, we investigated the cytotoxic effects of ECB in A549 and NCI-H358 human lung cancer cells. Culture of A549 and NCI-H358 cells with ECB induced apoptotic cell death, as revealed by an increase in annexin V staining. ECB treatment reduced mitochondrial membrane potential (MMP), disrupted the balance between pro-apoptotic and anti-apoptotic Bcl-2 proteins, and activated caspase-8, -9, and -3, as assessed by western blot analysis. Interestingly, pretreatment with a broad-spectrum caspase inhibitor (z-VAD-fmk) significantly attenuated ECB-induced apoptosis. Furthermore, gas chromatography–mass spectrometry (GC/MS) analysis of ECB identified six compounds. Among them, β-caryophyllene exhibited a potent anti-proliferative effect, and thus was identified as the major active compound. β- Caryophyllene induced $G_1$ cell cycle arrest by downregulating cyclin D1, cyclin E, cyclin-dependent protein kinase (CDK) -2, -4, and -6, and RB phosphorylation, and by upregulating p21[CIP1/WAF1] and p27[KIP1]. These results indicate that β-caryophyllene exerts cytotoxic activity in lung cancer cells through induction of cell cycle arrest.

**Keywords:** β-caryophyllene; essential oil from *Chrysanthemum boreale*; apoptosis; cell cycle; human lung cancer

---

## 1. Introduction

Lung cancer has the fourth highest incidence among cancers, and is the leading cause of cancer death worldwide [1]. The mortality rate is high due to signs of lung cancer generally not appearing until the disease is already at an advanced stage. The two main types of lung cancer consist of small cell lung cancer (SCLC), accounting for a minor percentage of all cases, and non-small cell lung cancer (NSCLC), representing more than 80 to 85% of cases [2]. When the cancer grade is higher than stage I, chemotherapeutic treatment is recommended for NSCLC [3]. Despite substantial progress in the oncology field as a whole, the outcomes following treatment for lung cancer are still poor [4]. However, chemotherapy is associated with significant side effects, and its use can lead to chemoresistance, which is often fatal. Hence, research and development of novel and less hazardous anti-cancer regimens for NSCLC are required [5–7].

Traditionally, botanical medicines have been used in developing countries as the major source of medical treatment [8]. Anti-cancer natural products or their derivatives included 49% of a total 175 FDA approved small molecules [9]. As an example, vinca alkaloids, isolated from *Catharanthus roseus* (Apocynaceae), cause microtubule disruption and induce cell cycle arrest at metaphase, resulting in apoptosis of cancer cells. SB365, a saponin D extracted from the roots of *Pulsatilla koreana* exhibits anti-proliferative effects in various cancer cell lines. In pancreatic cancer, SB365 induces apoptosis and inhibits angiogenesis, contributing to a rise in patient survival rate to 54%, with no reports of side effects [10]. Currently, the efficacy of intravenous SB365 treatment is being investigated in clinical trials [11].

*Chrysanthemum boreale* (Asteraceae) is a perennation. It is around 1–1.5 m tall, and has yellow flowers that are typically 1.5 cm in diameter. *C. boreale* is mainly distributed along the Korean Peninsula, and has spread to the Manchuria region. *C. boreale* and similar species, for instance *C. indicum* and *C. lavandulaefolium*, have been utilized in conventional eastern treatments for stomatitis, pneumonia, carbuncle, fever, and vertigo [12]. Moreover, the guaianolide derivative 8-acetoxy-4,10-dihydroxy-2,11(13)-guaiadiene-12,6-olide, isolated from *C. boreale*, has been shown to have cytotoxic activity against five human cancer cell lines [13]. However, the mechanisms that underlie the anti-proliferative effects of the essential oil from *C. boreale* (ECB) in NSCLC have not been thoroughly reported. A previous study described the isolation of 87 compounds from the ECB. Among them, the most represented were: monoterpenes camphor (17.93%), $\alpha$-thujone (13.13%), cis-chrysanthenol (12.80%), 1,8-cineole (3.95%), $\alpha$-pinene (3.83%), and sesquiterpene $\beta$-caryophyllene (3.04%) [12]. In this study, we investigated the cytotoxic potential of ECB in lung cancer cell lines. To determine the active ingredient of ECB, we tested six of its components (1,8-cineole, thujone, $\beta$-caryophyllene, camphor, endo-borneol, and 2-isopropyl-5-methyl-3-cyclohexen-1-one) for their anti-proliferative effects, and delineated the underlying molecular mechanisms associated with the cytotoxicity.

## 2. Results

### 2.1. ECB Induces Apoptosis in Human Lung Cancer Cells

To determine the anti-proliferative effects of ECB, we examined its cytotoxic potential in human lung carcinoma (A549), pancreatic adenocarcinoma (AsPC-1), and colon adenocarcinoma (HT-29) cell lines using 3-(4,5-dimethylthiazol-2-yl)-2,5-diphenyltetrazolium bromide (MTT) assay. Cells were treated with various concentrations of ECB for 48 h. The $IC_{50}$ values for A549, AsPC-1, and HT-29 were $28.18 \pm 1.96$, $30.86 \pm 2.32$, and $55.21 \pm 3.06$ µg/mL, respectively, indicating that ECB was more cytotoxic in A549 compared to other cell lines. For this reason, we decided to focus on lung carcinoma cells, using A549 and NCI-H358 as cell line models of NSCLC in our study. Similarly to A549, ECB showed dose-dependent cytotoxicity in NCI-H358 cells ($IC_{50}$: $31.19 \pm 2.01$ µg/mL,) compared to L132 normal lung epithelial cells ($IC_{50}$: > 100 µg/mL) (Figure 1a). Additionally, treatment with ECB induced a time-dependent increase in the sub-G1 cell death population in A549 and NCI-H358 cells (Figure 1b). In order to understand if ECB-induced cell death was apoptotic nature, we further examined whether ECB could induce exposure of phosphatidylserine (PS) in A549 and NCI-H358 cells by biparametric flow cytometry analysis, using PI and annexin V to stain DNA (dead cells) and PS (cells undergoing apoptosis), respectively. As shown in Figure 1c, treatment with ECB significantly increased the percentage of PI-annexin V double-positive cells in a concentration-dependent manner. These results suggest that ECB can induce A549 and NCI-H358 lung cancer cells death via apoptosis, rather than non-specific necrosis.

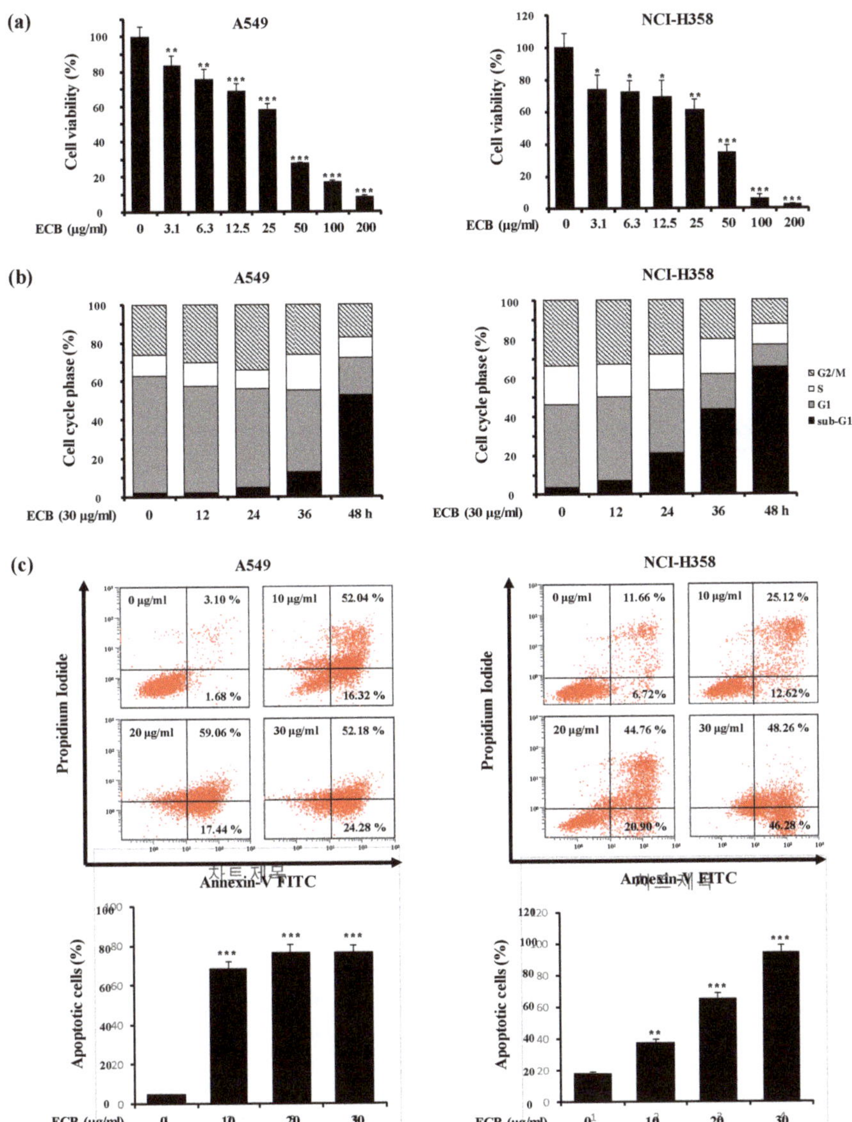

**Figure 1.** The effects of the essential oil from *C. boreale* (ECB) on cell viability and apoptosis in human lung cancer cells. (**a**) A549 cells and NCI-H358 cells were treated with increasing amounts of ECB for 48 h. To determine cell viability, MTT assay was performed. (**b**) Cells were treated with 30 μg/mL of ECB for the indicated times. The cell cycle progression was determined by staining with PI and flow cytometry. Results are representative of three independent experiments. (**c**) Cells treated with different concentrations of ECB (10, 20, or 30 μg/mL for 48 h) were double-stained with PI and annexin V, which specifically detects the externalization of phosphatidylserine (PS), and examined by flow cytometry. Data are presented as means ± SD of three independent experiments. * $p < 0.05$, ** $p < 0.01$, *** $p < 0.001$ vs. the control group.

## 2.2. ECB-Induced Apoptosis is Mediated by Caspase Activation and Mitochondrial Pathway in Human Lung Cancer Cells

The apoptotic process begins in response to intrinsic or extrinsic death signals, and several proteins are involved in this process, including caspases. Procaspases are the precursors of caspases. When cleaved, they become active, promoting apoptotic cues. Caspase-8 plays a pivotal role in the extrinsic apoptotic pathway [14]. By contrast, caspase-9 is activated as result of Bcl-2 proteins reducing the MMP in the intrinsic pathway. Finally, caspase-3 is activated through both the intrinsic and extrinsic pathways, and apoptosis occurs [15]. To examine the effect of ECB on the apoptotic process, we assessed the expression of apoptosis-related proteins by western blot analysis. Cancer cells treated with 10, 20, or 30 µg/mL ECB for 48 h exhibited reduced levels of procaspase-3, -8, and -9 (Figure 2a), suggesting activation of both the intrinsic and extrinsic apoptotic pathways, ultimately leading to apoptosis. Besides, analysis of PI-annexin V double-positive cells by flow cytometry showed that treatment with the pan-caspase inhibitor z-VAD-fmk could attenuate ECB-induced cell death in A549 and NCI-H358 cells (Figure 2b). Next, we measured the MMP of A549 and NCI-H358 cells after treatment with ECB, using the fluorescent dye DiOC6. As shown in Figure 2c, ECB increased mitochondrial membrane depolarization at 48 h compared to that at 0 h. Because Bcl-2 family proteins can regulate MMP [16], we evaluated levels of these proteins in the presence or absence of ECB. Western blot analysis revealed that the presence of ECB decreased anti-apoptotic Bcl-2 and Bcl-X$_L$ protein levels, and increased the expression of pro-apoptotic Bad (Figure 2d). Thus, these data indicate that ECB activates apoptosis by regulating the expression of Bcl-2 family proteins and reducing MMP, which leads to caspase activation in A549 and NCI-H358 cells.

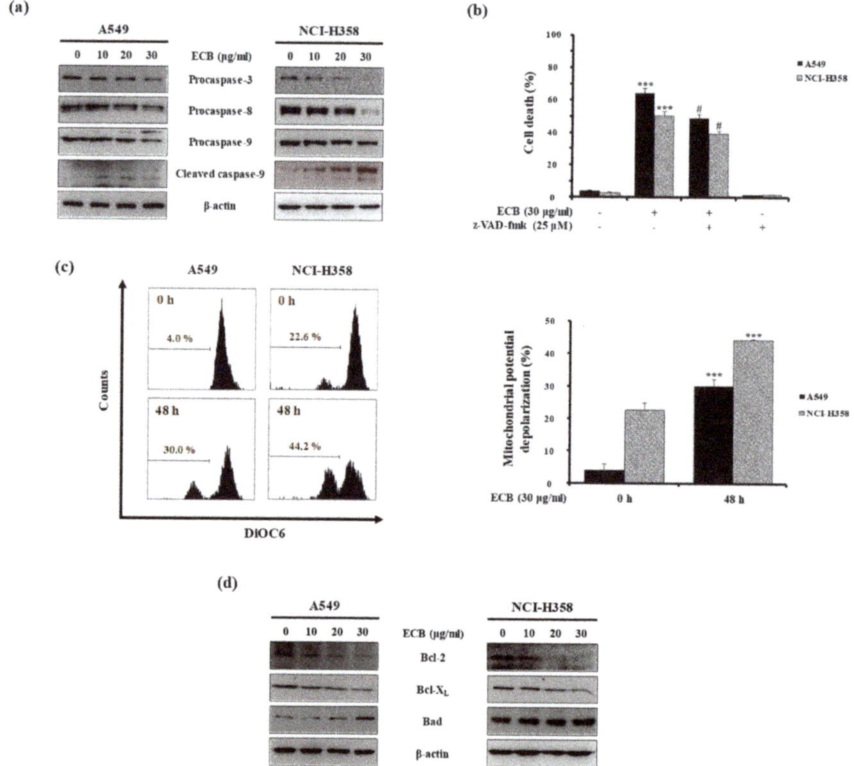

**Figure 2.** Effects of ECB on apoptosis-related proteins. (**a**) A549 and NCI-H358 cells were treated

with different doses of ECB for 48 h. Western blot analysis was conducted to determine the protein expression levels of procaspase-3, -8, -9, and cleaved caspase-9. (**b**) Cells were pretreated with or without z-VAD-fmk (25 µM) for 30 min, treated with ECB (30 µg/mL) for 48 h, and PI-annexin V double staining was performed to determine the fraction of apoptotic cells. (**c**) A549 and NCI-H358 cells were stained with DiOC6 at 0 and 48 h and flow cytometry was performed to measure mitochondrial membrane polarization. (**d**) A549 and NCI-H358 cells were treated with different concentrations of ECB for 48 h, and western blot analysis was performed to determine the protein expression levels of Bcl-2, Bcl-X$_L$, and Bad. Data are presented as means ± SD of three independent experiments. *** $p < 0.001$ vs. the control group; # $p < 0.001$ vs. the ECB-treated group.

## 2.3. β-Caryophyllene Regulates G₁ Cell Cycle Progression in Human Lung Cancer Cells

Next, we performed gas chromatography-mass spectrometry (GC/MS) of the ECB. This analysis revealed six major constituents (Figure 3), and their cytotoxicity was determined in A549 and NCI-H358 cells by MTT assay. As shown in Table 1, β-caryophyllene showed the strongest cytotoxicity among these compounds.

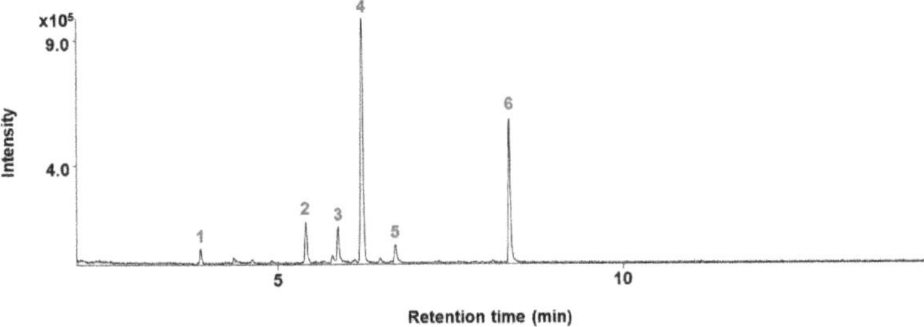


**Figure 3.** Total ion chromatogram of ECB. Peaks represent: 1. 1,8-cineole; 2. thujone; 3. β-caryophyllene; 4. camphor; 5. endo-borneol; 6. 2-isopropyl-5-methyl-3-cyclohexen-1-one.

**Table 1.** Cytotoxicity (IC$_{50}$) of major compounds isolated from ECB, assessed by MTT assay in A549 cells and NCI-H358 cells

| Compounds Isolated from ECB | Chemical Structure | IC$_{50}$ (µg/mL) | |
| --- | --- | --- | --- |
| | | A549 | NCI-H358 |
| 1,8-Cineole | | >200 | >200 |
| Thujone | | >200 | >200 |
| β-Caryophyllene | | 47.05 | 54.78 |
| Camphor | | >200 | >200 |
| Endo-borneol | | >200 | >200 |
| 2-Isopropyl-5 methyl-3-cyclohexen-1-one | | >200 | >200 |

In addition, flow cytometry analysis showed that β-caryophyllene treatment induced the accumulation of A549 and NCI-H358 cells in the $G_1$ phase in a time- and dose-dependent manner (Figure 4a,b). Based on these results, we examined the effect of β-caryophyllene on the expression levels of $G_1$ phase-regulatory proteins by western blot analysis. Our results revealed that β-caryophyllene decreased the expression of cyclin-dependent protein kinase (CDK) 2, CDK4, CDK6, cyclin D1, and cyclin E (Figure 4c,d). In addition, β-caryophyllene decreased the levels of phosphorylated retinoblastoma (p-RB) protein, but did not alter total RB levels. This is indicative of a reduced cyclin/CDK activity, which is consistent with the induction of cell cycle arrest. Moreover, β-caryophyllene increased the levels of p21$^{CIP1/WAP1}$ and p27$^{KIP1}$, two CDK inhibitors (CDKIs) that play key roles in the establishment of $G_1$ phase progression (Figure 4e). All together, these results indicate that β-caryophyllene promotes cell cycle arrest in $G_1$ phase in A549 and NCI-H358 cells.

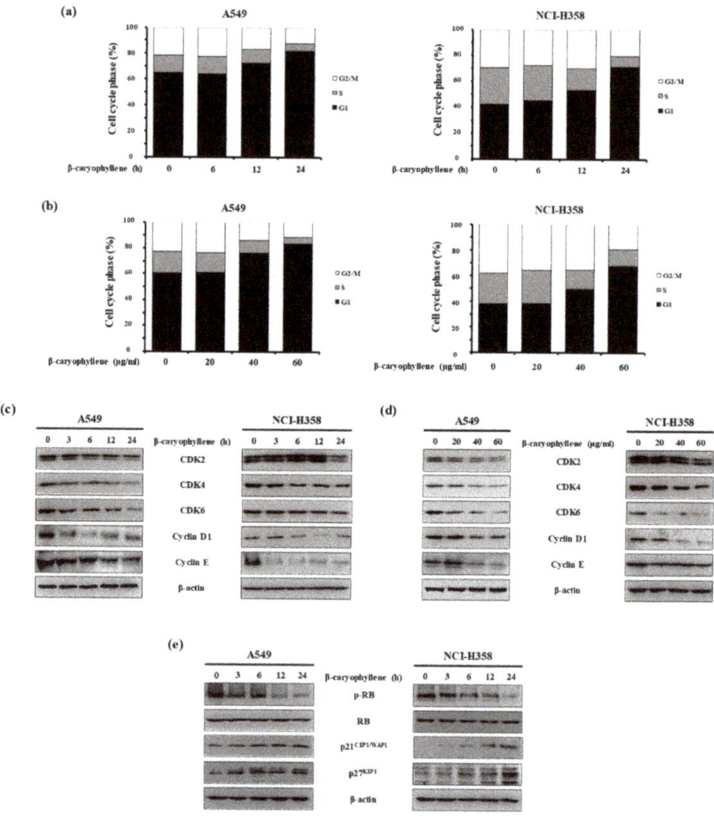

**Figure 4.** Effect of β-caryophyllene on G1 cell cycle arrest and cell cycle-related proteins in human lung cancer. (**a**) Cells were treated with β-caryophyllene (A549, 40 μg/mL; NCI-H358, 60 μg/mL) for the indicated times; (**b**) or with the indicated concentrations for 24 h. Cells were stained with PI for 30 min and then subjected to flow cytometry analysis to determine cell cycle progression. (**c** and **d**) A549 and NCI-H358 cells were treated with β-caryophyllene (A549, 40 μg/mL; NCI-H358, 60 μg/mL) for the indicated times and concentrations, and cells were collected for western blot analysis to measure cyclin/CDK protein expression. (**e**) Cells were treated with β-caryophyllene (A549, 40 μg/mL; NCI-H358, 60 μg/mL) for the indicated times, and then subjected to western blot analysis to determine the protein expression of p-RB, RB, p21$^{CIP1/WAP1}$, and p27$^{KIP1}$.

## 3. Discussion

Cancer chemoprevention is one of the crucial approaches to reduce or delay the occurrence of malignancy after the chronic administration of a synthetic, natural or biological agent [17]. The potential value of this approach has been demonstrated with trials in breast, prostate, and colon cancer [18]. Because of low cytotoxicity to normal cells, minimal side effects, and a wide margin of safety, medicinal herbs can be used to develop new pharmaceutical drugs. Fruits, leaves, and flowers are rich in essential oils, many of which contain important chemopreventive agents. Several essential oil extracts have been shown to exert anti-proliferative, anti-mutagenic, cytotoxic, anti-oxidant, pro-apoptotic, and anti-neoplastic effects. For example, the essential oil of *Eucalyptus benthamii* (Myrtaceae) contains limonene, which has cytotoxic and anti-proliferative effects in cancer cell lines [19].

*C. boreale* has been reported to possess anti-inflammatory [20] and anti-bacterial properties [21]; and β-caryophyllene, found in this and other plants (including the Asteraceae *Helichrysum gymnocephalum* and the Myrtaceae *Syzygium aromaticum*), retains anti-proliferative activities against various cancer cells through induction of apoptosis, anti-microbial, and antioxidant properties [22–24]. β-Caryophyllene is also contained in the essential oil of *Citrus aurantifolia* (Rutaceae), and studies have shown its ability to induce apoptosis in human colon cancer cells [25]. Furthermore, recent research has revealed an anti-carcinogenic effect for α-thujone, another major component of *C. boreale*. In particular, α-thujone treatment impaired the proliferation of glioblastoma (GBM) cells and angiogenesis, and reduced melanoma metastasis in a GBM rat model [26]. Though it was reported to exert an attenuating effect on the viability of the GBM [27], in the present study α-thujone did not show any cytotoxicity up to 500 μM against human lung cancer cells. Another compound in the ECB, endo-borneol, is a bicyclic organic terpene with the hydroxyl group located in an endo position. Many studies have reported that endo-borneol possesses metabolism-enhancing, anti-inflammatory, and antioxidant activities [28,29]. However, neither endo-borneol nor 2-isopropyl-5-methyl-3-cyclohexen-1-one showed any cytotoxicity in our lung cancer cytotoxicity assay. In the current study, we identified anti-proliferative factors that are required for ECB-induced cell cycle arrest and apoptosis of lung cancer cells. Our findings substantiate the results of published studies, and contribute to establishing novel natural compounds as anti-cancer therapeutics.

During the past decades, suppression of tumor cell proliferation through the induction of apoptosis has been recognized as a strategy for the identification of chemotherapeutic agents [27]. Apoptosis plays a major role in normal physiologic processes and is accompanied by blebbing of the cell membrane, DNA fragmentation, induction of apoptotic bodies, and activation of caspases [14]. Caspases are members of the cysteine-aspartic acid protease family and have crucial roles in triggering and executing apoptosis [30]. The extrinsic and intrinsic pathways form the two branches of caspase-dependent apoptotic signaling. The extrinsic pathway is activated by various receptors, such as TNF-α receptor, FasL receptor, toll-like receptor, and death receptor, which leads to the formation of the 'death-inducing signaling complex' (DISC) and activation of caspase-8 [31]. By contrast, the intrinsic pathway (or 'mitochondrial pathway') is activated by ROS, DNA damage, pro-apoptotic Bcl-2 family proteins, calcium, and some metals. Consequentially, activation of this pathway induces MMP and expedites the release of cytochrome *c*. Released cytochrome *c* forms part of the apoptosome, a complex that contains Apaf-1. The apoptosome triggers procaspase-9 cleavage and activation, which in turn triggers the activation of caspase-3, leading to apoptosis. In this study, we used PI-annexin V double staining to demonstrate that ECB induces apoptosis in human lung cancer cells. In addition, our data showed that ECB reduced MMP, likely due to the downregulation of Bcl-2 and Bcl-$X_L$, and upregulation of Bad protein expression.

Our results indicated that β-caryophyllene induces cell cycle arrest at the $G_1$ phase in human lung cancer cell lines. This assumption is based on the observation that β-caryophyllene promoted downregulation of $G_1$ cell cycle positive regulators, such as CDK2, CDK4, CDK6, cyclin $D_1$, and cyclin E, and upregulation of $G_1$ cell cycle negative regulators p21$^{CIP1/WAF1}$ and p27$^{KIIP1}$. Furthermore, β-caryophyllene decreased RB phosphorylation. In a previous study, β-caryophyllene has been shown to exert a selective anti-proliferative effect in HCT116 colon cancer cells, without affecting normal cell lines [22]. Interestingly, β-caryophyllene-treated HCT116 cells displayed several apoptotic features, such as DNA fragmentation, chromatin condensation, and MMP disruption. Thus, previous studies, together with our data, suggest that β-caryophyllene might have cell-type specific phenotypic effects, and that β-caryophyllene-induced cell cycle arrest might lead to apoptosis.

In summary, our study revealed that ECB exerts its anti-proliferative effect via caspase activation and mitochondria-dependent apoptosis in human lung cancer cells. The active constituent of ECB is β-caryophyllene, which promotes $G_1$ cell cycle arrest in these cell types. Taken together, these data suggest that ECB should be considered as a potential chemotherapeutic agent for the treatment of non-small cell lung cancer.

## 4. Materials and Methods

### 4.1. Materials

All kinds of cell culture-related agents were purchased from Life Technologies Inc (Grand Island, NY, USA). β-Caryophyllene, camphor, 1,8-cineole, 3-(4,5-dimethylthiazol-2-yl)-2,5-diphenyl-tetrazolium bromide (MTT), 3,3′-dihexyloxacarbocyanine iodide (DiOC6), propidium iodide (PI), bisacrylamide, sodium dodecyl sulfate (SDS), dimethyl sulfoxide (DMSO), RNase A, and other chemicals were purchased from Sigma Chemical Co. (St.Louis, MO, USA). $N,N,N',N'$-tetramethyl-ethylenediamine dihydrochloride (TEMED) were purchased from Bio-Rad Laboratories (Portland, ME, USA). The following antibodies for caspase-3, Bcl-2, Bcl-$X_L$, Bad, CDK2, CDK4, CDK6, Cyclin D1, Cyclin E, p21$^{CIP1/WAF1}$, p27$^{KIP1}$, RB, and β-actin were purchased from Santa Cruz Biotechnology Inc. (Santa cruz, CA, USA). p-RB antibody was purchased from Cell Signaling Technology (Danvers, MA, USA). The antibodies for caspase-8 and caspase-9 were purchased from BD Biosciences, Pharmingen (San Diego, CA, USA). z-VAD-fmk was purchased from Calbiochem (Bad Soden, Germany).

### 4.2. Preparation of Essential Oil

The aerial part of *Chrysanthemum boreale* (Asteraceae) was collected in October in Wonju city, Gangwon-do, Korea. This plant was identified by Sang-Cheol Lim (Department of Horticulture and Landscape, Sangji University, Korea). The voucher specimen (natchem# 38) was deposited at the Laboratory of Natural Products Chemistry, Department of Pharmaceutical Engineering, Sangji University, Korea. ECB was extracted using the method of steam distillation. The plant material (300 g) was distilled for 3 h by a steam distillation apparatus. The distilled liquid was fractionated three times with diethyl ether (each 400 mL). The diethyl ether layer was subjected to dehydration with anhydrous sodium sulfate, and then evaporated to give the essential oil (weight 14.1 g; volume 14.2 mL).

### 4.3. GC/MS Analysis

GC/MS analysis of ECB was performed in a GC-MS (GC: 6890 A, Agilent Technologies, Santa Clara, CA, USA; MS: 5973, Agilent Technologies) using a DB-5MS column (15 m, 0.25 mm i.d., 0.25 μm film thickness). ECB (0.1 μL) was injected on split mode at a 1:30 ratio. The oven temperature was set at 50 °C for 1 min, followed by a temperature gradient of 5 °C/min. When the temperature reached 160 °C, it was kept steady for 20 min. Then, a step of 5.0 °C/min was applied until oven temperature was 250 °C, where it was kept for 15 min. Helium was used as carrier gas with a flow rate of 1 mL/min. Injector and transfer line temperatures were both set at 280 °C. The mass spectrometer operated in the electron impact mode with the electron energy at 70 eV. Identification of volatile components was performed by matching their retention tmes and mass spectra with those of authentic standards.

### 4.4. Cell Culture

AsPC-1 human pancreatic adenocarcinoma, HT-29 human colon adenocarcinoma, A549 and NCI-H358 human lung adenocarcinoma, and human L132 normal lung epithelial cells were obtained from the Korean cell line bank (Seoul, Korea) and cultured in RPMI 1640 supplemented with 10% heat-inactivated fetal bovine serum (FBS), penicillin (100 units/mL), and streptomycin sulfate (100 µg/mL). Cells were maintained at 37 °C in a humidified atmosphere of 5% $CO_2$.

### 4.5. MTT Assay

To examine cell viability, the MTT assay was conducted using the previous modified method as described by Lee et al. [32]. Cells were harvested during the logarithmic growth phase and seeded in 96-well plates at a density of $2 \times 10^4$/mL in a final volume of 190 µL/well. After 24 h incubation, 10 µL ECB, or β-caryophyllene full-range concentration was added to 96-well plates. After 48 h, 50 µL of MTT (5 mg/mL stock solution in PBS) was added to each well for 4 h. Subsequently, the supernatant was removed, and MTT crystals were solubilized with 100 µL anhydrous DMSO each well. The optical density was measured at 540 nm.

### 4.6. PI Staining Analysis

After treatment with different concentrations of ECB or β-caryophyllene for various times, the cells were collected with trypsin and rinsed with ice-cold PBS twice. The pellets were resuspended and fixed in 70% EtOH at 4 °C overnight. Before detected by flow cytometry, the cells were washed twice with PBS and resuspended in a PI solution containing PI (1 µg/mL) and RNase A (10 µg/mL) for 30 min. The cells were evaluated by fluorescence-activated cell sorting (FACS) cytometer (Beckman Coulter, CA, USA).

### 4.7. PI and Annexin V Double Staining

Cells were harvested and washed twice with ice-cold PBS. Cells were resuspended with 1 × binding buffer (10 mM HEPES, pH 7.4, 140 mM NaCl, 2.5 mM $CaCl_2$), and then the cell suspension (100 µL) was incubated in dark place with 5 µL of annexin V-FITC and 10 µL of PI (50 µg/mL) for 15 min. Cells were analyzed for PI-annexin V double staining using FACS flow cytometry (Becton Dickinson Co, Heidelberg, Germany).

### 4.8. Determination of Mitochondria Membrane Potential (MMP)

Cells were collected and washed twice with ice-cold PBS. After treatment of 30 µg/mL ECB for 48 h, the cells were stained with DiOC6 at a final concentration of 30 nM for 30 min at 37 °C in the dark. The fluorescence intensity was analyzed with FACS cytometer (Beckman Coulter, CA, USA)

### 4.9. Western Blot Analysis

After the treatment of various concentrations of ECB or β-caryophyllene for different times, the cells were lyzed in protein lysis buffer (Intron, Seoul, Korea) for 30 min in ice. Cell debris was removed by microcentrifuge (4 °C, 15,000 rpm, 30 min) and then the protein concentration of supernatants was ascertained by bio-rad protein assay reagent, following the manufacturer's instruction. Measures of 20–30 µg of cell extracts were fractionated by 8–15 % SDS PAGE, and transferred to a PVDF. After incubating for 1 h with 5% skim milk in Tween 20/Tris-buffered saline (TBST) at 20 °C, the membranes were incubated with primary antibody at 4 °C for overnight. Membranes were washed three times with TBST and incubated with secondary antibody for 2 h at 25 °C, rewashed three times with TBST. Blots were developed using enhanced chemiluminescence detection agents (Amersham, Buckinghamshire, England).

### 4.10. Statistical Analysis

Data are expressed as mean ± S.D. Statistically significant values were compared using one-way ANOVA and Dunnett's post hoc test with the GraphPad Prism 5 statistical software, and $p$-values of less than 0.05 was considered statistically significant.

**Author Contributions:** K.-S.C., J.Y.H., and K.-T.L. were involved in the design of the study and analysis of data. J.-H.L. and H.-J.L. were involved in the technical support and conducted the experiments. K.-S.C., J.Y.H., J.Y.P., J.-H.C., J.H., and K.-T.L. were involved in the writing, reviewing, and revision of the manuscript. H.-J.P. and J.H. were involved in the material support. All authors have read and approved the final submitted manuscript.

**Funding:** This research was supported by Basic Science Research Program through the National Research Foundation of Korea funded by the Ministry of Science and ICT (NRF-2017R1A5A2014768).

**Conflicts of Interest:** The authors declare no conflict of interest.

## References


1. Schabath, M.B.; Cote, M.L. Cancer Progress and Priorities: Lung Cancer. *Cancer Epidemiol. Biomark. Prev.* **2019**, *28*, 1563–1579. [CrossRef] [PubMed]
2. Sher, T.; Dy, G.K.; Adjei, A.A. Small cell lung cancer. *Mayo Clin. Proc.* **2008**, *83*, 355–367. [CrossRef] [PubMed]
3. Hobbins, S.; West, D.; Peake, M.; Beckett, P.; Woolhouse, I. Patient characteristics, treatment and survival in pulmonary carcinoid tumours: An analysis from the UK National Lung Cancer Audit. *BMJ Open* **2016**, *6*, 012530. [CrossRef] [PubMed]
4. Polański, J.; Jankowska-Polanska, B.; Rosińczuk, J.; Chabowski, M.; Szymańska-Chabowska, A. Quality of life of patients with lung cancer. *OncoTargets Ther.* **2016**, *9*, 1023–1028.
5. Sun, G.; Fan, T.; Zhao, L.; Zhou, Y.; Zhong, R. The potential of combi-molecules with DNA-damaging function as anticancer agents. *Future Med. Chem.* **2017**, *9*, 403–435. [CrossRef]
6. Guohui, S.; Lijiao, Z.; Rugang, Z. The Induction and Repair of DNA Interstrand Crosslinks and Implications in Cancer Chemotherapy. *Anti-Cancer Agents Med. Chem.* **2015**, *16*, 221–246. [CrossRef]
7. Fan, T.; Sun, G.; Sun, X.; Zhao, L.; Zhong, R.; Peng, Y. Tumor Energy Metabolism and Potential of 3-Bromopyruvate as an Inhibitor of Aerobic Glycolysis: Implications in Tumor Treatment. *Cancers* **2019**, *11*, 317. [CrossRef]
8. Greenwell, M.; Rahman, P. Medicinal Plants: Their Use in Anticancer Treatment. *Int. J. Pharm. Sci. Res.* **2015**, *6*, 4103–4112.
9. Tewari, D.; Rawat, P.; Singh, P.K. Adverse drug reactions of anticancer drugs derived from natural sources. *Food Chem. Toxicol.* **2019**, *123*, 522–535. [CrossRef]
10. Moon, K.S.; Ji, J.Y.; Cho, Y.J.; Lee, J.H.; Choi, M.S.; Kim, E.E. Therapeutic Effects of SB Natural Anticancer Drug in 50 Patients with Stage IV Pancreatic Cancer. *J. Cancer Treat. Res.* **2015**, *3*, 42. [CrossRef]
11. Hong, J.-M.; Kim, J.-H.; Kim, H.; Lee, W.J.; Hwang, Y.-I. SB365, Pulsatilla Saponin D Induces Caspase-Independent Cell Death and Augments the Anticancer Effect of Temozolomide in Glioblastoma Multiforme Cells. *Molcules* **2019**, *24*, 3230. [CrossRef] [PubMed]
12. Kim, Y.-H.; Yu, H.-H.; Cha, J.-D.; You, Y.-O.; Kim, K.-J.; Jeong, S.-I.; Kil, B.-S. Antibacterial Activity and Chemical Composition of Essential Oil of Chrysanthemum boreale. *Planta Medica* **2003**, *69*, 274–277. [CrossRef] [PubMed]
13. Park, K.H.; Yang, M.S.; Park, M.K.; Kim, S.C.; Yang, C.H.; Park, S.J.; Lee, J.R. A new cytotoxic guaianolide from Chrysanthemum boreale. *Fitoterapia* **2009**, *80*, 54–56. [CrossRef] [PubMed]
14. Ashkenazi, A. Targeting the extrinsic apoptotic pathway in cancer: Lessons learned and future directions. *J. Clin. Investig.* **2015**, *125*, 487–489. [CrossRef]
15. Zielinski, R.R.; Eigl, B.J.; Chi, K.N. Targeting the Apoptosis Pathway in Prostate Cancer. *Cancer J.* **2013**, *19*, 79–89. [CrossRef]
16. Breckenridge, D.G.; Xue, D. Regulation of mitochondrial membrane permeabilization by BCL-2 family proteins and caspases. *Curr. Opin. Cell Biol.* **2004**, *16*, 647–652. [CrossRef]
17. Ahmad, N.; Adhami, V.M.; Afaq, F.; Feyes, D.K.; Mukhtar, H. Resveratrol causes WAF-1/p21-mediated G(1)-phase arrest of cell cycle and induction of apoptosis in human epidermoid carcinoma A431 cells. *Clin. Cancer Res.* **2001**, *7*, 1466–1473.
18. Downey, K.; Jafar, M.; Attygalle, A.D.; Hazell, S.; A Morgan, V.; Giles, S.L.; A Schmidt, M.; Ind, T.E.J.; Shepherd, J.H.; DeSouza, N.M. Influencing surgical management in patients with carcinoma of the cervix using a T2- and ZOOM-diffusion-weighted endovaginal MRI technique. *Br. J. Cancer* **2013**, *109*, 615–622. [CrossRef]

19. Döll-Boscardin, P.M.; Sartoratto, A.; Maia, B.H.L.D.N.S.; De Paula, J.P.; Nakashima, T.; Farago, P.V.; Kanunfre, C.C. In Vitro Cytotoxic Potential of Essential Oils of Eucalyptus benthamii and Its Related Terpenes on Tumor Cell Lines. *Evid.-Based Complement. Altern. Med.* **2012**, *2012*, 1–8. [CrossRef]

20. Pyee, Y.; Chung, H.J.; Choi, T.J.; Park, H.J.; Hong, J.Y.; Kim, J.S.; Kang, S.S.; Lee, S.K. Suppression of inflammatory responses by handelin, a guaianolide dimer from chrysanthemum boreale, via downregulation of nf-kappab signaling and pro-inflammatory cytokine production. *J. Nat. Prod.* **2014**, *77*, 917–924. [CrossRef]

21. Kim, B.-S.; Park, S.-J.; Kim, M.-K.; Kim, Y.-H.; Lee, S.-B.; Lee, K.-H.; Choi, N.-Y.; Lee, Y.-R.; Lee, Y.-E.; You, Y.-O. Inhibitory Effects of Chrysanthemum boreale Essential Oil on Biofilm Formation and Virulence Factor Expression of Streptococcus mutans. *Evid.-Based Complement. Altern. Med.* **2015**, *2015*, 1–11.

22. Dahham, S.S.; Tabana, Y.M.; Iqbal, M.A.; Ahamed, M.B.; Ezzat, M.O.; Majid, A.S.; Majid, A.M. The anticancer, antioxidant and antimicrobial properties of the sesquiterpene beta-caryophyllene from the essential oil of aquilaria crassna. *Molecules* **2015**, *20*, 11808–11829. [CrossRef] [PubMed]

23. Ramachandhiran, D.; Sankaranarayanan, C.; Murali, R.; Babukumar, S.; Vinothkumar, V. Beta-caryophyllene promotes oxidative stress and apoptosis in kb cells through activation of mitochondrial-mediated pathway—An in-vitro and in-silico study. *Arch. Physiol. Biochem.* **2019**, 1–15. [CrossRef] [PubMed]

24. Fidyt, K.; Fiedorowicz, A.; Strzadala, L.; Szumny, A. Beta-caryophyllene and beta-caryophyllene oxide-natural compounds of anticancer and analgesic properties. *Cancer Med.* **2016**, *5*, 3007–3017. [CrossRef]

25. Patil, J.R.; Jayaprakasha, G.; Murthy, K.C.; Tichy, S.E.; Chetti, M.B.; Patil, B.S. Apoptosis-mediated proliferation inhibition of human colon cancer cells by volatile principles of Citrus aurantifolia. *Food Chem.* **2009**, *114*, 1351–1358. [CrossRef]

26. Torres, A.; Vargas, Y.; Uribe, D.; Carrasco, C.; Torres, C.; Rocha, R.; Oyarzun, C.; San Martin, R.; Quezada, C. Pro-apoptotic and anti-angiogenic properties of the alpha /beta-thujone fraction from thuja occidentalis on glioblastoma cells. *J. Neurooncol.* **2016**, *128*, 9–19. [CrossRef]

27. Pudelek, M.; Catapano, J.; Kochanowski, P.; Mrowiec, K.; Janik-Olchawa, N.; Czyz, J.; Ryszawy, D. Therapeutic potential of monoterpene alpha-thujone, the main compound of thuja occidentalis l. Essential oil, against malignant glioblastoma multiforme cells in vitro. *Fitoterapia* **2019**, *134*, 172–181. [CrossRef]

28. Slameňová, D.; Horváthová, E. Cytotoxic, anti-carcinogenic and antioxidant properties of the most frequent plant volatiles. *Neoplasma* **2013**, *60*, 343–354. [CrossRef]

29. Cheng, C.; Liu, X.-W.; Du, F.-F.; Li, M.-J.; Xu, F.; Wang, F.-Q.; Liu, Y.; Li, C.; Sun, Y. Sensitive assay for measurement of volatile borneol, isoborneol, and the metabolite camphor in rat pharmacokinetic study of Borneolum (Bingpian) and Borneolum syntheticum (synthetic Bingpian). *Acta Pharmacol. Sin.* **2013**, *34*, 1337–1348. [CrossRef]

30. Logue, S.E.; Martin, S.J. Caspase activation cascades in apoptosis. *Biochem. Soc. Trans.* **2008**, *36*, 1–9. [CrossRef]

31. Matthews, G.M.; Newbold, A.; Johnstone, R.W. Intrinsic and Extrinsic Apoptotic Pathway Signaling as Determinants of Histone Deacetylase Inhibitor Antitumor Activity. *Mol. Cell. Basis Metastasis Road Ther.* **2012**, *116*, 165–197.

32. Lee, K.-W.; Chung, K.-S.; Seo, J.-H.; Yim, S.-V.; Park, H.-J.; Choi, J.-H.; Lee, K.-T. Sulfuretin from heartwood of Rhus verniciflua triggers apoptosis through activation of Fas, Caspase-8, and the mitochondrial death pathway in HL-60 human leukemia cells. *J. Cell. Biochem.* **2012**, *113*, 2835–2844. [CrossRef] [PubMed]

**Sample Availability:** Samples of the compounds are not available from the authors.

 *molecules*

*Article*

# Chemical Composition of Two Different Lavender Essential Oils and Their Effect on Facial Skin Microbiota

**Marietta Białoń [1,*], Teresa Krzyśko-Łupicka [2], Ewa Nowakowska-Bogdan [3] and Piotr P. Wieczorek [1]**

[1] Faculty of Chemistry, University of Opole, Oleska 48, 45-052 Opole, Poland
[2] Independent Department of Biotechnology and Molecular Biology, Faculty of Natural and Technical Science, University of Opole, Kominka 6A, 45-035 Opole, Poland
[3] The Institute of Heavy Organic Synthesis "Blachownia", Energetyków 9, 47-225 Kędzierzyn-Koźle, Poland
* Correspondence: Marietta.Bialon@uni.opole.pl

Academic Editor: Francesca Mancianti
Received: 23 July 2019; Accepted: 5 September 2019; Published: 8 September 2019

**Abstract:** Lavender oil is one of the most valuable aromatherapy oils, its anti-bacterial and anti-fungal activities can be explained by main components such as linalool, linalyl acetate, lavandulol, geraniol, or eucalyptol. The aim of the study was to assess the anti-microbial effects of two different lavender oils on a mixed microbiota from facial skin. The commercial lavender oil and essential lavender oil from the Crimean Peninsula, whose chemical composition and activity are yet to be published, were used. Both oils were analysed by gas chromatography coupled to mass spectrometry. The composition and properties of studied oils were significantly different. The commercial ETJA lavender oil contained 10% more linalool and linalyl acetate than the Crimean lavender oil. Both oils also had different effects on the mixed facial skin microbiota. The Gram-positive bacilli were more sensitive to ETJA lavender oil, and Gram-negative bacilli were more sensitive to Crimean lavender oil. However, neither of the tested oils inhibited the growth of Gram-positive cocci. The tested lavender oils decreased the cell number of the mixed microbiota from facial skin, but ETJA oil showed higher efficiency, probably because it contains higher concentrations of monoterpenoids and monoterpenes than Crimean lavender oil does.

**Keywords:** facial skin microbiota; gas chromatography with mass spectrometry; lavender essential oil

---

## 1. Introduction

The generic name "lavender" dates back to ancient times and derives from the Latin word *lavare*, which means washing and bathing. According to Romans, not only the aromatic qualities but also antiseptic properties were important [1]. Therefore, lavender oil was used as a panacea even in the case of wounds associated with tissue loss [2].

Today, as a preservative and skin regenerator, it is widely used in the cosmetic industry to produce safe tonics, lotions, creams, shampoos, conditioners, shower gels, and soaps by specialist cosmetic companies, such as Dr. Beta-Pollena Aroma, Farmona Organique, Sanoflore, and Yves Rocher.

Because of the mild climate, adequate sunshine, alkaline soil, and natural wind protection, lavender is naturally present in Mediterranean countries.

Lavender belongs to the *Labiatae* family, which includes ~30 species of *Lavandula*. However, only three species with lavender fragrance are of industrial importance. These are as follows [3]:

1.  Narrow-leaved lavender (real, medical) *Lavandula officinalis* Chaix, syn. *L. vera* DC, *L. angustifolia* Mill.;
2.  Broad-leaved lavender (spike lavender) *Lavandula latifolia* Vill. (Syn. *L. spica* DC);

3.   Lavandin, a hybrid of the two preceding species.

In the cosmetics industry, the most popular essential oils are those derived from these plants. Nevertheless, both their odour and chemical composition are determined by a number of factors: plant species or varieties, climatic conditions, growth method and conditions, harvesting, transport, storage, and oil preparation techniques. Lavender oil obtained from *L. angustifolia* is the most valuable and the most expensive because its efficiency is two-fold lower than that of spike oil, and four-fold lower than that of lavandin oil [4]. Therefore, lavender oil can be falsified with cheaper oils (lavandin or spike). Sometimes, synthetic products such as linalyl acetate are added [4].

Although the main active ingredients are monoterpenes (linalool, linalyl acetate, lavandulol, geraniol, bornyl acetate, borneol, terpineol, and eucalyptol or lavandulyl acetate), these oils may have different anti-bacterial and anti-fungal effects, depending on their chemical composition [5,6]. A high and almost equal content of linalool and linalyl acetate (a ratio above one) is required for good anti-microbial properties of lavender essential oil [5,6]. The high concentration of lavandulol and its acetate is also desirable, giving the oil a rosaceous, sharp floral aroma. Good anti-microbial activity requires that the ratio of the content of the sum of linalyl acetate with linalool to the content of terpinen-4-ol in lavender essential oil is more than 13 [3,4]; moreover, this ratio can help to determine the type of oil and its applicability, as presented in Table 1 [7–9]. On the other hand, high concentrations of ocimene, cineole, camphor, or terpinen-4-ol adversely affect the quality of this oil [3,4]. The chemical composition of different lavender species depends on the geographic region of origin (Tables 1 and 2). On the basis of the presented observations, it is obvious that *L. angustifolia*, *Lavandula stoechas*, and *Lavandula dentata* contain large amounts of eucalyptol and camphor (Tables 1 and 2). It should also be noted that lavender oil from Brazil [10] contains borneol at a concentration of 22.4%, which is much higher than that in other *L. angustifolia* oils (Table 1). Although *L. abrialis* oil from France [11] and *L. bipinnata* from Algeria [12] contain large amounts of camphor, they do not contain eucalyptol (Table 2).

**Table 1.** The components of lavender oil from *Lavandula angustifolia* described in the literature.

| Name | *Lavandula angustifolia*—Place and Area (%) | | | | | | | | |
| | Italy | | | Australia | | China | Greece | Brazil | Iran |
| | (a) | (b) | (c) | (a) | (b) | | (a) | | |
|---|---|---|---|---|---|---|---|---|---|
| tricyclene | | | - | 0.02–0.04 | - | - | - | - | - |
| α-thujene | 0.13 | 0.11 | 0.05 | 0.02–0.17 | - | - | - | - | - |
| α-pinene | 0.51 | 0.93 | 0.26 | 0.08–0.73 | - | - | - | 0.70 | 1.41 |
| camphene | 0.26 | 0.59 | 0.19 | 0.02–0.54 | 0.19 | 3.98 | - | 0.50 | - |
| β-phellandrene | - | - | - | 0.02–0.11 | - | - | - | 3.40 | - |
| β-pinene | 0.14 | 0.30 | 1.58 | 0.03–1.21 | - | - | - | 1.10 | 1.52 |
| octen-3-ol | 0.22 | 0.20 | 0.16 | 0.03–1.14 | 0.28 | 0.35 | - | - | - |
| 3-octanone | 0.36 | 0.68 | - | 0.33–3.49 | - | - | - | - | - |
| myrcene | 0.12 | - | trace | 0.26–1.22 | 0.56 | 0.87 | - | 0.70 | 1.02 |
| 3-carene | - | - | - | 0.05–0.30 | - | 0.45 | - | 0.90 | 0.76 |
| sabinene | 0.31 | 0.75 | 0.31 | 0.04–0.39 | - | - | - | - | - |
| *p*-cymene | 0.29 | 0.20 | - | 0.10–0.45 | - | - | - | - | - |
| *o*-cymene | 0.04 | 0.09 | 0.30 | 0.03–0.12 | - | - | - | - | - |
| limonene | 1.10 | 2.36 | - | 0.18–3.92 | 0.24 | 0.19 | - | - | - |
| eucalyptol | 3.98 | 10.89 | 6.75 | 0.1–10.87 | 1.51 | 2.30 | 4.80 | 7.90 | 3.93 |
| (*E*)-β-ocimene | 1.24 | 0.42 | 0.56 | 0.34–2.36 | 0.09 | 0.52 | - | - | - |
| (*Z*)-β-ocimene | 1.02 | 1.32 | 0.77 | 0.95–6.17 | 0.44 | | - | - | - |
| *trans*-linalool oxide | - | 0.07 | 1.57 | 0.26–0.99 | - | - | - | - | - |
| *cis*-linalool oxide | 0.09 | 0.07 | 1.48 | 0.34–1.09 | 0.24 | - | - | - | 1.67 |
| *trans*-sabinene hydrate | 0.39 | 0.34 | - | 0.04–0.20 | - | - | - | - | - |
| linalool | 35.96 | 36.51 | 35.31 | 23.03–57.48 | 52.59 | 44.54 | 44.50 | - | 4.91 |
| *cis*-*p*-menth-2-en-1-ol | - | - | - | 0.02–0.04 | - | - | - | - | - |
| *trans*-pinocarveol | 0.18 | 0.10 | - | 0.01–0.34 | - | - | - | - | - |
| camphor | 5.56 | 11.76 | 7.81 | 0.09–7.10 | 8.79 | - | - | 3.50 | 2.83 |
| myrtenol | - | - | - | 0.04–0.21 | - | - | - | 0.40 | - |

**Table 1.** *Cont.*

| Name | *Lavandula angustifolia*—Place and Area (%) | | | | | China | Greece | Brazil | Iran |
| | Italy | | | Australia | | | | | |
| | (a) | (b) | (c) | (a) | (b) | | (a) | | |
|---|---|---|---|---|---|---|---|---|---|
| borneol | 2.71 | 4.21 | 2.98 | 0.30–4.04 | 7.50 | 2.45 | 3.90 | 22.40 | 8.57 |
| *p*-cymen-8-ol | 0.33 | 0.55 | - | 0.12–0.27 | - | - | - | - | - |
| lavendulol | 0.05 | 0.05 | 0.55 | 0.05–0.86 | - | - | - | - | - |
| terpinen-4-ol | 6.57 | 2.10 | 3.34 | 0.11–8.07 | 2.45 | - | 6.90 | 0.90 | - |
| sabine ketone | - | - | - | 0.02–0.10 | - | - | - | - | - |
| *m*-cymen-8-ol | 0.03 | 0.09 | - | 0.02–0.18 | - | - | - | - | - |
| *α*-terpineol | 0.06 | 0.07 | 4.39 | 0.12–6.02 | 3.03 | 6.75 | 3.50 | 1.20 | 1.98 |
| hexyl butyrate | - | - | 0.43 | 0.12–1.72 | - | - | - | - | - |
| isobornyl formate | 0.06 | 0.10 | - | 0.10–0.52 | - | - | - | - | - |
| geraniol | - | - | 0.67 | - | - | 11.02 | - | - | 1.24 |
| cumin aldehyde | - | - | - | 0.04–0.53 | 0.16 | - | - | - | - |
| carvone | - | - | - | 0.02–0.19 | - | - | - | 0.40 | - |
| linalyl acetate | 21.74 | 14.42 | 12.09 | 4.01–35.39 | 9.27 | - | 32.70 | - | - |
| *α*-bisabolol | 1.12 | 0.89 | 3.76 | 0.02–0.71 | 0.78 | - | - | 13.10 | 2.32 |
| dihydrocarveol | - | - | - | 0.03–0.51 | - | - | - | - | - |
| bornyl acetate | - | - | - | 0.03–0.32 | 1.11 | - | - | - | 1.67 |
| lavendulyl acetate | - | - | - | 0.70–6.16 | 1.32 | 10.78 | - | - | - |
| neryl acetate | 0.06 | - | 1.31 | 0.07–1.23 | 1.21 | - | - | - | 2.16 |
| *β*-bourbonene | 0.17 | 0.09 | - | 0.02–0.09 | - | - | - | - | - |
| *α-trans*-bergamotene | 0.07 | 0.05 | - | 0.02–0.15 | 0.07 | - | - | - | - |
| *α*-cederene | - | - | - | 0.01–0.09 | - | - | - | - | - |
| *β*-caryophellene | - | - | - | - | - | - | - | 3.20 | 1.60 |
| caryophyllene | 2.87 | 2.42 | 1.30 | 0.45–2.83 | 1.00 | 0.50 | 0.30 | - | - |
| *α*-santalene | - | - | - | - | 0.05 | - | - | - | - |
| *α-cis*-bergamotene | 0.07 | 0.06 | - | 0.02–0.09 | - | - | - | - | - |
| *β*-farnesene | 4.02 | 1.07 | 1.00 | 0.17–1.69 | 1.83 | - | - | - | 0.71 |
| germacene D | 0.77 | 1.50 | - | 0.16–0.94 | 0.37 | - | - | - | - |
| *β*-bisabolene | 0.03 | - | - | - | 0.26 | - | - | 0.80 | - |
| *γ*-cadinene | 0.26 | 0.39 | - | 0.03–0.39 | 0.28 | - | - | 2.90 | - |
| caryophyllene oxide | - | - | - | - | 0.22 | - | - | 4.50 | 2.73 |
| spathulenol | 0.06 | 0.31 | - | 0.01–0. 06 | - | - | - | - | - |
| *γ*-muurolol | - | - | - | 0.02–0.48 | - | - | - | - | - |
| *τ*-cadinol | - | - | - | 0.02–0.42 | - | - | - | - | 1.85 |
| *γ*-terpinene | - | - | - | 0.05–0.22 | - | - | - | - | - |
| octen-3-yl acetate | - | - | - | 0.19–4.16 | - | - | - | - | - |
| norborneol acetate | - | - | - | 0.04–0.51 | - | - | - | - | - |

Italy (a)—low Friuli–Venezia Giulia (northeast Italy) [7]; Italy (b)—high Friuli–Venezia Giulia (northeast Italy) [7]; Italy (c)—Betulla srl (Italy) [9]; Australia (a)—Australian Botanical Products (Hallam, Australia) [13]; Australia (b)—The Lavender Patch lavender farm, Victoria, Australia [8]; China (a)—Xinjiang, China [14]; Greece (a)—Crete (Greece) [15]; Brazil—Franca, State of Sao Paulo, Brazil [10]; Iran—Astara, north of Iran [16].

**Table 2.** The components of other lavender oils described in the literature.

| Name | Place and Area (%) ± SD | | | | | | | | |
| | *Lavandin abrialis* | *stoechas* | | *dentata* | *Lavandula bipinnata* | *gibsoni* | *canarien-sis* | *multifida* | |
| | France | Turkey | Greece (b) | Algeria | India (a) | India (b) | Australia (c) | Portugal (a) | (b) |
|---|---|---|---|---|---|---|---|---|---|
| tricyclene | 0.03 | - | 0.20 | 0.40 | - | - | - | - | - |
| *α*-thujene | - | - | - | trace | - | - | - | - | - |
| *α*-pinene | 0.40 | 1.31 | 2.52 | trace | 1.01 | 1.47 | trace | 0.80 | 0.60 |
| camphene | 0.30 | 1.40 | 1.30 | - | 0.35 | - | - | - | - |
| *β*-phellandrene | - | - | 0.10 | - | - | - | - | - | - |
| *β*-pinene | 0.30 | - | 1.30 | 0.20 | - | - | trace | - | - |
| octen-3-ol | 0.30 | - | - | - | - | 2.20 | - | 0.60 | 0.50 |
| 3-octanone | 1.00 | - | - | - | - | - | - | 0.40 | 0.30 |
| myrcene | 0.30 | - | 0.10 | trace | 0.25 | 2.78 | trace | 5.70 | 5.50 |

**Table 2.** *Cont.*

| Name | Place and Area (%) ± SD | | | | | | | | |
|---|---|---|---|---|---|---|---|---|---|
| | *Lavandin abrialis* | *stoechas* | | *dentata* | *Lavandula bipinnata* | *gibsoni* | *canarien-sis* | *multifida* | |
| | France | Turkey | Greece (b) | Algeria | India (a) | India (b) | Australia (c) | Portugal (a) | (b) |
| 3-carene | 0.02 | - | - | - | 0.37 | 1.52 | - | 0.50 | 0.50 |
| sabinene | 0.10 | - | - | 1.40 | 0.39 | - | - | - | - |
| *p*-cymene | 0.04 | - | 4.90 | - | - | 0.82 | trace | 0.30 | 0.20 |
| *o*-cymene | - | - | - | - | - | - | - | - | - |
| limonene | 0.70 | - | - | - | - | 2.30 | trace | 0.60 | 0.30 |
| eucalyptol | - | 8.03 | 16.30 | 38.40 | - | - | - | - | - |
| (*E*)-*β*-ocimene | 2.60 | - | trace | 0.10 | - | trace | trace | 27.40 | 27.00 |
| (*Z*)-*β*-ocimene | 3.00 | - | - | - | - | 0.30 | 0.60 | 1.70 | 1.50 |
| *trans*-linalool oxide | 0.20 | - | - | - | - | - | - | - | - |
| *β*-citral | | | | | | - | 0.30 | - | - |
| *cis*-linalool oxide | 0.1 | - | - | - | - | - | - | - | - |
| *trans*-sabinene hydrate | - | - | - | 0.10 | - | - | - | - | - |
| linalool | 35.00 | 0.29 | 1.20 | trace | 0.94 | 2.65 | 0.90 | 0.30 | 0.20 |
| *cis-p*-menth-2-en-1-ol | - | - | - | trace | - | - | - | - | - |
| *trans*-pinocarveol | - | - | - | - | - | - | - | - | - |
| camphor | 8.90 | 18.18 | 9.90 | 1.60 | 7.09 | - | - | - | - |
| myrtenol | - | - | 1.10 | 1.70 | - | - | - | - | - |
| borneol | 2.90 | - | 0.90 | - | - | - | - | - | - |
| *p*-cymen-8-ol | - | - | - | 3.80 | - | - | 0.20 | - | - |
| lavandulol | 0.60 | - | 0.40 | - | 0.38 | - | - | - | - |
| terpinen-4-ol | - | - | 0.80 | 0.70 | 0.73 | - | - | - | - |
| sabine ketone | - | - | - | 0.50 | - | - | - | - | - |
| *m*-cymen-8-ol | - | - | 0.20 | - | - | - | - | - | - |
| *α*-terpineol | 0.50 | - | 0.50 | 1.80 | - | 0.77 | 0.30 | - | - |
| hexyl butyrate | - | - | - | - | - | - | - | - | - |
| isobornyl formate | - | - | - | - | - | - | - | - | - |
| geraniol | - | - | - | - | - | - | - | - | - |
| cumin aldehyde | - | - | - | 1.10 | - | - | - | - | - |
| carvone | - | - | 0.10 | - | - | - | - | - | - |
| linalyl acetate | 27.00 | - | 0.20 | trace | 3.37 | - | - | - | - |
| *α*-bisabolol | - | - | trace | - | - | - | 1.50 | 0.20 | 0.20 |
| dihydrocarveol | - | - | - | - | - | - | - | - | - |
| bornyl acetate | - | 1.32 | trace | - | 0.21 | - | - | - | - |
| lavendulyl acetate | 1.00 | - | 3.21 | - | 1.79 | - | - | - | - |
| verbenone | - | - | 0.60 | 0.40 | - | - | - | - | - |
| neryl acetate | 0.70 | - | - | - | - | - | - | - | - |
| *β*-bourbonene | - | - | - | trace | - | - | trace | - | - |
| *α-trans*-bergamotene | - | - | - | trace | - | - | - | - | - |
| *α*-cederene | - | - | - | trace | - | - | - | - | - |
| *β*-caryophellene | - | - | - | - | 0.18 | - | 7.60 | 0.80 | 0.90 |
| caryophyllene | 0.70 | - | - | 0.50 | - | trace | - | - | - |
| *α*-santalene | 0.20 | - | - | - | - | - | - | - | - |
| *α-cis*-bergamotene | - | - | - | - | - | - | - | - | - |
| *β*-farnesene | 0.30 | - | - | - | - | - | 0.30 | trace | 0.10 |
| germacene D | - | - | 0.10 | - | 1.66 | - | 2.20 | 0.50 | 0.30 |
| *β*-bisabolene | - | - | | | - | | 20.80 | 5.60 | 5.00 |
| *γ*-cadinene | trace | 0.80 | 0.20 | 0.40 | - | - | - | 0.20 | 0.20 |
| caryophyllene oxide | 0.30 | 0.33 | 0.90 | - | 3.68 | 1.21 | 2.00 | 0.30 | 0.20 |
| spathulenol | - | - | - | 0.30 | - | - | 2.20 | 0.60 | 0.80 |
| cubenol | - | - | trace | - | - | - | - | - | - |
| *γ*-muurolol | - | - | - | - | - | - | - | - | - |
| *τ*-cadinol | - | - | 4.20 | - | - | - | - | 0.20 | 0.30 |
| *γ*-terpinene | trace | - | 0.20 | 0.50 | - | - | - | 0.20 | 0.10 |
| octen-3-yl acetate | 0.03 | - | - | - | - | - | - | - | - |
| norborneol acetate | - | - | - | - | - | - | - | - | - |

France—southern France [11]; Turkey—district of Alahan (Hatay) [17]; Greece (b)—north part of Greece at Chalkidiki peninsula [18]; Algeria—Cherchel (northwest of Algiers region, Algeria) [12]; India (a)—Asangihal village in Sindagi taluk of Bijapur district, India [19]; India (b)—Purandar Fort region [20]; Australia (c)—Randwick, Sydney (Australia) [21]; Portugal (a)—region Sesimbra/Arrábida, south of Portugal [22]; Portugal (b)—region Mértola, south of Portugal [17].

Because each lavender oil has a quantitatively and qualitatively distinct profile of chemical compounds, it is necessary to determine the quantity and identity of its individual components (Tables 1 and 2). These data will allow researchers to determine the effects of essential lavender oils on the autochthonous microbiota of the skin. From the literature [23,24], it is known that the chemical composition of oils and their macerates has a decisive influence on their microbiological properties.

Lavender essential oil contains several anti-microbial compounds, such as eucalyptol, linalool, terpinen-4-ol, and α-terpineol. Among them, linalool was demonstrated to be the strongest active ingredient against a wide range of microorganisms [8]. Borneol and eucalyptol were identified also as the main compounds in the many essential oils exhibiting anti-parasitic activity [10]. Terpinen-4-ol, α-pinene, β-pinene, 1,8-cineol, linalool, and 4-terpineol also showed high anti-fungal activity against Gram-positive and Gram-negative strains [25,26]. Linalool and linalyl acetate have local anaesthetic effects, proven in animal tests (in vivo and in vitro) [25]. Various monoterpenoids, such as α-terpineol, terpinen-4-ol, eucalyptol, and linalool, have antiviral activity against influenza strains [26]. Eucalyptol, terpinen-4-ol, thymol, and carvacrol also have extensive anti-inflammatory effects [26].

Composition of the facial skin microbiota varies and depends on many factors, such as proper hygiene, state of health, oiliness, skin hydration, pH, local temperature, and the reduction potential [27]. The skin microbiota contains persistent indigenous microorganisms (so-called residents) that are located on its surface for almost the entire lifespan of the individual, including transitory microorganisms from the environment, animals, food, or water. The microbiota of adult skin [27–29] is mainly formed by Gram-positive cocci (*Staphylococcus epidermidis, S. haemolyticus, S. hominis, S. aureus* (carrier), *Enterococcus faecalis, Micrococcus* spp., and *Streptococcus*), Gram-positive bacilli (*Corynebacterium* spp., *Propionibacterium acnes, P. granulosum, P. avidum,* and *Bacillus* spp.), Gram-negative bacilli (*Acinetobacter* spp. and *Escherichia coli*), and yeast-like fungi (*Pityrosporum ovale* and *Candida* spp.). Cosmetics containing active anti-bacterial substances of natural origin (essential oils) help to control the growth of microorganisms and additionally have a beneficial effect on the processes taking place on the surface and in the skin. Essential oils accelerate the regeneration and development of skin cells, and, for this reason, the skin becomes stronger and regenerates faster by supporting the processes of granulation of a wounded epidermis [30]. As a result, skin ageing processes are delayed.

The chemical composition and anti-microbial activity of different lavender oils depend on the geographic region of origin, which is why the aim of the present study was to analyse the correlation of the chemical composition of two lavender oils of different origins with their anti-microbial effects on the mixed microbiota of facial skin and on dominant bacterial isolates extracted from the surface of facial skin. The chemical composition and activity of essential lavender oil from the Crimean Peninsula are yet to be published.

## 2. Results

### 2.1. Chemical Analysis

A complex chromatogram was obtained as a result of this analysis, where we identified 101 compounds. Among these identified components, there were 64 oil compounds that were already described in the literature and 37 other compounds, which are yet to be reported (Table 3).

Based on the results, it was found that Crimean lavender oil contains several-fold larger quantities of monoterpenes such as 3-carene, *o*-cymene, or bicyclic sesquiterpenes (bergamotene isomers, caryophyllene, or γ-cadinene) as compared with the literature data. In the extract of Crimean lavender, *p*-cymene-1-ol, 3-octanone, and terpenes were detected at much smaller concentrations, e.g., sabinene or ocimene and sesquiterpene germacrene D.

**Table 3.** The components of ETJA and Crimean lavender oils.

| Name | Abbreviation | RI | | Oil, Column and Area (%) ± SD | | |
|---|---|---|---|---|---|---|
| | | Literat. | Exper. | ETJA ZB-5HT | Crimean HP-5MS | SupelcoWAX |
| tricyclene | B MO | 923 | 920 | - | 0.04 ± 0.00 | 0.01 ± 0.00 |
| $\alpha$-thujene | B M | 928 | 928 | - | 0.18 ± 0.01 | 0.04 ± 0.00 |
| $\alpha$-pinene | B M | 936 | 933 | 0.7 ± 0.01 | 0.36 ± 0.01 | - |
| camphene | B M | 950 | 947 | 0.31 ± 0.02 | 0.27 ± 0.01 | 0.05 ± 0.01 |
| $\beta$-phellandrene | M M | 973 | 973 | 0.09 ± 0.00 | 0.09 ± 0.01 | - |
| $\beta$-pinene | B M | 978 | 974 | 0.22 ± 0.01 | 0.11 ± 0.01 | - |
| octen-3-ol | OT | 980 | 983 | 0.04 ± 0.01 | 0.61 ± 0.01 | 0.16 ± 0.01 |
| 3-octanone | OT | 985 | 988 | - | 0.10 ± 0.01 | 0.01 ± 0.01 |
| $\beta$-myrcene | A M | 989 | 991 | 0.38 ± 0.02 | 0.25 ± 0.01 | 0.08 ± 0.00 |
| 3-carene | B M | 1011 | 1005 | 0.36 ± 0.01 | 0.86 ± 0.01 | 0.20 ± 0.01 |
| sabinene | B M | 1004 | 1009 | - | 0.02 ± 0.01 | 0.02 ± 0.01 |
| $p$-cymene | M M | 1024 | 1020 | 0.58 ± 0.01 | 0.22 ± 0.02 | - |
| $o$-cymene | M M | 1041 | 1022 | | 1.03 ± 0.06 | 0.18 ± 0.00 |
| limonene | M M | 1029 | 1026 | 19.02 ± 0.07 | 0.55 ± 0.02 | 0.15 ± 0.00 |
| eucalyptol | B MO | 1031 | 1027 | - | 5.00 ± 0.10 | 1.66 ± 0.02 |
| (E)-$\beta$-ocimene | A M | 1048 | 1040 | 0.02 ± 0.01 | 0.99 ± 0.01 | 0.44 ± 0.01 |
| (Z)-$\beta$-ocimene | A M | 1037 | 1049 | | 0.75 ± 0.02 | 0.41 ± 0.01 |
| *trans*-linalool oxide | A MO | 1083 | 1070 | 0.05 ± 0.01 | 0.53 ± 0.03 | 0.08 ± 0.00 |
| $\beta$-citral | A M | 1245 | 1084 | - | 0.08 ± 0.01 | - |
| *cis*-linalool oxide | A MO | 1075 | 1091 | 0.15 ± 0.01 | 0.52 ± 0.02 | 0.07 ± 0.01 |
| *trans*-sabinene hydrate | B MO | 1098 | 1097 | - | 0.10 ± 0.04 | 0.04 ± 0.00 |
| linalool | A MO | 1099 | 1105 | 41.84 ± 0.10 | 34.13 ± 0.25 | 52.71 ± 0.33 |
| *cis*-$p$-menth-2-en-1-ol | M MO | 1123 | 1116 | - | 0.01 ± 0.00 | - |
| *trans*-pinocarveol | B MO | 1140 | 1135 | - | 0.04 ± 0.01 | - |
| camphor | B MO | 1143 | 1141 | 0.15 ± 0.01 | 0.54 ± 0.01 | 0.09 ± 0.01 |
| myrtenol | B MO | 1150 | 1146 | 0.19 ± 0.01 | 0.08 ± 0.04 | - |
| borneol | B MO | 1166 | 1168 | - | 1.50 ± 0.01 | - |
| $p$-cymen-8-ol | B MO | 1184 | 1171 | - | 0.04 ± 0.01 | 0.03 ± 0.00 |
| lavandulol | A MO | 1168 | 1175 | 0.18 ± 0.01 | 0.54 ± 0.01 | 0.11 ± 0.02 |
| terpinen-4-ol | M MO | 1177 | 1181 | 0.29 ± 0.01 | 6,66 ± 0.04 | 2.29 ± 0.02 |
| sabine ketone | B MO | 1194 | 1190 | - | 0.50 ± 0.02 | 0.08 ± 0.00 |
| $m$-cymen-8-ol | B MO | 1180 | 1192 | 0.03 ± 0.01 | 0.16 ± 0.01 | 0.03 ± 0.00 |
| $\alpha$-terpineol | M MO | 1190 | 1197 | 0.07 ± 0.00 | 1.54 ± 0.03 | - |
| hexyl butyrate | OT | 1191 | 1203 | - | 0.63 ± 0.02 | - |
| isobornyl formate | B MO | 1240 | 1230 | - | 0.13 ± 0.01 | 0.61 ± 0.01 |
| geraniol | A MO | 1255 | 1234 | 0.02 ± 0.01 | 0.08 ± 0.01 | 0.03 ± 0.01 |
| cumin aldehyde | M MO | 1238 | 1242 | - | 0.18 ± 0.00 | 0.03 ± 0.00 |
| carvone | M MO | 1242 | 1245 | 0.04 ± 0.01 | 0.05 ± 0.01 | 0.01 ± 0.00 |
| linalyl acetate | A MO | 1255 | 1259 | 32.70 ± 0.08 | 23.29 ± 0.30 | 36.56 ± 0.34 |
| $\alpha$-bisabolol | M SO | 1282 | 1266 | - | 0.03 ± 0.01 | - |
| dihydrocarveol | M MO | 1194 | 1277 | - | 0.05 ± 0.07 | - |
| bornyl acetate | B MO | 1283 | 1278 | - | 0.15 ± 0.01 | 0.04 ± 0.01 |
| lavendulyl acetate | A MO | 1289 | 1285 | 0.06 ± 0.01 | 2.45 ± 0.02 | 0.51 ± 0.01 |
| verbenone | B MO | 1206 | 1296 | - | 0.04 ± 0.01 | - |
| neryl acetate | A MO | 1363 | 1359 | 0.39 ± 0.02 | 0.22 ± 0.01 | 0.05 ± 0.01 |
| $\beta$-bourbonene | B S | 1384 | 1373 | - | 0.05 ± 0.00 | 0.01 ± 0.00 |
| $\alpha$-*trans*-bergamotene | B S | 1434 | 1382 | - | 0.17 ± 0.01 | - |
| $\alpha$-cedrene | B S | 1412 | 1398 | - | 0.06 ± 0.00 | - |
| $\beta$-caryophellene | B S | 1406 | 1402 | - | 0.11 ± 0.10 | - |
| caryophyllene | B S | 1420 | 1408 | 0.5 ± 0.01 | 4.19 ± 0.10 | 1.63 ± 0.02 |
| $\alpha$-santalene | B S | 1421 | 1411 | - | 1.05 ± 0.03 | 0.23 ± 0.01 |
| $\alpha$-*cis*-bergamotene | B S | 1414 | 1429 | - | 0.28 ± 0.00 | 0.02 ± 0.00 |
| $\beta$-farnesene | A S | 1456 | 1454 | 0.1 ± 0.01 | 0.96 ± 0.01 | 0.28 ± 0.02 |
| germacene D | A S | 1481 | 1496 | - | 0.03 ± 0.02 | - |
| $\beta$-bisabolene | M S | 1508 | 1504 | - | 0,02 ± 0.01 | - |
| $\gamma$-cadinene | B S | 1513 | 1507 | - | 0.53 ± 0.01 | - |
| caryophyllene oxide | B SO | 1581 | 1572 | - | 2.59 ± 0.01 | 0.29 ± 0.00 |
| spathulenol | B SO | 1576 | 1601 | - | 0.02 ± 0.00 | - |

**Table 3.** *Cont.*

| Name | Abbreviation | RI | | Oil, Column and Area (%) ± SD | | |
| --- | --- | --- | --- | --- | --- | --- |
| | | Literat. | Exper. | ETJA ZB-5HT | Crimean HP-5MS | SupelcoWAX |
| cubenol | B SO | 1636 | 1603 | - | 0.04 ± 0.01 | - |
| γ-muurolol | B S | 1645 | 1622 | - | 0.02 ± 0.01 | 0.07 ± 0.00 |
| τ-cadinol | B SO | 1640 | 1628 | - | 0.25 ± 0.01 | 0.03 ± 0.01 |
| γ-terpinene | M M | 1060 | 1060 | 0.08 ± 0.01 | - | 0.02 ± 0.00 |
| octen-3-yl acetate | OT | 1110 | 1091 | - | - | 0.21 ± 0.01 |
| norborneol acetate | B MO | 1114 | 1129 | - | - | 0.08 ± 0.00 |
| α-terpinene | M M | 1017 | 1016 | 0.53 ± 0.01 | - | - |
| **Not described in literature** | | | | | | |
| *m*-cymene | M M | 999 | 970 | - | 0.03 ± 0.01 | - |
| butanoic acid, butyl ester | OT | 990 | 997 | - | 0.15 ± 0.01 | 0.13 ± 0.01 |
| acetic acid, hexyl ester | OT | 1004 | 1016 | - | 0.22 ± 0.00 | 0.07 ± 0.00 |
| bicyclo[3.1.0]hexan-2-ol | B MO | | 1064 | - | 0.20 ± 0.01 | - |
| isopulegone | M MO | 1157 | 1068 | - | 0.01 ± 0.00 | - |
| eucarvone | M MO | 1048 | 1077 | - | 0.01 ± 0.00 | - |
| 2-carene | B M | 1006 | 1079 | 0.02 ± 0.0 | 0.01 ± 0.01 | - |
| 6-camphenol | B MO | 1110 | 1081 | - | 0.01 ± 0.00 | - |
| cinerone | M MO | 1084 | 1088 | - | 0.04 ± 0.01 | - |
| octen-1-ol acetate | OT | 1192 | 1112 | - | 1.06 ± 0.04 | - |
| isopulegol isomer | M MO | 1146 | 1139 | - | 0.03 ± 0.01 | - |
| butanoic acid, hexyl ester | OT | 1190 | 1156 | - | 0.11 ± 0.01 | 0.02 ± 0.01 |
| Z-farnesol | A SO | | 1160 | - | 0.02 ± 0.01 | - |
| 3,7 octadiene-2,6 diol 2,6-dimethyl isomer | A MO | 1189 | 1200 | - | 0.22 ± 0.01 | 0.02 ± 0.00 |
| isopulegol isomer | M MO | 1156 | 1214 | - | 0.02 ± 0.01 | - |
| *p*-menthane-1,2,3-triol | M MO | | 1251 | - | 0.16 ± 0.01 | - |
| E-farnesol | A SO | | 1269 | - | 0.02 ± 0.01 | - |
| α-limonene diepoxide | M MO | 1297 | 1271 | - | 0.13 ± 0.01 | - |
| 3,7-octadiene-2,6-diol-2,6-dimethyl isomer | A MO | 1229 | 1332 | - | 0.12 ± 0.01 | - |
| hydroxy linalool | A MO | 1367 | 1345 | - | 0.15 ± 0.01 | - |
| limonene oxide | M MO | 1336 | 1348 | - | 0.15 ± 0.00 | - |
| epicubenol | B SO | 1627 | 1364 | - | 0.03 ± 0.00 | - |
| linalool formate | A MO | 1219 | 1379 | - | 0.27 ± 0.01 | - |
| nerolidyl acetate | A SO | 1687 | 1385 | - | 0.01 ± 0.00 | - |
| β-cedrene | B S | 1419 | 1439 | - | 0.09 ± 0.01 | - |
| epi-β-santalene | B S | 1431 | 1441 | - | 0.04 ± 0.01 | 0.01 ± 0.00 |
| santalol | B SO | 1617 | 1442 | - | 0.01 ± 0.00 | - |
| humulene | B S | 1455 | 1444 | - | 0.13 ± 0.01 | 0.03 ± 0.00 |
| limonen-6-ol pivalate | M SO | | 1452 | - | 0.13 ± 0.01 | - |
| β-cubenene | B S | 1527 | 1473 | - | 0.09 ± 0.00 | |
| β-carryophyllene isomer | B S | 1455 | 1479 | - | 0.10 ± 0.02 | - |
| α-santanol | B SO | 1671 | 1514 | - | 0.06 ± 0.01 | - |
| calamenene | B S | 1543 | 1515 | - | 0.03 ± 0.01 | - |
| tricyclo[7.2.0.0(2,6)]undecan-5-ol,2,6,10-tetramethyl | B SO | | 1548 | - | 0.07 ± 0.01 | - |
| zingiberene | M S | 1495 | 1495 | - | - | 0.01 ± 0.00 |
| benzeneethanol | OT | 1121 | 1120 | - | | 0.06 ± 0.01 |
| cyclofenchene | B M | 882 | 892 | 0.34 ± 0.01 | - | - |
| β-copaene | B M | 1433 | 1433 | 0.04 ± 0.01 | - | - |
| 2-bornanone | B MO | 1143 | 1136 | 0.05 ± 0.01 | - | - |
| isoborneol | B MO | 1158 | 1145 | 0.46 ± 0.01 | - | - |

Abbreviations: A M—aliphatic monoterpenes; M M—monocyclic monoterpenes; B M—bi- and tricyclic monoterpenes; A MO—aliphatic monoterpenoids; M MO—monocyclic monoterpenoids; B MO—bi- and tricyclic monoterpenoids; A S—aliphatic sesquiterpenes; M S—monocyclic sesquiterpenes; B S—bi- and tricyclic sesquiterpenes; A SO—aliphatic sesquiterpenoids; M SO—monocyclic sesquiterpenoids; B SO—bi- and tricyclic sesquiterpenoids. SD—standard deviation; RI—retention indexes; literat.—literature data; exper.—determined experimentally for the non-polar columns: HP-5MS for CRIMEA oil and ZB-5HT for ETJA oil.

Analysis of hexane solutions of lavender essential oil was performed too (on the SupelcoWAX column), and the average retention parameters and peak areas are listed in Table 3. Less complex chromatograms and worse separation of the sample components were obtained in this assay; specifically, the peaks corresponding to the main components of oil were close to one another and were not completely separated.

GC–MS analysis on the SupelcoWAX column allowed us to identify 50 compounds contained in Crimean lavender essential oil. Among these compounds, 42 were already described in the literature. Three components of lavender oil—which were already described in the literature—could not be identified by means of the HP-5MS column but could be identified using the SupelcoWAX column. These compounds included $\gamma$-terpinene, octene-3-yl acetate, and norborneol acetate.

Furthermore, eight previously undescribed components were identified, including the five already identified via the HP-5MS column. In addition, the presence of three small peaks corresponding to benzoic acid, butyric acid, zingiberene, and 2-phenylethanol was detected.

Moreover, sharp, clear-cut, and completely separated peaks of limonene (1) and eucalyptol (2) for 1:10 dilutions were obtained using this column. They could not be analysed by means of the HP-5MS column. In the case of samples with the dilution of 1:10, overlapping limonene (1) and eucalyptol (2) peaks were observed. Some improvement of separation of these components of lavender oil was achieved by greater dilution, but only the use of polar columns yielded satisfactory results. The comparison of separation of these two terpenes at different dilutions on both columns is shown in Figure 1.

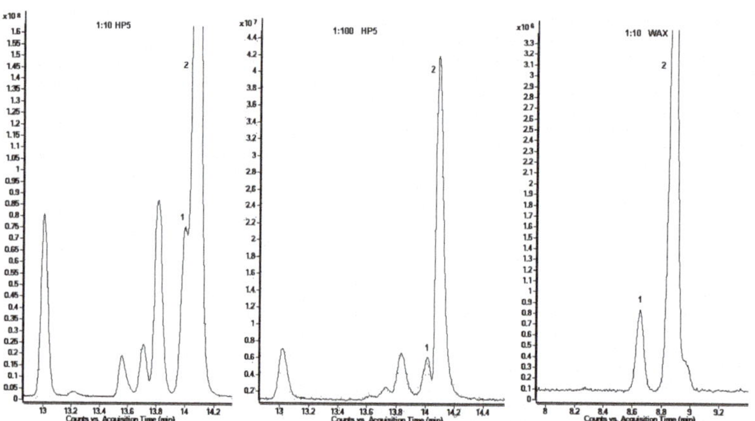

**Figure 1.** Comparison of limonene (1) and eucalyptol (2) peak separation, between the HP-5MS column at a dilution of 1:10 (*v/v*) or 1:100 (*v/v*) and the SupelcoWAX$^{\mathrm{TM}}$ 10 column at dilution 1:10 (*v/v*).

From the presented data, it can be concluded that the separation of Crimean lavender oil on the polar column was not satisfactory; better results were obtained on the non-polar column. Therefore, only this column was employed to determine the chemical composition of ETJA lavender oil.

On the basis of the conducted studies, it was demonstrated that the tested lavender oils differ in chemical composition and anti-microbial activity both quantitatively and qualitatively. In ETJA oil (Table 3), 33 components were identified, including 28 already described in the literature, as well as five compounds not described previously (2-carene, cyclofenchene, $\beta$-copaene, 2-bornanone, and isoborneol).

ETJA lavender oil turned out to contain higher concentrations of linalool (41.8%), linalyl acetate (32.7%), and limonene (19.0%), whereas Crimean lavender oil contained linalool (34.1% according to HP-5MS column analysis or 52.7% in accordance with the SupelcoWAX column analysis), linalyl acetate (23.3% according to the HP-5MS column analysis or 36.6% according to the SupelcoWAX column analysis), and eucalyptol (5.0% in accordance with the HP-5MS column analysis or 1.7% judging by the SupelcoWAX column analysis; Table 3). ETJA lavender oil was composed mainly of monoterpenoids (76.7%) and monoterpenes (22.7%), whereas Crimean lavender oil was found to be composed mainly of monoterpenoids (80.1% according to the HP-5MS column analysis or 95.1% in accordance with the SupelcoWAX column analysis), much less monoterpenes (5.8% according to the HP-5MS column analysis or 1.6% judging by the SupelcoWAX column analysis), and some sesquiterpenes (8.0% in accordance with the HP-5MS column analysis or 2.3% judging by the SupelcoWAX column analysis; Table 4).

**Table 4.** The list of terpenes in the tested lavender oils.

| | Abbreviation | Oil, Column and Area (%) | | |
| --- | --- | --- | --- | --- |
| | | ETJA | Crimean | |
| | | ZB-5HT | HP-5MS | SupelcoWAX |
| Aliphatic monoterpenes | A M | 0.40 | 2.07 | 0.93 |
| Monocyclic monoterpenes | M M | 20.30 | 1.92 | 0.35 |
| Bi- and tricyclic monoterpenes | B M | 1.99 | 1.81 | 0.31 |
| **Monoterpenes** | **M** | **22.69** | **5.80** | **1.59** |
| Aliphatic monoterpenoids | A MO | 75.39 | 62.52 | 90.14 |
| Monocyclic monoterpenoids | M MO | 0.40 | 9.04 | 2.33 |
| Bi- and tricyclic monoterpenoids | B MO | 0.88 | 8.53 | 2.67 |
| **Monoterpenoids** | **MO** | **76.67** | **80.09** | **95.14** |
| Aliphatic sesquiterpenes | A S | 0.10 | 0.99 | 0.28 |
| Monocyclic sesquiterpenes | M S | - | 0.02 | 0.01 |
| Bi- and tricyclic sesquiterpenes | B S | 0.50 | 6.94 | 2.00 |
| **Sesquiterpenes** | **S** | **0.60** | **7.95** | **2.29** |
| Aliphatic sesquiterpenoids | A SO | - | 0.05 | - |
| Monocyclic sesquiterpenoids | M SO | - | 0.16 | - |
| Bi- and tricyclic sesquiterpenoids | B SO | - | 3.07 | 0.32 |
| **Sesquiterpenoids** | **SO** | **-** | **3.28** | **0.32** |
| **Others** | **OT** | **0.04** | **2.88** | **0.66** |

### 2.2. Biological Analysis

The effect of lavender oils on the mixed microbiota of the face skin without signs of lesions depended on the origin of the oil and the concentration used. ETJA lavender oil at all concentrations tested reduced the number of skin microbial cells 1000–10,000-fold, compared to the control. The strongest microbial cell number reduction was observed after application of 70 $\mu$L/cm$^3$ oil and slightly less at 50 $\mu$L/cm$^3$ (Figure 2). On the other hand, Crimean lavender oil exerted much weaker anti-microbial activity, and only at the highest concentration did it suppress the growth of the microbiota hundred-fold (Figure 2).

Lavender oils, depending on their origin, also had a different influence on the qualitative changes in the facial skin microbiota. In the presence of the highest concentration of ETJA lavender oil tested, bacteria of the following species survived: *Micrococcus luteus*, *E. coli*, *Staphylococcus warneri*, and *Enterococcus faecium*. Crimean lavender oil, however, did not inhibit the growth of *Bacillus* (*B. cereus*, *B. subtilis*, and *B. mycoides*), *Corynebacterium* spp., *E. faecium*, and *S. warneri*. The most sensitive to ETJA lavender oil were Gram-positive bacilli, and Gram-negative bacilli were the most sensitive to Crimean lavender oil. On the other hand, none of the tested oils inhibited the growth of Gram-positive cocci.

Therefore, an attempt was made to determine those oil concentrations which would effectively decrease the growth of individual isolates. Inhibitory effects on the growth of bacterial isolates that survived in a mixed microbial population from facial skin were exerted by the tested oils only at concentrations between 40 and 80 $\mu L/cm^3$ and growth inhibition zones between 12.5 and 44 mm, with higher effectiveness of oils at the highest concentrations. ETJA lavender oil was more effective because it limited the growth of most bacteria under study, including *Bacillus.* Neither of the oils tested inhibited the growth of *E. faecium* (Table 5, Figure 3).

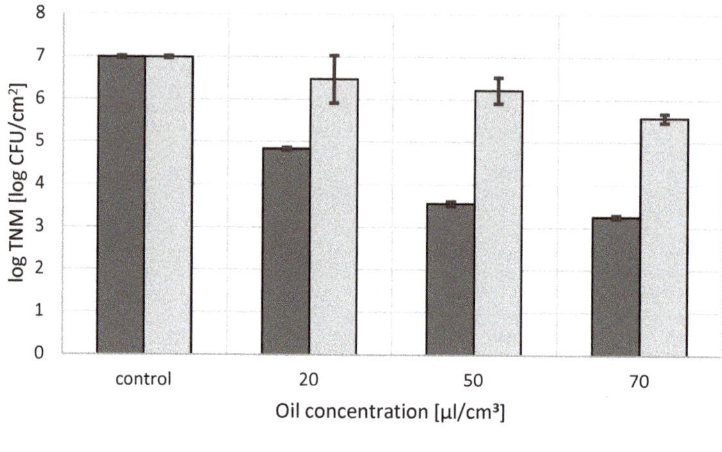


**Figure 2.** The influence of concentrations of the lavender oils under study on the number of microbiota cells from facial skin.

A.                                                                                          B.

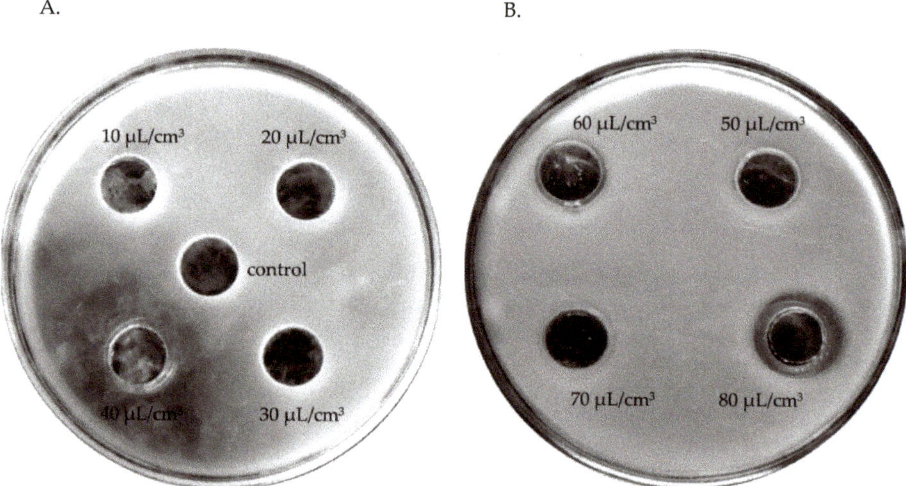


**Figure 3.** Zones of growth inhibition of a *Bacillus cereus* isolate in the presence of tested concentrations (10–80 $\mu L/cm^3$) of lavender oils: (**A**) ETJA, 40 $\mu L/cm^3$; (**B**) Crimean lavender oil, 80 $\mu L/cm^3$.

At lower concentrations (10–40 $\mu L/cm^3$), lavender oils manifested neutral effects (Figure 4) or, as in the case of Crimean lavender oil, stimulated bacterial growth (Figure 5).

**Figure 4.** Neutral effects of lavender oils on the growth of the bacterial species *Enterococcus faecium*.

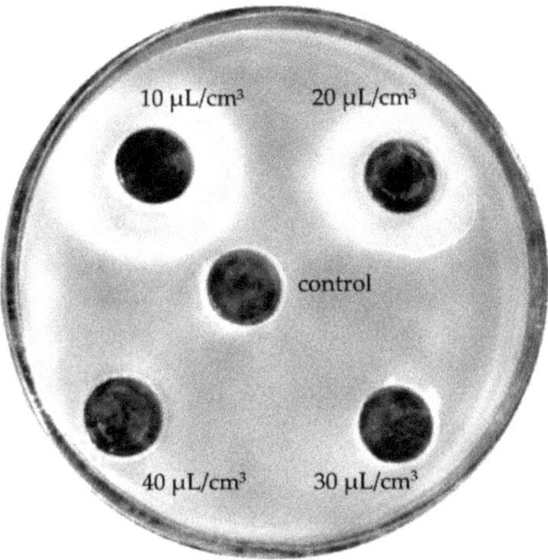

**Figure 5.** Stimulation of *Bacillus mycoides* growth in the presence of a low concentration (10 or 20 µL/cm$^3$) of Crimean lavender oil.

**Table 5.** Zones of growth inhibition of dominant bacterial isolates.

| Species of Isolates | Oil Concentration (µL/cm$^3$) | Zones of Inhibition (mm) ± SD | |
|---|---|---|---|
| | | ETJA | Crimean |
| *Bacillus cereus* | 40 | 47.0 ± 4.2 | 0 |
| | 80 | 40.0 ± 3.5 | 19.5 ± 0.7 |
| *Bacillus* subtilis | 80 | 23.5 ± 0.7 | 0 |
| *Bacillus mycoides* | 60 | 33.0 ± 1.4 | 0 |
| *Staphylococcus warneri* | 80 | 22.8 ± 0.4 | 0 |
| *Micrococcus luteus* | 80 | 19.0 ± 1.4 | 12.5 ± 0.7 |
| *Enterococcus faecium* | 80 | 0 | 0 |
| *Corynebacterium* spp. | 50 | 18.5 ± 0.7 | 13.0 ± 1.4 |
| *Escherichia coli* | 80 | 0 | 16.0 ± 1.4 |

To sum up, the tested lavender oils reduced the mixed population of microbes from facial skin, but ETJA lavender oil with higher amounts of monoterpenoids (linalool and linalyl acetate) and monoterpenes (limonene) was characterised by higher effectiveness than Crimean lavender oil.

## 3. Discussion

Shortly after birth, the human skin is immediately colonised by microorganisms, such as bacteria, yeasts, fungi, and viruses. Nonetheless, the composition of the skin microbiota varies quantitatively and qualitatively, and it depends on humidity, temperature, pH, and body area [27]. A child's skin is mainly colonised by bacteria of genus *Staphylococcus*, *Enterococcus*, *Corynebacterium*, and *Escherichia*, and, in the teenage period, by *Sarcina*. At an elderly age, however, an increase in the number of fungal cells is observed, mainly *Candida albicans* yeast [27–29].

The challenge in skincare is oily skin because it has to be properly cleaned and moisturised, but comedogenic agents (which block sebaceous glands, resulting in blackheads) cannot be used [27,31,32]; therefore, biological substances effective at low concentrations are sought for care for this type of skin. Lavender oil is a strong antiseptic. Therefore, it is an additive to pharmaceuticals (salve and lotions for hard-to-heal wounds, eczema, and anti-rheumatic preparations), as well as cosmetics. It is used in mouth, throat, upper respiratory tract, and lung infectious diseases, as well as in dermatology to treat difficult-to-heal wounds, ulcers, and burns, and in cosmetology. However, the anti-microbial effect of lavender oil depends on the species and the variety of lavender from which it is obtained. Cavanagh and Wilkinson [33] and Sienkiewicz et al. [34] showed that the anti-microbial activity of essential oils depends on their chemical composition. According to literature data, *Lavandula angustifolia* oil has the most variable chemical composition. Bulgarian lavender oil contains ocimene (6.8–7.7%), linalool (30–34%), and linalyl acetate (35–38%), while it does not contain lavandulol and lavandulol acetate. The main ingredients in oils from China and India were linalool, linalool acetate, and lavandulol, all found in various amounts; however, ocimene was not identified [14,35]. Adaszyńska et al. [36] showed that the highest content of linalool was found in essential oils from the variety "Lavender Lady" and "Elegance Purple" (23.9% and 22.4%). At the same time, these oils contained small amounts of *cis*-β-ocimene. The best anti-bacterial properties against *S. aureus* and *Pseudomonas aeruginosa* were found in oils obtained from varieties "Blue River" and "Munstead". Essential oil obtained from *Lavandula angustifolia* Mill. has strong bactericidal properties against methicillin-resistant *Staphylococcus aureus* (MRSA) and vancomycin-resistant *Enterococcus* sp. (VRE) [33]. Essential oil from *Lavandula heterophylla* "Avonview" inhibits growth of *Streptococcus pyogenes*, *Enterobacter aerogenes*, *Staphylococcus aureus* MRSA, *Pseudomonas aeruginosa*, *Citrobacter freundii*, *Proteus vulgaris*, *Escherichia coli* VRE, *Shigella sonnei*, and *Propionibacterium acnes* [33,37]. A suitable substance may be lavender oil, which, according to the manufacturer, can be added even at a concentration of ~5% to cosmetic preparations and pharmaceuticals. At this concentration, lavender oil has a strong effect on some types of skin without irritation; however, more sensitive skin requires preparations with lower lavender oil content, and, during a disease with relevant symptoms, preparations with a higher concentration of lavender oil.

The scientific literature mainly contains the results of studies on the effects of lavender oils (obtained from many varieties of lavender by various research techniques) on selected individual isolates of microorganisms. Sabara and Kunicka-Styczyńska [38] reported that lavender oil (from *Lavandula angustifolia*) at concentrations of 100–200 µL/cm$^3$ inhibited the growth of all tested microorganisms—*E. coli*, *Bacillus subtilis*, *Candida mycoderma*, and *Aspergillus niger*—and the inhibitory effect on *Bacillus* bacteria and *Aspergillus* fungi growth was obtained at a 10-fold lower dose (10 µL/mL). Roller et al. [39], when comparing the anti-microbial efficacy of several lavender oils (from different varieties of the plant), tested them individually, as well as in mixtures, against methicillin-resistant and non-methicillin-resistant *S. aureus* and noted that the best anti-microbial effects were obtained by combining several oils. In other studies, lavender oil at concentrations below 2000 ppm (parts per million) was less active against bacteria of the genera *Bacillus*, *Lactobacillus*, *Clostridium*, and *Bifidobacterium* [40].

The obtained results showed a significant reduction in the number of microbial cells in the mixed population from the skin at the dose of 50 µL/cm$^3$ lavender oil, but the most effective was lavender oil at the concentration of 70 µL/cm$^3$, although no complete inhibition of the growth of the mixed microbiota from the skin was observed.

After application of ETJA lavender oil to the mixed microbiota from the skin, bacteria *M. luteus*, *E. coli*, *S. warneri*, and *E. faecium* survived, whereas, after the application of Crimean lavender oil, *Bacillus* (*B. cereus*, *B. subtilis*, and *B. mycoides*), *Corynebacterium* sp., *E. faecium*, and *S. warneri* survived. Only oils at concentrations between 40 and 80 µL/cm$^3$ inhibited the growth of individual bacterial isolates, whereby oils used at the highest concentrations showed higher effectiveness. ETJA lavender oil inhibited the growth of most bacteria tested, including *Bacillus*, but neither oil inhibited the growth of *E. faecium*.

The most sensitive to ETJA lavender oil were Gram-positive bacilli, and Gram-negative bacilli were the most sensitive to Crimean lavender oil. On the other hand, neither of the tested oils inhibited the growth of Gram-positive cocci.

The essence of the effect of lavender oils on skin microbiota depends on the quantitative and qualitative chemical composition. Essential oils have an affinity for lipid cell structures; therefore, they destroy the cell wall and membranes of bacteria, mainly Gram-positive ones (less often Gram-negative) and fungi, and, as a consequence, there is leakage and coagulation of the cytoplasm. In addition, lavender oils inhibit the synthesis of RNA, DNA, proteins, and polysaccharides, while, in fungi, they act as anti-mycotics and inhibit the production of enzymes [40].

Monoterpenes, especially linalool, have an anti-microbial effect on bacteria. The mechanism of action consists of disturbing the lipid structure of cell membranes and increasing the permeability of these membranes to monoterpenes, which—by penetrating bacterial cells—block their metabolism, thereby leading to cell death [41].

According to the literature, the spectrum of action of lavender oil is broad because it has anti-viral and anti-fungal properties, in addition to bactericidal activity. Studies conducted by Minami et al. [42] revealed that narrow-leaved lavender (*L. latifolia*) spike oil at a concentration of 1% suppressed the replication of a herpes virus in vitro. According to those researchers, this outcome can be explained by the impact of oil components on the areola and virus glycoprotein [42].

Our study indicates that the main components of the tested Crimean lavender oil are similar to those of oils extracted from two species of lavender: *L. angustifolia* from Australia [13] and *Lavandin abrialis* from France [11]. Much greater differences from the literature data were observed here in the components present in the tested oils in quantities not exceeding 1%. Crimean lavender oil is characterised by several-fold higher concentration of terpenes such as 3-carene, *o*-cymene, caryophyllene, and bergamotene as compared to other lavender oils. Nonetheless, sabinene, ocimene, and germacrene D were present in much smaller quantities. In addition, we were able to identify other terpene compounds not yet reported as components of lavender oil. These include terpene alcohols (geraniol, farnesol, santalol, isopulegol, and cubenol), terpene aldehydes (citral), and terpene ketones (verbenone, cinerone, and eucarvone).

## 4. Materials and Methods

### 4.1. Materials

The research experiments consisted of two oils: commercial lavender essential oil (*Lavandula angustifolia* Oil) from ETJA (produced by ETJA, Elbląg, Poland) and Crimean lavender oil (*Lavandula angustifolia* Oil) from lavender grown in the gardens of the Institute of Essential Oil of the Ukrainian Academy of Agricultural Sciences in Simferopol, Crimea, Ukraine. Both oils were obtained by steam distillation.

### 4.2. Gas Chromatography with Mass Spectrometry (GC–MS)

The analysis of Crimean lavender oil was performed at the Institute of Heavy Organic Synthesis "Blachownia" in Kędzierzyn-Koźle on an Agilent Technologies Gas Chromatograph 7890 GC (Agilent, Santa Clara, CA, USA) system coupled with a mass spectrometer, GC–MS 7000—Triple Quad (Agilent, Santa Clara, CA, USA). Two types of capillary columns of different polarity—non-polar HP-5MS (5% diphenyl, 95% dimethylpolysiloxane, Agilent J&W, Palo Alto, CA, USA) and SupelcoWAX™ 10 polar (polyethylene glycol Carbowax®20M, Merck KGaA, Darmstadt, Germany)—were employed; both columns had a length of 30 m, internal diameter of 0.25 mm, and film thickness of 0.25 microns. Helium served as the carrier gas, and its flow rate was 1.5 mL/min. Analyses were performed in the temperature range 45–250 °C; the initial temperature was maintained for 6 min, and the heating rate was 3 °C/min. Samples with a volume of 0.5 mL were prepared by means of an auto-sampler. The gas chromatograph was equipped with a split injector; the split ratio was 100:1. Injector temperature was 250 °C. The test solutions were prepared by diluting an oil sample with *n*-hexane at a volume ratio of 1:10 or 1:100.

ETJA lavender oil was analysed in the Faculty of Chemistry, University of Opole, on a Hewlett Packard HP 6890 series GC system chromatograph (Hewlett Packard, Waldbronn, Germany), which was coupled with a Hewlett Packard 5973 mass selective detector (Hewlett Packard, Waldbronn, Germany). The chromatograph was equipped with the non-polar, high-temperature ZB-5HT capillary column (length, 30 m; inner diameter, 0.32 mm; film thickness, 0.25 μm, Phenomenex Inc., Torrance, CA, USA). Helium served as the carrier gas, and its flow rate was 2 mL/min. Assays were performed in the temperature range 60–280 °C, and the heating rate was 10 °C/min; the auxiliary temperature was 300 °C. Samples with a volume of 1 mL were manually dosed. The gas chromatograph was equipped with an on-column injector with programmable temperature (the same as the analysis temperature). The test solutions were prepared by diluting an oil sample with dichloromethane at a volume ratio of 1:10 or 1:100.

Components were identified by comparison of their mass spectra with the spectrometer database of the NIST 11 Library (National Institute of Standards and Technology, Gaithersburg, MD, USA) and by comparison of their retention index calculated against *n*-alkanes ($C_9$–$C_{20}$). Each chromatographic analysis was repeated three times. The average values of relative composition of essential oil (percentages) were calculated from the peak areas.

### 4.3. Biological Experiment

The object of this experiment was the microbiota of oily facial skin without signs of lesions; the microbiota was isolated by a surface swab, and two lavender oils of various origins—from the ETJA company (ETJA, Elbląg, Poland) and oil extracted from Crimean lavender (not yet described in the literature)—were employed at concentrations 10–80 μL/cm$^3$.

The biological material was collected from five areas of facial skin, i.e., the cheeks, nose, forehead, and chin (i.e., from a total area of 20 cm$^2$) and was resuspended in broth (control) and in broth with the addition of tested lavender oils at concentrations of 20, 50, and 70 μL/cm$^3$ (a concentration of 50 μL/cm$^3$ is recommended by the manufacturers when this oil serves as an additive in cosmetic and pharmaceutical preparations). Because the effectiveness of an oil in suppressing the growth of the skin microbiota depends on the concentration and origin of the oil, lower and higher concentrations than those recommended by the manufacturer were tested. The samples were incubated for 24 h at a temperature of 35 °C.

The anti-microbial effects of these oils on the mixed microbiota from facial skin were evaluated by the surface culture method (10-fold dilutions in water containing 0.05% Tween-80) in parallel with the Nutrient LAB Agar™ medium by the BIOCORP company (BIOCORP, Warszawa, Poland) for determination of the bacterial cell count and in addition to selective media (Braid–Parker, *Enterococcus* agar, Hektoena, ENDO, and *Pseudomonas* agar) of the BTL company (BTL sp. z o.o., Łódź, Poland) for determination of a cell count of potentially pathogenic bacteria. After incubation, the total number of lavender oil-non-sensitive bacteria was determined, and the results were expressed in log colony-forming units (CFU)/cm$^2$ of the facial skin surface. The dominant bacterial isolates were identified by API tests from BIOMERIEUX company (BIOMERIEUX SSC Europe Sp. z o.o., Warszawa, Poland; ID32GN: Gram-negative bacilli, 50CHB: Gram-positive bacilli, ID32 STAPH: Gram-positive cocci). In the presence of the highest concentration of ETJA lavender oil used, bacteria of the following species survived: *Micrococcus luteus*, *Escherichia coli*, *Staphylococcus warneri*, and *Enterococcus faecium*. In the presence of the highest concentration of Crimea oil used, bacteria of the following species survived: *Bacillus cereus*, *B. subtilis*, *B. mycoides*, *Corynebacterium* spp., and *Enterococcus faecium*.

Next, the bactericidal activities of the tested oils on these dominant bacterial isolates were evaluated by the diffusion cylinder plate method on Nutrient LAB Agar$^{TM}$ medium [43]. The media were inoculated with 1 cm$^3$ of a standard bacterial suspension with the optical density of $\zeta = 2$ at a wavelength of 550 nm. The results were presented as a mean value of the growth inhibition diameter (in mm). The inhibitory effect was assumed to be the lack of growth around wells, whereas growth stimulation intensified growth around wells, and the neutral effect caused growth inhibition at the edges of the wells. The control was water containing 0.05% Tween-80. The essential oils and extracts were used at the following concentrations: 10, 20, 30, 40, 60, 70, or 80 µL/cm$^3$ (*v/v*). Each experiment was repeated three times.

## 5. Conclusions

Lavender oils from ETJA and Crimea most effectively reduced the number of mixed microbiota cells from facial skin at a concentration of 70 µL/cm$^3$, although no complete bactericidal activity was observed. The most sensitive to ETJA lavender oil were Gram-positive bacilli, and Gram-negative bacilli were the most sensitive to Crimean lavender oil. On the other hand, neither of the tested oils inhibited the growth of Gram-positive cocci. The tested lavender oils differed in their chemical composition quantitatively and qualitatively; 33 ingredients were identified in ETJA oil, including five compounds not described before (e.g., cyclofenchene and isoborneol); 101 components were identified in Crimean lavender oil, including 37 compounds not described before (e.g., octen-1-ol acetate and linalool formate). Two types of columns of different polarity allowed for better separation and identification of essential oil components such as limonene and eucalyptol. ETJA lavender oil was composed mainly of monoterpenoids (76.7%) and monoterpenes (22.7%), whereas Crimean lavender oil consisted mainly of monoterpenoids (80%), much less monoterpenes (5.8%), and some sesquiterpenes (8.0%; Table 5). Such differences in chemical composition were most likely due to the different geographical origins of the plant material. The analysed lavender oils differed in their bactericidal effect; ETJA lavender oil with higher monoterpenoid content (linalool and linalyl acetate) and monoterpene content (limonene) was characterised by higher efficiency than Crimean lavender oil.

**Author Contributions:** Conceptualization, M.B., E.N.-B. and T.K.-Ł.; methodology, M.B., E.N.-B. and T.K.-Ł.; formal analysis, M.B., E.N.-B. and T.K.-Ł.; data curation, M.B.; writing—original draft preparation, M.B., E.N.-B. and T.K.-Ł.; writing—review and editing, M.B.; visualization, M.B.; supervision, P.P.W.

**Funding:** This research received no external funding.



## References

1. Jabłońska-Trypuć, A.; Farbiszewski, R. *Sensory and Basics of Perfumery*; MedPharm: Wroclaw, Poland, 2008; pp. 114–115, 121–129.
2. Lazzara, M.V. *Aromatherapy. Healing Baths*; Bauer-Weltbild Media Sp. z.o.o.: Warszawa, Poland, 2003; pp. 19–29.
3. Góra, J.; Lis, A. *The Most Valuable Essential Oils*; UMK Publishing: Torun, Poland, 2005; pp. 165–175.
4. Góra, J.; Lis, A. The most valuable oils–Lavender oil. *Aromaterapia PTA* **1995**, *2*, 5–11.
5. Glinka, R.; Glinka, M. *Cosmetic Recipe with Elements of Cosmetology*; MA Publishing: Lodz, Poland, 2008; pp. 70–73.

6.  Janeczko, Z.; Pisulewska, E. *Domestic Oil Plants. Occurrence, Cultivation, Chemical Composition, Application;* "Know-How" Publishing: Kraków, Poland, 2008; pp. 7–11, 43–47.
7.  Da Porto, C.; Decorti, D.; Kikc, I. Flavour compounds of *Lavandula angustifolia* L. to use in food manufacturing: Comparison of three different extraction methods. *Food Chem.* **2009**, *112*, 1072–1078. [CrossRef]
8.  Danh, L.T.; Han, L.N.; Triet, N.D.A.; Zhao, J.; Mammucari, R.; Foster, N. Comparison of chemical composition, antioxidant and antimicrobial activity of lavender (*Lavandula angustifolia* L.) Essential oils extracted by supercritical $CO_2$, hexane and hydrodistillation. *Food Bioprocess Technol.* **2013**, *6*, 3481–3489. [CrossRef]
9.  Reverchon, E.; Della Porta, G. Supercritical $CO_2$ extraction and fractionation of Lavender essential oil and waxes. *J. Agric. Food Chem.* **1995**, *43*, 1654–1658. [CrossRef]
10. Mantovani, A.L.L.; Vieira, G.P.G.; Cunha, W.R.; Groppo, M.; Santos, R.A.; Rodrigues, V.; Magalhães, L.G.; Crotti, A.E.M. Chemical composition, antischistosomal and cytotoxic effects of the essential oil of *Lavandula angustifolia* grown in Southeastern Brazil. *Rev. Bras. Farmacogn.* **2013**, *23*, 877–884. [CrossRef]
11. Steltenkamp, R.J.; Casazza, W.T. Composition of the essential oil of Lavandin. *J. Agric. Food Chem.* **1967**, *15*, 1063–1069. [CrossRef]
12. Dob, T.; Dahmane, D.; Berramdane, T.; Chelghoum, C. Chemical composition of the essential oil of *Lavandula dentata* L. from Algeria. *Int. J. Aromather.* **2005**, *15*, 110–114. [CrossRef]
13. Shellie, R.; Mondello, L.; Marriott, P.; Dugo, G. Characterisation of lavender essential oils by using gas chromatography-mass spectrometry with correlation of linear retention indices and comparison with comprehensive two-dimensional gas chromatography. *J. Chromatogr. A.* **2002**, *970*, 225–234. [CrossRef]
14. Cong, Y.; Abulizi, P.; Zhi, L.; Wang, X. Chemical composition of the essential oil of *Lavandula angustifolia* from Xinjiang, China. *Chem. Nat. Compd.* **2008**, *44*, 810–815. [CrossRef]
15. Daferera, D.J.; Ziogas, B.N.; Polissou, M.G. GC-MS analysis of essential oils from some Greek aromatic plants and their fungitoxicity on *Penicillium digitatum*. *J. Agric. Food Chem.* **2000**, *48*, 2576–2581. [CrossRef]
16. Yazdani, E.; Sendi, J.J.; Aliakbar, A.; Senthil-Nathan, S. Effect of *Lavandula angustifolia* essential oil against lesser mulberry pyralid Glyphodes pyloalis Walker (Lep: *Pyralidae*) and identification of its major derivatives. *Pest. Bioch. Physiol.* **2013**, *107*, 250–257. [CrossRef]
17. Dadalioglu, I.; Evrendilek, G.A. Chemical compositions and antibacterial effects of essential oils of Turkish Oregano (*Origanum minutiflorum*), Bay Laurel (*Laurus nobilis*), Spanish Lavender (*Lavandula stoechas* L.), and Fennel (*Foeniculum vulgare*) on common foodborne pathogens. *J. Agric. Food Chem.* **2004**, *52*, 8255–8260. [CrossRef] [PubMed]
18. Hassiotis, C.N. Chemical compounds and essential oil release through decomposition process from *Lavandula stoechas* in Mediterranean region. *Biochem. Sys. Ecol.* **2010**, *38*, 493–501. [CrossRef]
19. Hanamanthagouda, M.S.; Kakkalameli, S.B.; Naik, P.M.; Nagella, P.; Seetharamareddy, H.R.; Murthy, H.N. Essential oils of *Lavandula bipinnata* and their antimicrobial activities. *Food Chem.* **2010**, *118*, 836–839. [CrossRef]
20. Kulkarni, R.R.; Pawar, P.V.; Joseph, M.P.; Akulwad, A.K.; Sen, A.; Joshi, S.P. *Lavandula gibsoni* and *Plectranthus mollis* essential oils: Chemical analysis and insect control activities against *Aedes aegypti*, *Anopheles stephensi* and *Culex quinquefasciatus*. *J. Pest. Sci.* **2013**, *86*, 713–718. [CrossRef]
21. Palá-Paúl, J.; Brophy, J.J.; Goldsack, R.J.; Fontaniella, B. Analysis of the volatile components of *Lavandula canariensis* (L.) Mill., a Canary Islands endemic species, growing in Australia. *Biochem. Sys. Ecol.* **2004**, *32*, 55–62. [CrossRef]
22. Zuzarte, M.; Vale-Silva, L.; Gonçalves, M.J.; Cavaleiro, C.; Vaz, S.; Canhoto, J.; Pinto, E.; Salgueiro, L. Antifungal activity of phenolic-rich *Lavandula multifida* L. essential oil. *Eur. J. Clin. Microbiol. Infect. Dis.* **2012**, *31*, 1359–1366. [CrossRef]
23. Białoń, M.; Krzyśko-Łupicka, T.; Pik, A.; Wieczorek, P.P. Chemical composition of herbal macerates and corresponding commercial essential oils and their effect on bacteria *Escherichia coli*. *Molecules* **2017**, *22*, 1887. [CrossRef]
24. Białoń, M.; Krzyśko-Łupicka, T.; Koszałkowska, M.; Wieczorek, P.P. The influence of chemical composition of commercial lemon essential oils on the growth of *Candida* strains. *Mycopathologia* **2014**, *177*, 29–39. [CrossRef]
25. Adaszyńska-Skwirzyńska, M.; Swarcewicz, M. Chemical composition and biological activity of medical lavender. *Wiad. Chem.* **2014**, *68*, 11–12.

26. Król, S.K.; Skalicka-Woźniak, K.; Kandefer-Szerszeń, M.; Stepulak, A. The biological and pharmacological activity of essential oils in the treatment and prevention of infectious diseases. *Postepy Higieny Medycyny Doswiadczalnej* **2013**, *67*, 1000–1007.

27. Martini, M.C. *Cosmetology and Pharmacology of the Skin*; PZWL Publishing: Warszawa, Poland, 2007; pp. 37–44, 52, 105–112, 122.

28. Grzybowski, J.; Zaborowski, P. *Theoretical and Practical Basis of Infectology*; Institute of Health Protection Problems, Borgis Publishing: Warszawa, Poland, 2007; pp. 34–55.

29. Szewczyk, E.M.; Dudkiewicz, B.; Lisiecki, P.; Różalska, M.; Sobiś-Glinkowska, M.; Szarapińska-Kwaszewska, J. *Bacteriological Diagnosis*; PWN Publishing: Warszawa, Poland, 2005; pp. 11–159, 205–207, 297–342.

30. Brud, W.; Konopacka-Brud, I. Essential oils as active substances in cosmetics. *Herb. Messages* **1998**, *7/8*, 8–10.

31. Jaroszewska, B. *Cosmetology*; Atena Publishing: Warszawa, Poland, 2004; pp. 93–94.

32. Peters, B.; Kerkhoff, E.; Kuska, S.; Schweig, W.; Wulfhorst, B. *Cosmetology*; Stam REA Publishing: Warszawa, Poland, 2006; pp. 42–43, 233.

33. Cavanagh, H.M.A.; Wilkinson, J.M. Lavender essential oil: Review. *Austr. Infect. Contr.* **2005**, *10*, 35–37. [CrossRef]

34. Sienkiewicz, M.; Denys, P.; Kowalczyk, E. Antibacterial and immunostimulatory effect of essential oils. *Int. Rev. Allergol. Clin. Immunol.* **2011**, *17*, 40–44.

35. Boelens, M.H. Chemical and sensory evaluation of *Lavandula* oils. *Perf. Flav.* **1995**, *20*, 23–51.

36. Adaszyńska, M.; Swarcewicz, M.; Markowska-Szczupak, A. Comparison of chemical composition and antimicrobial activity of Lavender varieties from Poland. *Post. Fitoter.* **2013**, *2*, 90–96.

37. Gören, A.C.; Topçu, G.; Bilsel, G.; Bilsel, M.; Aydoğmuş, Z.; Pezzuto, J.M. The chemical constituents and biological activity of essential oil of *Lavandula stoechas* ssp. *stoechas*. *Zeitschrift Für Naturforschung C* **2002**, *15*, 797–800. [CrossRef] [PubMed]

38. Sabara, D.; Kunicka-Styczyńska, A. Lavender oil–Flavouring or active cosmetic ingredient? *Food Chem. Biotechnol.* **2009**, *73*, 37–42.

39. Roller, S.; Ernest, N.; Buckle, J. The antimicrobial activity of high-necrodane and other lavender oils on methicillin-sensitive and -resistant *Staphylococcus aureus* (MSSA and MRSA). *J. Altern. Complement. Med.* **2009**, *15*, 275–279. [CrossRef]

40. Kalemba, D. Antibacterial and antifungal properties of essential oils. *Post. Mikrobiol.* **1998**, *38*, 185–203. [CrossRef]

41. Trytek, M.; Paduch, R.; Fiedurek, J.; Kandefer-Szerszeń, M. Monoterpenes–Old compounds, new applications, and biotechnological methods of their production. *Biotechnologia* **2007**, *76*, 135–155.

42. Grabowska, K.; Janeczko, Z. Essential oils with antiviral activity. *Aromaterapia PTA* **2009**, *58*, 9–20.

43. Johnson, O.O.; Ayoola, G.A.; Adenipekun, T. Antimicrobial activity and the chemical composition of the volatile oil blend from *Allium sativum* (Garlic Clove) and *Citrus reticulate* (Tangerine fruit). *Int. J. Pharm. Sci. Drug Res.* **2013**, *5*, 187–193.

**Sample Availability:** Samples of the compounds are not available from the authors.

*Article*

# Identification and Allelopathy of Green Garlic (*Allium sativum* L.) Volatiles on Scavenging of Cucumber (*Cucumis sativus* L.) Reactive Oxygen Species

Fan Yang [†], Xiaoxue Liu [†], Hui Wang, Rui Deng, Hanhan Yu and Zhihui Cheng *

College of Horticulture, Northwest A&F University, Taicheng Road No.3, Yangling, Shaanxi 712100, China
* Correspondence: chengzh@nwsuaf.edu.cn; Tel.: +86-151-2918-3300
† These authors contributed equally to the article.

Academic Editor: Igor Jerković
Received: 13 August 2019; Accepted: 5 September 2019; Published: 7 September 2019

**Abstract:** Garlic and formulations containing allicin are used widely as fungicides in modern agriculture. However, limited reports are available on the allelopathic mechanism of green garlic volatile organic compounds (VOCs) and its component allelochemicals. The aim of this study was to investigate VOCs of green garlic and their effect on scavenging of reactive oxygen species (ROS) in cucumber. In this study, green garlic VOCs were collected by HS-SPME, then analyzed by GS-MS. Their biological activity were verified by bioassays. The results showed that diallyl disulfide (DADS) is the main allelochemical of green garlic VOCs and the DADS content released from green garlic is approximately 0.08 mg/g. On this basis, the allelopathic effects of green garlic VOCs in vivo and 1 mmol/L DADS on scavenging of ROS in cucumber seedlings were further studied. Green garlic VOCs and DADS both reduce superoxide anion and increase the accumulation of hydrogen peroxide of cucumber seedlings. They can also regulate active antioxidant enzymes (SOD, CAT, POD), antioxidant substances (MDA, GSH and ASA) and genes (*CscAPX, CsGPX, CsMDAR, CsSOD, CsCAT, CsPOD*) responding to oxidative stress in cucumber seedlings.

**Keywords:** green garlic volatile organic compounds; volatiles isolation and analysis; allelochemicals; GC-MS; biological activity; reactive oxygen species

---

## 1. Introduction

Plants synthesize and release various volatile organic compounds (VOCs) [1]. VOCs play essential roles in attracting pollinators and seed-dispersers, defense against herbivores and pathogens, interplant signaling and allelopathy [2,3]. Garlic (*Allium sativum* L.), an economically important vegetable [4], contains various volatile components, including diallyl, dimethyl, and allyl methyl sulfides, disulfides, and trisulfides, as well as some other minor components, all of which are formed by the decomposition of allicin and are released upon crushing garlic. These organosulfur compounds can inhibit carcinogen activation, boost detoxifying processes, cause cell cycle arrest, stimulate the mitochondrial apoptotic pathway and increase the acetylation of histones [5]. Garlic also has potential allelopathic effects and is widely used in crop rotation and intercropping with many other crops, e.g., cucumber [6], tomato [7], eggplant [8], pepper [9]. In this kind of intercropping system, garlic has been confirmed to alleviate continuous cropping issues from the aspects of reducing plant diseases and improving the physical and chemical properties of soil [10]. Green garlic is young garlic with tender leaves that harvested at the early stage before the bulb is formed and it is consumed by people of many countries. The green garlic-cucumber intercropping system has been shown to have the benefits of improving soil fertility, promoting the activity of soil enzymes and increasing cucumber biomass [6,11]. However, there is a lack of identification of the main allelochemicals for green garlic volatiles, and knowledge about what kind of allelochemicals regulated the increased biomass of cucumber, which are the subjects in this study.

Diallyl disulfide (DADS) is a major allelochemical of the VOCs in garlic [7]. In animals, DADS has been shown having the effect of reducing cellular toxins and inhibiting the proliferation of cancer cells through several actions which include the activation of metabolizing enzymes that detoxify carcinogens, suppression of the formation of DNA adducts, antioxidant effects; regulation of cell-cycle arrest, induction of apoptosis and differentiation, histone modification and inhibition of angiogenesis and invasion [12]. Kubota et al. found that the active substances from garlic that can break the bud dormancy of grapevines are sulfur-containing compounds, among which the most effective one is DADS [13]. Previous studies found that DADS can also regulate tomato root growth by affecting cell division, endogenous phytohormone levels, expansin gene expression and the pathways of tomato root sulphate assimilation and glutathione (GSH) metabolism [7,14]. Besides, a low concentration DADS is able to promote cucumber root growth and induce main root elongation by up-regulating the expression of *CsCDKA* and *CsCDKB* genes and regulating adjusting the hormone balance of roots [15]. However, the effect of DADS on scavenging of cucumber reactive oxygen species (ROS) has not been investigated.

It is important to choose a suitable method to collect as much as possible the volatile compounds of plants in their natural state. For determination of garlic flavor components by gas chromatography-mass spectrometry (GC-MS), several sampling techniques including steam distillation (SD), simultaneous distillation and solvent extraction (SDE), microwave assisted hydrodistillation extraction (MWHD), ultrasound-assisted extraction (USE), solid-phase trapping solvent extraction (SPTE) and headspace solid-phase microextraction (HS-SPME) were applied and compared in previous studies [16,17]. Compared with SD, SDE, and SPTE, the HS-SPME method had several advantages which are rapid solvent-free extraction, no apparent thermal degradation, less laborious manipulation and less sample requirement. Five different fiber coatings were evaluated to select a suitable fiber for HS-SPME of garlic flavor components, among which the divinyl benzene/carboxen/ polydimethylsiloxane (DVB/CAR/PDMS) one was the most efficient among the investigated fibers [16]. Warren investigated the analysis of thiol compounds using a needle trap device and HS-SPME coupled to GC-MS, comparing the advantages and disadvantages of two methods [18]. Based on the above studies, the HS-SPME coupled to GC-MS method seems to be a good choice for determining the allelochemicals of green garlic. We chose a fiber-HS-SPME-GC-MS method to determine green garlic allelochemicals based on these previous studies.

Similar to other biological active factors, allelochemicals derived from plants can target some specific biological processes, which include destroying membrane permeability, influencing photosynthetic and respiratory chain electron transport, influencing cell division and ultrastructure, changing enzyme activity, altering reactive oxygen species (ROS) levels and effecting expression of related genes [19]. ROS are involved in many biological processes, such as growth, development, response of biotic and environmental stresses and programmed cell death [20,21]. It has been reported that higher ROS levels might cause oxidative damage to plants. In order to reduce this damage, plants have developed efficient antioxidant defense systems to scavenging ROS, such as superoxide dismutase (SOD), catalase (CAT), peroxidase (POD), ascorbic acid (ASA), GSH, monodehydroascorbate (MDA), the expressions of their synthetic genes are related to oxidation reactions [22,23].

In this study, we will explore the regulation mechanism of VOCs derived from green garlic in scavenging ROS of cucumber from the following three aspects. Firstly, the major effective allelochemicals of VOCs that collected from green garlic will be identified; secondly, the different ROS scavengers induced by VOCs of green garlic and the ROS levels will be investigated among treated cucumber seedlings; thirdly, the expression levels of six genes (*CsPOD*, *CsCAT*, *CscAPX*, *CsMDAR*, *CsGPX*, *CsSOD*) related to oxidation reactions will be checked following the treatment of cucumber seedlings with VOCs and DADS. This study will provide more information for understanding the mechanism of green garlic VOCs in scavenging ROS of vegetables.

## 2. Results

### 2.1. Identification of Volatile Allelochemicals in Green Garlic

Identification of the effective components of VOCs collected from green garlic is the basis to reveal its mechanism in scavenging ROS. For the VOCs collected from the green garlic segments that were contained in the SPME vial, only two compounds including DADS and diallyl sulfide (DAS) were identified (Figure 1A and Table 1). The retention time and relative peak area of DADS were 13.83 min and 99.52%, respectively, while, DAS had a retention time of 7.04 min and a relative peak area of 0.48%. Among the VOCs collected using green garlic segments that were put in a sealed desiccator, 17 compounds were detected (Table S1), in which there were three sulfur compounds including DADS, methyl propenyl disulfide and DAS, with relative peak areas of 86.33%, 1.53% and 0.35%, respectively (Figure 1B and Table 1). All the other compounds were heterocyclic compounds and long chain hydrocarbon compounds, which are impurities in air. There were 42 compounds identified in the VOCs obtained from the whole green garlics that were placed in a sealed desiccator (Table S2). Among the 42 detected compounds there was only one sulfur compound, which was DADS with a relative peak area of 15.3% (Figure 1C and Table 1), while the others were also heterocyclic compounds and long chain hydrocarbon compounds. The qualitative identification result showed that DADS is the main volatile compound of green garlic. Compared with the whole green garlic, the cut green garlic segments produced more DADS with about a 5-fold difference.

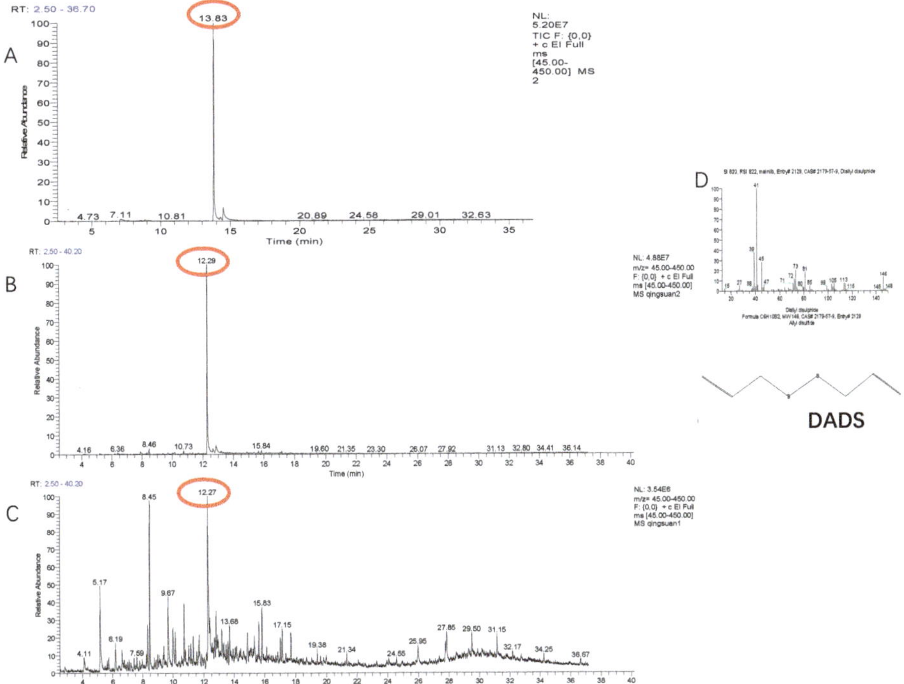

**Figure 1.** Identification of the collected volatiles from green garlic volatiles by GC-MS. (**A**) Cut (SPME vial); (**B**) Cut (sealed desiccator); (**C**) Whole (sealed desiccator); (**D**) MS spectrum and compound structure of DADS. *n* = 3, three biological replicates.

**Table 1.** Total sulfur compounds of green garlic volatile compounds among different collecting methods.

| Treatment | Total Sulfur Compounds | 1 | | | 2 | | | 3 | | |
|---|---|---|---|---|---|---|---|---|---|---|
| | | Compound Name | Area% | RT | Compound Name | Area% | RT | Compound Name | Area% | RT |
| Cut (SPME vial) | 2 | Diallyl disulphide | 99.52 | 13.83 | Methyl propenyl disulfide | 0 | – | Diallyl sulfide | 0.48 | 7.04 |
| Cut (sealed desiccator) | 3 | Diallyl disulphide | 86.33 | 12.29 | Methyl propenyl disulfide | 1.53 | 7.87 | Diallyl sulfide | 0.35 | 6.36 |
| Whole (sealed desiccator) | 1 | Diallyl disulphide | 15.30 | 12.27 | Methyl propenyl disulfide | 0 | – | Diallyl sulfide | 0 | – |

The result of an external standard method for quantitative analysis of volatiles showed that the linear relation between DADS concentration (mmol/L) (y) and peak area (x) fits a simple linear regression equation (Equation (1)):

$$y = 5E + 09x + 3E + 07, R^2 = 0.9903 \tag{1}$$

Based on this equation, the DADS that released by green garlics has a concentration of about 0.08 mg/g.

Our biological tests showed that the green garlic treatment had a consistent effect with the DADS treatment (Figure 2). When the concentration of DADS was higher than 1 mmol/L, cucumber leaves turned yellowish (Figure 2). Inversely, when the green garlic treatment was less than 2 g or the concentration of DADS was lower than 1 mmol/L, the cucumber leaves had no morphology changes (Figure 2). The amount of DADS volatilized by 2 g of green garlic approached 1 mmol/L DADS treatment based on Equation (1). In addition, higher concentrations of green garlic volatiles and DADS can cause cucumber leaves turn to soft and rot (Figure 2). These results confirmed that DADS is the main allelochemical of green garlic volatiles.

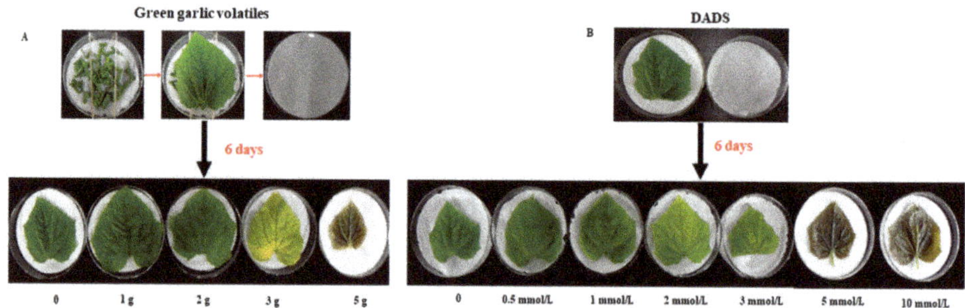


**Figure 2.** The biological test and its results of green garlic volatiles and DADS on cucumber leaf. (**A**) effect of green garlic (0.0, 1.0, 2.0, 3.0, 5.0 g) volatiles on cucumber leaf color and structure. (**B**) effect of DADS (0, 0.5, 1, 2, 3, 5, and 10 mmol/L) on cucumber leaf color and structure. Five leaves from 5 plants were used for each treatment.

*2.2. Effect of Green Garlic Volatiles and DADS on Cucumber ROS and Antioxidant Activity*

2.2.1. Effect of Green Garlic Volatiles and DADS on Cucumber ROS ($O_2^{\bullet-}$ and $H_2O_2$)

Figure 3A showed the results of histological observation of $H_2O_2$ in cucumber leaves stained by DAB dye. Compared with the control groups, the colors of dyed cucumber leaves gradually turned much darker with the increasing concentrations of green garlic volatiles and DADS (Figure 3A).

This suggested that green garlic volatiles and DADS led to an increasing $H_2O_2$ content at histological levels. After 10-days of co-culture with green garlic, the $O_2^{\bullet-}$ contents of treated cucumber leaves were significantly lower that of control ($p < 0.05$). Similarly, the $O_2^{\bullet-}$ contents of cucumber leaves under 1 mmol/L DADS treatment was also lower than that of control group (Figure 3B). The $H_2O_2$ content in cucumber leaves exhibited a rising trend with the increase of the concentration of green garlic volatiles. Besides, the $H_2O_2$ contents in cucumber leaves of all green garlic treatments and DADS were significantly higher than that of control ($p < 0.05$), which were consistent with the histological observations of $H_2O_2$ in cucumber leaves (Figure 3A).

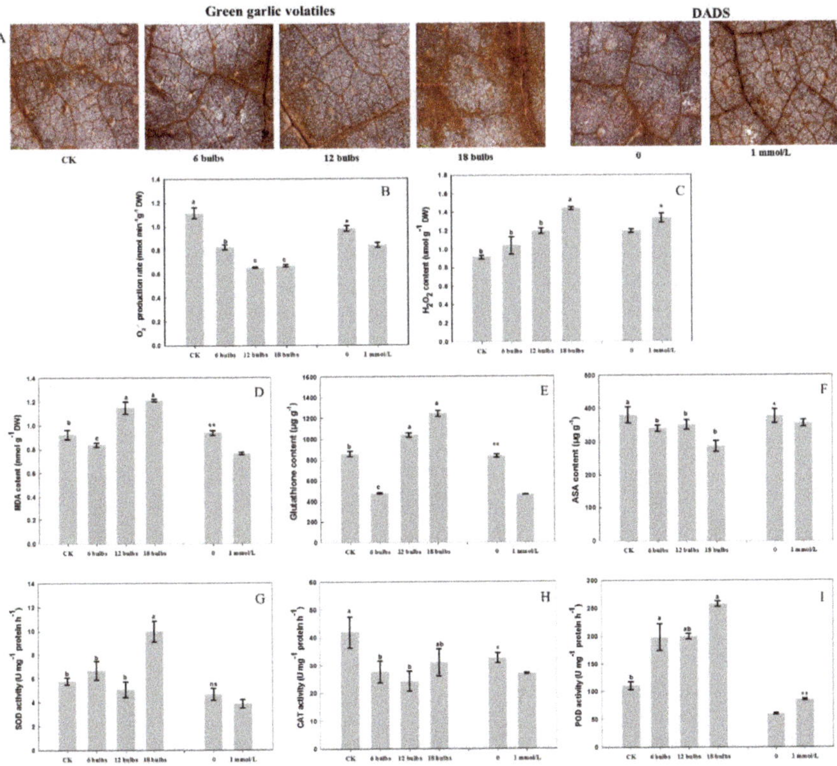


**Figure 3.** Effect of green garlic volatiles and DADS on cucumber ROS, antioxidant substance and antioxidant enzymes activity. (**A**) Histological observation of $H_2O_2$ in cucumber leaves that stained by DAB dye. (**B,C**), effect of green garlic volatiles and DADS on cucumber ROS ($O_2^{\bullet-}$ and $H_2O_2$); (**D–F**), effect of green garlic volatiles and DADS on cucumber antioxidant substance content (MDA, GSH and ASA); (**G–I**), effect of green garlic volatiles and DADS on cucumber antioxidant enzyme activity (SOD, CAT and POD). * $p < 0.05$; ** $p < 0.01$; ANOVA, followed by Tukey test and *t*-test. Data are means ± standard errors ($n = 3$ for (**B,C**), three biological replicates).

### 2.2.2. Effect of Green Garlic Volatiles and DADS on Cucumber Antioxidant Substances (MDA, GSH and ASA)

The MDA content of cucumber leaves co-cultured with 6-bulbs of green garlic was significantly lower than that of control ($p < 0.05$) (Figure 3D), while, that for co-cultures with 12- and 18-bulbs of green garlic the MDA content was significantly higher than the control group ($p < 0.05$) (Figure 3D). When cucumber seedlings were treated with DADS, the MDA content of cucumber leaves was also significantly lower than that of control ($p < 0.01$) which was consistent with the treatment co-cultured with 6-bulbs of green garlic (Figure 3D). The GSH content had an increasing trend with the increasing number of bulbs of green garlic. The GSH contents of 6-bulb green garlic and DADS treatment were both significantly lower than that of control ($p < 0.05$), but the treatments with 12- and 18-bulbs of green garlic were higher than that of control (Figure 3E). The ASA content in cucumbers treated with green garlic bulbs and DADS were consistently lower than that of control (Figure 3F).

### 2.2.3. Effect of Green Garlic Volatiles and DADS on Cucumber Antioxidant Enzymes (SOD, CAT and POD)

In this study, the SOD activity of cucumber that was co-cultured with 18-bulbs of green garlic was significantly higher than other treatments ($p < 0.05$) (Figure 3G). There were no significant difference among the treatments co-cultured with 6- and 12-bulbs of green garlic, the treatment of DADS and the control groups ($p < 0.05$) (Figure 3G). All the green garlic volatiles treatments had lower CAT activity than the control ($p < 0.01$). For the CAT activity, the treatments of DADS exhibited similar results to the treatments of green garlic volatiles (Figure 3H). Both the green garlic volatiles and DADS treatments had higher POD activity than that of control ($p < 0.01$) (Figure 3I).

### 2.3. Gene Expression Changes in Response to Green Garlic Volatiles and DADS

In order to further clarify the potential mechanism of action of green garlic volatiles and DADS treatments on cucumber ROS and antioxidant enzyme activities, the expression levels of six related genes (*CscAPX*, *CsGPX*, *CsMDAR*, *CsSOD*, *CsCAT*, *CsPOD*) were examined with real-time quantitative polymerase chain reaction (RT-qPCR). The expression levels of these six genes were significantly changed in cucumber leaves after the treatments (Figure 4).

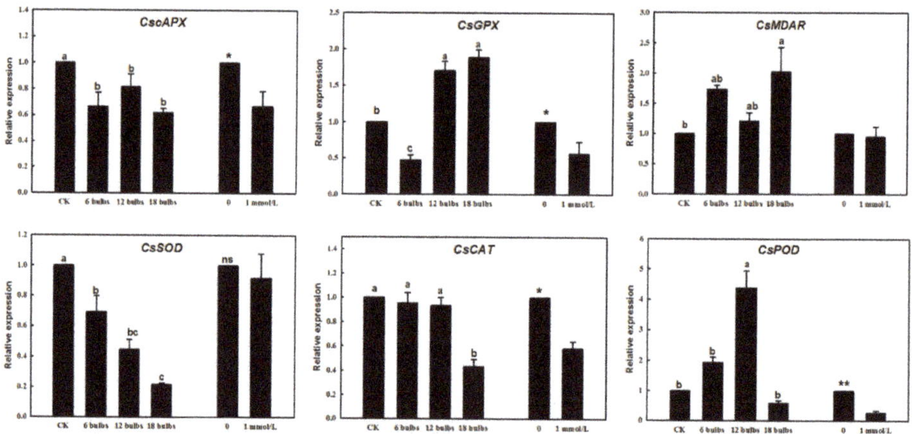

**Figure 4.** Changes in gene relative expression in response to green garlic volatiles and DADS. *CscAPX*, cucumber cytosol ascorbic acid peroxidase gene; *CsGPX*, cucumber glutathione peroxidase gene; *CsMDAR*, cucumber gene of scavenge MDA; *CsSOD*, *CsCAT* and *CsPOD*, cucumber genes of SOD, CAT, POD. * $p < 0.05$; ** $p < 0.01$; ANOVA, followed by Tukey test and *t*-test. Data are means ± standard errors ($n = 3$, three biological replicates).

The green garlic volatiles and DADS treatments increased the expression levels of *CsGPX*, *CsMDAR*, and *CsPOD* genes and decreased the expression levels of *CscAPX*, *CsCAT* and *CsSOD* (Figure 4). Compared with the control group, the expression level of *CsGPX* was significantly decreased by 2-fold ($p < 0.05$) in the treatments with 12- and 18-bulbs of green garlic. The expression level of *CsPOD* in cucumber co-cultured with 12-bulbs of green garlic was increased almost 4-fold when compared with the control ($p < 0.01$) (Figure 4). However, the expression level of *CsCAT* was decreased in cucumbers that were co-cultured with green garlic, especially after 18-bulb treatment, and DADS treatment (Figure 4). The *CscAPX* expression level was decreased by green garlic volatiles and the DADS treatments (Figure 4). The expression of *CsMDAR* was increased by 2-fold when compared with the control ($p < 0.05$) (Figure 4).

## 3. Discussion

Garlic organosulfur compounds are biosynthesized for defensive purposes such as protection against abiotic stressors and are formed quickly once the plant tissues are damaged [24]. Fresh garlic only contains alliin, a derivative of cysteine. When the cells of fresh garlic cell are crushed, allinase can convert alliin to allicin. Similarly, the DADS content of the cut green garlic segments were higher than that of the whole garlic plants (Table 1), which suggested that the damaged green garlics can volatilize more DADS. In general, DATS, DADS, and VDTs are the major sulfides derived from the garlic extracts. When garlic oil is extracted by steam distillation or simultaneous distillation, the resulting extracts mainly contain DADS and DATS [25]. When garlic was treated by high temperature aging, crushing, and roasting, the compounds identified by SPME-GC/MS were also DADS and DATS [26]. DADS (97.85%), DAS (0.01%), and DATS (0.01%) were found to be the predominant flavor components of garlic samples extracted by HS-SPME using a 50/30-micron DVB/CAR/PDMS fiber [16]. In cut green garlic segments, we can detect two sulfur compounds, DADS and DAS, while in whole green garlic, only DADS can be detected (Figure 1). The green garlic allelochemicals were determined by fiber-HS-SPME-GC-MS in this study. The analytical performances of needle trap micro-extraction (NTME) coupled with GC-MS were evaluated by analyzing a mixture of twenty-two representative breath volatile organic compounds (VOCs) belonging to different chemical classes (i.e., hydrocarbons, ketones, aldehydes, aromatics and sulfurs), which confirmed the reliability of this method [27,28]. Previous studies were mainly focused on the identification of garlic compounds produced from garlic cloves, which might not be identical to those seen this study, because we collected the volatile compounds of green garlic plants

By calculation and comparison, in general, the amount of DADS released by green garlic was approximately 0.08 mg/g. The four treatments of 0, 6, 12 and 18 bulbs of green garlic in the mini-greenhouse, corresponded to the DADS concentrations of 0, 0.025, 0.046 and 0.057 µmol/L per cucumber plant (Table S1). This means that the DADS concentration of a 6-bulb treatment in a mini-greenhouse is approximately equivalent to 50 mL of 1 mmol/L DADS treatment (0.024 µmol/L per cucumber plant). It has been reported that lower concentrations (0.01–0.62 mmol/L) of DADS can significantly promoted root growth, whereas higher levels (6.20–20.67 mmol/L) will have inhibitory effects [6]. Therefore, considering the effect and cost, 1 mmol/L DADS concentration is suitable for use.

Many studies have shown that allelochemicals significantly inhibit the activities of antioxidant enzymes, increase the level of free radicals, lead to membrane lipid peroxidation and membrane potential changes, thereby reducing the scavenging effect on ROS and destroying the entire membrane system of plants [9]. ROS play an important role in plants' signal transduction pathways, as key regulators of processes such as growth, development, response to biotic and environmental stimuli, and plant metabolism, especially $H_2O_2$ which is important for programmed cell death to resist disease [20]. Green garlic volatiles and DADS significantly decreased the $O_2^{\bullet-}$ contents (Figure 3B). However, they significantly increased the $H_2O_2$ content in cucumber leaves (Figure 3C). Plants have complex antioxidant systems to deal with ROS damage, including enzymatic systems (such as SOD, CAT and POD) and non-enzymatic systems (such as MDA, GSH and ASA) [23]. Previous studies have shown that the activities of POD, CAT, SOD and MDA contents in the leaves and roots of tomato were increased after the treatment of DADS [14]. In this study, the activity of POD was increased, while that of CAT was decreased, and the MDA content was also decreased after when treated with DADS. SOD is the most important ROS scavenger by converting $O_2^{\bullet-}$ into molecular oxygen and $H_2O_2$ [23]. $H_2O_2$ can be decomposed into molecular oxygen and $H_2O$ by APX GPX, CAT, or POD [20,29]. Green garlic volatiles and DADS promoted the activities of POD and reduced the contents of GSH and ASA and the activity of CAT, then led to the change of $H_2O_2$ content (Figure 3). The gene expression changes were consistent with the changes of antioxidant enzyme and antioxidant substances. The green garlic volatiles decreased the expression of *CscAPX, CsGPX, CsCAT* and *CsPOD. SOD, CAT,* and *POD* genes play key roles in plant tissue antioxidant defenses [30]. *SOD* gene is the defense first line to against the ROS. Genes like *GPX, cAPX, MDAR, CAT* and *POD* work to further convert $H_2O_2$ into $H_2O$ and $O_2$ through different reactions [23]. *GPX* utilizes glutathione (GSH) as an electron donor to reduce ROS [31]. The *POD* gene is another reported defense-related enzyme gene and 24 *POD* genes have been identified in transcriptome analysis after DADS treatment [14]. *CAT* gene directly decompose $H_2O_2$ into $H_2O$ and $O_2$. *CAT* gene is indispensable for ROS detoxification [32]. APX enzymes catalyze the conversion of $H_2O_2$ into $H_2O$ and MDA using ascorbate [33]. *MDAR* helps to scavenge the MDA and generate dehydroascorbate (DHA) [34].

In summary, a scheme (Figure 5) showing green garlic volatiles' effect on cucumber ROS, antioxidant enzymes, antioxidant substances and genes can be proposed. Firstly, green garlic volatiles decreased the expression of *CsSOD* gene, which made the $O_2^{\bullet-}$ turn into molecular oxygen and $H_2O_2$. Then green garlic volatiles decreased the expression of *CscAPX, CsGPX, CsCAT* and *CsPOD* gene, which can promote the activity of POD and reduce the contents of GSH and ASA and the activity of CAT. Finally, green garlic volatiles increased the accumulation of $H_2O_2$. This increased accumulation of $H_2O_2$ might increase the disease resistance of cucumber. We will continue to work on the effects of green garlic volatiles on cucumber disease resistance in the future.

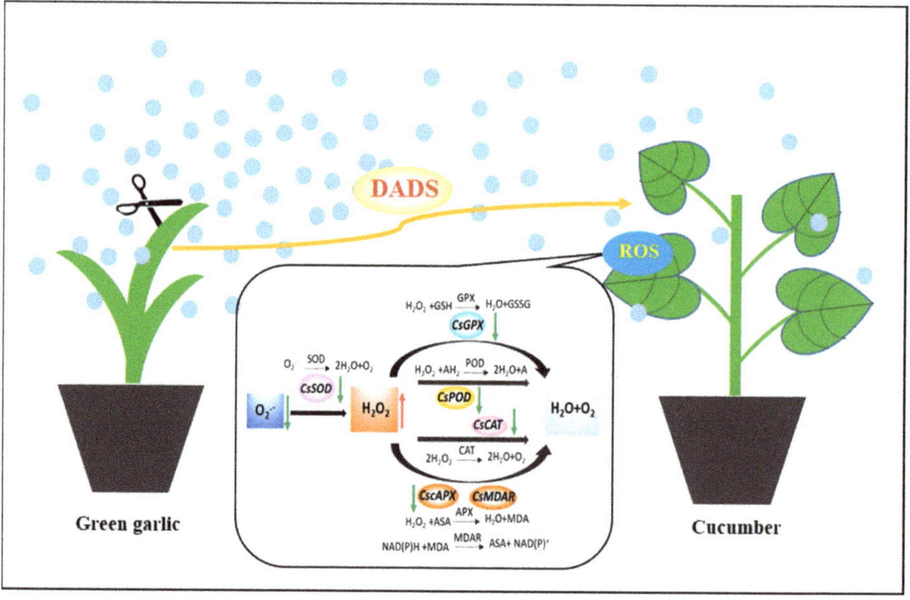


**Figure 5.** A scheme of green garlic volatiles effect on cucumber ROS, antioxidant enzymes, antioxidant substances and genes.

## 4. Materials and Methods

### 4.1. Materials and Equipment

Cucumber seeds of Jinyou 40, a North China fresh market type variety, were obtained from Tianjin Kernel Cucumber Research Institute (Tianjin, China). Garlic bulbs of Gailiang were provided by the College of Horticulture, Northwest A&F University (Yangling, Shaanxi Province, China). To prepare the DADS stock solutions, laboratory grade DADS (purity 80%) was purchased from Sigma-Aldrich Co. (St. Louis, MO, USA) and was firstly dissolved in Tween-80 with a ratio of 1:2 (*w/w*), then distilled water was added to get a 10 mmol/L stock solution which was stored at 4 °C for further use [35]. SPME fibers (50/30 μm DVB/CAR/PDMS, Sigma-Aldrich Co.) were used to absorb the VOCs and GC-MS (ISQ, Thermo Fisher, Waltham, MA, USA) was used to identify the collected VOCs. A ultraviolet spectrophotometer (UNIC 3200, UNIC, Shanghai, China) was used to measure the enzyme activities and ROS content. A microscope (BX63, Olympus, Japan) was used for histological observation of $H_2O_2$ effects.

### 4.2. Plant Growth and Experimental Design

Garlic bulbs of uniform size (approximately diameter = 4 cm, height = 4 cm and weight = 35 g) were selected and sown in plastic pots (diameter = 21 cm) filled with sterilized culture substrate. The green garlic plants that grew to 25 cm in plant height (about 30 days after sowing) were used for collecting VOCs and identification of green garlic allelopathic compounds and co-culturing with cucumber seedlings.

Germinated cucumber seeds were planted in plastic pots filled with sterilized culture substrate and cultivated in growth chamber with a day/night temperature of 25/18 °C, a day/night regime of 16/8 h and a relative humidity of 80%. The cucumber seedlings at two-true-leaf stage were used for following experiments.

To verify whether DADS is the main allelochemical of green garlic VOCs, bioassay analysis was performed. The detached cucumber leaves were treated with different weights of green garlic (0.0, 1.0, 2.0, 3.0, 5.0 g) that was cut into 1 cm segments and different concentrations of DADS (0, 0.5, 1, 2, 3, 5, and 10 mmol/L) in Petri dishes (see details in Figure 2). To ensure the treatment effect, the green garlic pieces were replaced with fresh ones and DADS was added again after 3 days of treatment. The morphology changes of these detached cucumber leaves were observed at the end of the sixth day.

To explore the effect of VOCs that derived from green garlic on scavenging ROS of cucumber, green garlic plants and cucumber seedlings were co-cultured in a greenhouse of Northwest A&F University, Yangling (N 34°16′, E 108°4′), China. Green garlic plants and cucumbers were co-cultured for 10 days in a plastic mini-greenhouse (60 × 60 × 58 cm), in which 12 cucumber seedlings were surrounded with green garlic seedlings. Four treatments including different number of green garlic seedlings (derived from 0, 6, 12 and 18 garlic bulbs, respectively) were used in this experiment (see details in Figure S1). Each treatment was replicated three times. To investigate the effect of DADS on ROS scavenging in cucumber, 12 cucumber seedlings were also planted in the plastic mini- greenhouse, and 5 mL of DADS solution with a concentration of 1 mmol/L were sprayed on each cucumber seedling, while an equal volume of distilled water was sprayed as a control (see details in Figure S1). During the 10-day cultivation of these cucumber seedlings, DADS and distilled water were sprayed twice, on the first and sixth day. For all treatments, the second true leaves of cucumber were sampled. One part of the leaves were put in a stationary liquid for histological observation of $H_2O_2$, and the other parts of the leaves were stored at −80 °C for ROS and genes assay.

### 4.3. Identification of Green Garlic Allelochemicals by Fiber-HS-SPME-GC-MS

The volatiles derived from green garlic samples were collected by three methods. For the first one, we used the headspace solid phase microextraction (HS-SPME) method, in which the green garlic plants were firstly cut into approximately 1 cm segments and 2 g samples were weighed and put into a 40 mL SPME vial (see details in Figure S2) which was pre-incubated for 10 min at 45 °C, and then the SPME fiber was exposed for 45 min at the same temperature to adsorb the volatiles. For the second and the third method, a hermetic glass container with 8 L in volume was used to collect the volatiles, the difference being that the green garlic seedlings were cut into segments or not (see details in Figure S2). Eight uniform green garlic seedlings were selected and put into these containers for three hours, then the SPME fibers were used to absorb the produced volatiles. All the collected volatile compounds were directly measured by a GC-MS system. Moreover, an external standard method was used for quantitative analysis of these volatiles, in which 2 mL DADS standard solutions with concentrations of

0.1, 0.5, 1.0, 1.5 and 2.0 mmol/L were prepared and added into a 20 mL SPME vial. Therefore, the final concentrations of DADS in SPME vials that used for the external standard method were 0.01, 0.05, 0.10, 0.15, 0.20 mmol/L, respectively.

The GC was equipped with a DB-5MS column (30 m × 0.25 mm × 0.25 µm, Agilent Technologies, Santa Clara, CA, USA). UHP helium was used as the carrier gas at a flow rate of 1 mL min$^{-1}$. The desorption time was 2.5 min. The injector temperature was set at 230 °C and injection was performed in a split mode and split ratio of 30:1. The gas flow was set at 0.8 mL/min. The temperature was programed as follows: 40 °C for 2 min, then rising at 6 °C /min to 170 °C and held for 5 min, where it then was increased at 10 °C /min to 250 °C, where it was held for 1 min, running for 40 min. For MS, the electron multiplier was operating at 70 eV and all samples were analyzed in electron impact mode. Mass ranges of $m/z$ 40–450 u for green garlic analysis were acquired. The temperatures of the transfer line and ion trap were both held at 250 °C. All the treatments were done in triplicates.

The compound TIC was used to identify the allelochemicals of VOCs. The Xcalibar workstation NIST standard spectrum database was used to search for each detected component, and the related literatures and standard spectra were used to check and confirm the inspection results. Only components with a matching degree greater than 800 were reported, and the relative content of each component was calculated based on the area normalization method.

### 4.4. H₂O₂ Histological Analysis: 3,3-Diamino-Benzidine (DAB) Staining

DAB-vascular uptake staining method was used to measure the H₂O₂ production. The cucumber leaves (1 cm sections) were immersed in a solution containing 1 mg/mL DAB that was dissolved in HCl-acidified (pH 3.8) distilled water, then incubated under light for 8 h, then the treated leaves were examined under a microscope [36].

### 4.5. ROS Assay

The H₂O₂ content in cucumber was measured according to the method described by Gong et al. using potassium iodide [37]. Superoxide anion ($O_2^{\bullet-}$) content was measured according to the method of Wang et al. [38].

### 4.6. Antioxidative Enzyme and Antioxidant Substances Analysis

To assay antioxidative enzymes, 0.5 g of cucumber leaves were weighed and ground together with 6 mL of 200 mmol/L phosphate buffer (pH 7.8) which contains 1% (*w/v*) soluble polyvinyl pyrrolidone (PVP) under ice bath conditions. The homogenates were centrifuged at 11,000× g for 20 min at 4 °C and the supernatant were collected to measure the activities of SOD, CAT, POD and the contents of soluble protein and MDA [8]. For the assay of antioxidant substances, 1.0 g of cucumber leaves were homogenized in 5 mL of 1% (*w/v*) saline buffer (1% (*w/v*) PVP, 1 mmol/L ethylenediaminetetraacetic acid (EDTA)). Homogenates were centrifuged at 11,000 g for 15 min at 4 °C and the supernatants were collected to check the content of ASA and GSH [39].

### 4.7. Total RNA Extraction and Antioxidant Enzyme Gene Expression Analysis

Total RNA was isolated from cucumber leaves using Trizol total RNA Extraction Reagent (Bioer, Hangzhou, China). First-strand cDNA was synthesized by a HiFiScript cDNA Synthesis Ki (CoWin Biosciences, Beijing, China). The expression of selected genes was analyzed by real-time quantitative polymerase chain reaction (RTq-PCR). The corresponding primer sequences were selected from Zhao et al. [23]. The qPCR was performed with the iCycleriQTM 5 multicolor real-time PCR detection system (Bio-Rad, Hercules, CA, USA) following the manufacturer's instructions. Gene expression levels were calculated on the basis of the $2^{-\Delta\Delta Ct}$ method [40]. Three biological and technical replications were performed, respectively.

*4.8. Statistical Analysis*

The statistical analyses were performed using SPSS 13.0 (IBM, Armonk, NY, USA). The significant differences between control and experimental groups that were treated with green garlic was tested by one-way ANOVA followed by Tukey's test. The significant differences between DADS treatment and control was determined by a t-test, with * $p$ values < 0.05 and ** $p$ values < 0.01, respectively.

## 5. Conclusions

This study's results showed that diallyl disulfide (DADS) is the main allelochemical of green garlic volatiles and it is released naturally at a level of approximately 0.08 mg/g green garlic. Green garlic volatiles and DADS showed obvious concentration effects. They both can reduce $O_2^-$ and they increase the accumulation of $H_2O_2$ by decreasing the expression of *CscAPX*, *CsGPX*, *CsCAT* and *CsPOD*, which may increase the disease resistance of cucumber. We will continue to study the effect of green garlic volatiles on cucumber disease resistance in the future.

**Supplementary Materials:** The following are available online. Figure S1: Plant growth and experimental design; Figure S2: Process of collection and identification of main allelochemicals of green garlic volatiles by GC/MS; Table S1: Total 17 compounds of green garlic volatile compounds of cutting segments (in sealed desiccator); Table S2: Total 42 compounds of whole green garlic (in sealed desiccator); Table S3: The green garlic weight and DADS concentration of different treatments.

**Author Contributions:** Formal analysis, F.Y. and X.L.; Funding acquisition, Z.C.; Investigation, H.W., R.D. and H.Y.; Methodology, F.Y., X.L. and H.W.; Project administration, Z.C.; Resources, Z.C.; Writing – original draft, F.Y.; Writing – review & editing, F.Y., X.L., H.W., R.D., H.Y. and Z.C.

**Funding:** This research was funded by the National Natural Science Foundation of China (No. 31772293).

**Acknowledgments:** We thanks to key laboratory of Northwest A&F University. Thanks to Yupeng Pan, Cheng F., Liu HQ and Muhamad A. give valuable advices on the revision.

**Conflicts of Interest:** The authors declare no conflict of interest.

## References

1. Adebesin, F.; Widhalm, J.R.; Boachon, B.; Lefevre, F.; Pierman, B.; Lynch, J.H.; Alam, I.; Junqueira, B.; Benke, R.; Ray, S.; et al. Emission of volatile organic compounds from petunia flowers is facilitated by an ABC transporter. *Science* **2017**, *356*, 1386–1388. [CrossRef] [PubMed]
2. Dudareva, N.; Klempien, A.; Muhlemann, J.K.; Kaplan, I. Biosynthesis, function and metabolic engineering of plant volatile organic compounds. *New Phytol.* **2013**, *198*, 16–32. [CrossRef] [PubMed]
3. Pichersky, E.; Gershenzon, J. The formation and function of plant volatiles: Perfumes for pollinator attraction and defense. *Curr. Opin. Plant. Biol.* **2002**, *5*, 237–243. [CrossRef]
4. Martins, N.; Petropoulos, S.; Ferreira, I.C. Chemical composition and bioactive compounds of garlic (*Allium sativum* L.) as affected by pre- and post-harvest conditions: A review. *Food Chem.* **2016**, *211*, 41–50. [CrossRef] [PubMed]
5. Jae, J.H.; Dong, S.J.; Mi, K.K.; Hoon, C.Y.; Ki Hyung, K.; Young Chul, P. Diallyl disulfide induces reversible G2/M phase arrest on a p53-independent mechanism in human colon cancer HCT-116 cells. *Oncol. Rep.* **2008**, *19*, 275–280.
6. Xiao, X.M.; Cheng, Z.H.; Meng, H.W.; Khan, M.A.; Li, H.Z. Intercropping with garlic alleviated continuous cropping obstacle of cucumber in plastic tunnel. *Acta. Agr. Scand. B-S P.* **2012**, *62*, 696–705. [CrossRef]
7. Cheng, F.; Cheng, Z.H.; Meng, H.W.; Tang, X.W. The garlic allelochemical diallyl disulfide affects tomato root growth by influencing cell division, phytohormone balance and expansin gene expression. *Front. Plant. Sci.* **2016**, *7*. [CrossRef] [PubMed]
8. Wang, M.Y.; Wu, C.N.; Cheng, Z.H.; Meng, H.W.; Zhang, M.R.; Zhang, H.J. Soil chemical property changes in eggplant/garlic relay intercropping systems under continuous cropping. *PLoS ONE* **2014**, *9*. [CrossRef]
9. Ding, H.Y.; Cheng, Z.H.; Liu, M.L.; Hayat, S.; Feng, H. Garlic exerts allelopathic effects on pepper physiology in a hydroponic co-culture system. *Biol. Open.* **2016**, *5*, 631–637. [CrossRef]
10. Cheng, F.; Cheng, Z.H. Research progress on the use of plant allelopathy in agriculture and the physiological and ecological mechanisms of allelopathy. *Front. Plant. Sci.* **2015**, *6*. [CrossRef]

11. Xiao, X.M.; Cheng, Z.H.; Meng, H.W.; Liu, L.H.; Li, H.Z.; Dong, Y.X. Intercropping of green garlic (*Allium sativum* L.) induces nutrient concentration changes in the soil and plants in continuously cropped cucumber (*Cucumis sativus* L.) in a plastic tunnel. *PLoS ONE* **2013**, *8*. [CrossRef] [PubMed]

12. Yi, L.; Su, Q. Molecular mechanisms for the anti-cancer effects of diallyl disulfide. *Food Chem. Toxicol.* **2013**, *57*, 362–370. [CrossRef] [PubMed]

13. Kubota, N.; Yamane, Y.; Toriu, K.; Kawazu, K.; Higuchi, T.; Nishimura, S. Identification of active substances in garlic responsible for breaking bud dormancy in grapevines. *J. Jpn. Soc. Hortic. Sci.* **1999**, *68*, 1111–1117. [CrossRef]

14. Cheng, F.; Cheng, Z.H.; Meng, H.W. Transcriptomic insights into the allelopathic effects of the garlic allelochemical diallyl disulfide on tomato roots. *Sci Rep.-Uk.* **2016**, *6*. [CrossRef] [PubMed]

15. Ren, K.L.; Hayat, S.; Qi, X.F.; Liu, T.; Cheng, Z.H. The garlic allelochemical DADS influences cucumber root growth involved in regulating hormone levels and modulating cell cycling. *J. Plant. Physiol.* **2018**, *230*, 51–60. [CrossRef] [PubMed]

16. Lee, S.N.; Kim, N.S.; Lee, D.S. Comparative study of extraction techniques for determination of garlic flavor components by gas chromatography-mass spectrometry. *Anal. Bioanal Chem.* **2003**, *377*, 749–756. [CrossRef] [PubMed]

17. Kimbaris, A.C.; Siatis, N.G.; Daferera, D.J.; Tarantilis, P.A. Pappas CS, Polissiou MG: Comparison of distillation and ultrasound-assisted extraction methods for the isolation of sensitive aroma compounds from garlic (*Allium sativum* L.). *Ultrason. Sonochem.* **2006**, *13*, 54–60. [CrossRef] [PubMed]

18. Warren, J.M.; Parkinson, D.R.; Pawliszyn, J. Assessment of thiol compounds from garlic by automated headspace derivatized in-needle-NTD-GC-MS and derivatized in-fiber-SPME-GC-MS. *J. Agr. Food Chem.* **2013**, *61*, 492–500. [CrossRef] [PubMed]

19. Godinez-Vidal, D.; Rocha-Sosa, M.; Sepulveda-Garcia, E.B.; Lara-Reyna, J.; Rojas-Martinez, R.; Zavaleta-Mejia, E. Phenylalanine ammonia lyase activity in chilli CM-334 infected by Phytophthora capsici and Nacobbus aberrans. *Eur. J. Plant. Pathol.* **2008**, *120*, 299–303. [CrossRef]

20. Rio D, L.A. ROS and RNS in plant physiology: An overview. *J. Exp. Bot.* **2015**, *66*, 2827–2837. [CrossRef]

21. Gill, S.S.; Tuteja, N. Reactive oxygen species and antioxidant machinery in abiotic stress tolerance in crop plants. *Plant. Physiol. Bioch.* **2010**, *48*, 909–930. [CrossRef] [PubMed]

22. Talukdar, D. Allelopathic effects of Lantana camara L. on Lathyrus sativus L.: Oxidative imbalance and cytogenetic consequences. *Allelopath. J.* **2013**, *31*, 71–90.

23. Zhao, L.J.; Hu, Q.R.; Huang, Y.X.; Fulton, A.N.; Hannah-Bick, C.; Adeleye, A.S.; Keller, A.A. Activation of antioxidant and detoxification gene expression in cucumber plants exposed to a Cu(OH)$_2$ nanopesticide. *Env. Sci-Nano.* **2017**, *4*, 1750–1760. [CrossRef]

24. Putnik, P.; Gabrić, D.; Roohinejad, S.; Barba, F.J.; Granato, D.; Mallikarjunan, K.; Lorenzo, J.M.; Bursać Kovačević, D. An overview of organosulfur compounds from Allium spp.: From processing and preservation to evaluation of their bioavailability, antimicrobial, and anti-inflammatory properties. *Food Chem.* **2019**, *276*, 680–691. [CrossRef] [PubMed]

25. Chyau, C.C.; Mau, J.L. Release of volatile compounds from microwave heating of garlic juice with 2,4-decadienals. *Food Chem.* **1999**, *64*, 531–535. [CrossRef]

26. Kim, N.Y.; Park, M.H.; Jang, E.Y.; Lee, J. Volatile distribution in garlic (*Allium sativum* L.) by solid phase microextraction (SPME) with different processing conditions. *Food Sci. Biotechnol.* **2011**, *20*, 775–782. [CrossRef]

27. Biagini, D.; Lomonaco, T.; Ghimenti, S.; Bellagambi, F.G.; Onor, M.; Scali, M.C.; Barletta, V.; Marzilli, M.; Salvo, P.; Trivella, M.G. Determination of volatile organic compounds in exhaled breath of heart failure patients by needle trap micro-extraction coupled with gas chromatography-tandem mass spectrometry. *J. Breath Res.* **2017**, *11*. [CrossRef]

28. Ghimenti, S.; Lomonaco, T.; Bellagambi, F.G.; Tabucchi, S.; Onor, M.; Trivella, M.G.; Ceccarini, A.; Fuoco, R.; Francesco, F.D. Comparison of sampling bags for the analysis of volatile organic compounds in breath. *J. Breath Res.* **2015**, *9*. [CrossRef]

29. Roxas, V.P.; Lodhi, S.A.; Garrett, D.K.; Mahan, J.R.; Allen, R.D. Stress tolerance in transgenic tobacco seedlings that overexpress glutathione S-transferase/glutathione peroxidase. *Plant. Cell Physiol.* **2000**, *41*, 1229–1234. [CrossRef]

30. Liu, J.G.; Zhang, X.L.; Sun, Y.H.; Lin, W. Antioxidative capacity and enzyme activity in Haematococcus pluvialis cells exposed to superoxide free radicals. *Chin. J. Oceanol. Limn.* **2010**, *28*, 1–9. [CrossRef]

31. Tanaka, T.; Izawa, S.; Inoue, Y. GPX2, encoding a phospholipid hydroperoxide glutathione peroxidase homologue, codes for an atypical 2-Cys peroxiredoxin in Saccharomyces cerevisiae. *J. Biol. Chem.* **2005**, *280*, 42078–42087. [CrossRef] [PubMed]

32. Garg, N.; Manchanda, G. ROS generation in plants: Boon or bane? *Plant. Biosyst.* **2009**, *143*, 81–96. [CrossRef]

33. Caverzan, A.; Passaia, G.; Rosa, S.B.; Ribeiro, C.W.; Lazzarotto, F.; Margis-Pinheiro, M. Plant responses to stresses: Role of ascorbate peroxidase in the antioxidant protection. *Genet. Mol. Biol.* **2012**, *35*, 1011–1019. [CrossRef] [PubMed]

34. Hossain, Z.; Komatsu, S. Contribution of proteomic studies towards understanding plant heavy metal stress response. *Front. Plant. Sci.* **2013**, *3*. [CrossRef] [PubMed]

35. Zho, Y.J.; Su, J.; Shi, L.; Liao, Q.J.; Su, Q. DADS downregulates the Rac1-ROCK1/PAK1-LIMK1-ADF/cofilin signaling pathway, inhibiting cell migration and invasion. *Oncol. Rep.* **2013**, *29*, 605–612. [CrossRef] [PubMed]

36. Zhang, H.C.; Wang, C.F.; Cheng, Y.L.; Chen, X.M.; Han, Q.M.; Huang, L.L.; Wei, G.R.; Kang, Z.S. Histological and cytological characterization of adult plant resistance to wheat stripe rust. *Plant. Cell Rep.* **2012**, *31*, 2121–2137. [CrossRef] [PubMed]

37. Gong, H.J.; Randall, D.P.; Flowers, T.J. Silicon deposition in the root reduces sodium uptake in rice (*Oryza sativa* L.) seedlings by reducing bypass flow. *Plant. Cell Env.* **2006**, *29*, 1970–1979. [CrossRef] [PubMed]

38. Wang, B.; Zhu, S.J. Pre-storage cold acclimation maintained quality of cold-stored cucumber through differentially and orderly activating ROS scavengers. *Postharvest Biol. Tec.* **2017**, *129*, 1–8. [CrossRef]

39. Wang, G.; Zhang, H.; Lai, F.; Wu, H. Germinating peanut (*Arachis hypogaea* L.) seedlings attenuated selenite-induced toxicity by activating the antioxidant enzymes and mediating the ascorbate-glutathione cycle. *J. Agric. Food Chem.* **2016**, *64*, 1298–1308. [CrossRef]

40. Livak, K.J.; Schmittgen, T.D. Analysis of relative gene expression data using real-time quantitative PCR and the $2^{-\Delta\Delta Ct}$ method. *Methods* **2001**, *25*, 402–408. [CrossRef]

**Sample Availability:** Samples of the compounds are not available from the authors.

*Article*

# The Profile of Urinary Headspace Volatile Organic Compounds After 12-Week Intake of Oligofructose-Enriched Inulin by Children and Adolescents with Celiac Disease on a Gluten-Free Diet: Results of a Pilot, Randomized, Placebo-Controlled Clinical Trial


Natalia Drabińska [1,*], Elżbieta Jarocka-Cyrta [2], Norman Mark Ratcliffe [3] and Urszula Krupa-Kozak [1,*]

[1]  Department of Chemistry and Biodynamics of Food, Institute of Animal Reproduction and Food Research of Polish Academy of Sciences, Tuwima 10 Str., 10-748 Olsztyn, Poland
[2]  Department of Pediatrics, Gastroenterology, and Nutrition, Collegium Medicum, University of Warmia & Mazury, Oczapowskiego 2 Str., 10-719 Olsztyn, Poland; ejarocka@op.pl
[3]  Institute of Biosensor Technology, the University of the West of England, University of the West of England, Coldharbour Lane, Frenchay, Bristol BS16 1QY, UK; norman.ratcliffe@uwe.ac.uk
*   Correspondence: n.drabinska@pan.olsztyn.pl or natalia_drabinska@wp.pl (N.D.); u.krupa-kozak@pan.olsztyn.pl (U.K.-K.); Tel.: +48-895-234-639 (N.D); +48-895-234-618 (U.K.-K.)



Academic Editor: Derek J. McPhee
Received: 20 March 2019; Accepted: 4 April 2019; Published: 5 April 2019



**Abstract:** The concentration of volatile organic compounds (VOCs) can inform about the metabolic condition of the body. In the small intestine of untreated persons with celiac disease (CD), chronic inflammation can occur, leading to nutritional deficiencies, and consequently to functional impairments of the whole body. Metabolomic studies showed differences in the profile of VOCs in biological fluids of patients with CD in comparison to healthy persons; however, there is scarce quantitative and nutritional intervention information. The aim of this study was to evaluate the effect of the supplementation of a gluten-free diet (GFD) with prebiotic oligofructose-enriched inulin (Synergy 1) on the concentration of VOCs in the urine of children and adolescents with CD. Twenty-three participants were randomized to the group receiving Synergy 1 (10 g per day) or placebo for 12 weeks. Urinary VOCs were analyzed using solid-phase microextraction and gas chromatography–mass spectrometry. Sixteen compounds were identified and quantified in urine samples. The supplementation of GFD with Synergy 1 resulted in an average concentration drop (36%) of benzaldehyde in urine samples. In summary, Synergy 1, applied as a supplement of GFD for 12 weeks had a moderate impact on the VOC concentrations in the urine of children with CD.



**Keywords:** volatile organic compounds; celiac disease; gluten-free diet; gas chromatography–mass spectrometry; solid-phase microextraction; prebiotic


---

## 1. Introduction

Volatile organic compounds (VOCs) are carbon-based molecules that are volatile at ambient temperature [1]. Hundreds of VOCs are secreted by cells of the human body, as a result of metabolic processes. The qualitative and quantitative profile of VOCs in biological fluids can vary depending on the metabolic changes; therefore, the pattern of volatile metabolites may reflect the presence of disease [2]. Several studies showed an association between the pattern of volatile biomarkers and the presence of gastrointestinal diseases [3–9]. Gas chromatography coupled with mass spectrometry (GC–MS), a "gold standard" in VOC analysis, was applied to distinguish patients with diarrhea-predominant irritable bowel syndrome, Crohn's disease, ulcerative colitis, and healthy controls [8], as well as celiac disease (CD) and irritable bowel syndrome [6]. Moreover, the effect of a gluten-free diet (GFD) on the exhaled breath was evaluated [10,11]. The great success of previous studies contributed to the tremendous progress in the development of new analytical techniques for VOC detection, such as field-asymmetric ion mobility spectrometry and selected ion flow tube mass spectrometry, successfully applied in the analysis of VOCs in gastrointestinal diseases [5,12]. Recently, volatolomics was established as a new scientific domain with significant diagnostic potential [13]. The application of the VOC analysis can be an innovative and non-invasive tool for the diagnosis of diseases, as well as for the monitoring of the effectiveness of treatment [14].

It is believed that changes in VOCs observed in gastrointestinal diseases are the result of the impaired fermentation activity of the gut microbiota [6]. In many clinical trials, the changes in the metabolism of bacteria were suggested as more informative than the microbiota composition itself [15–17]. Moreover, many of the nutritional interventions had moderate or no effect on qualitative and/or quantitative changes of intestinal microbiota; however, they had a much more prominent effect on their metabolism [14,16,17]. In the intestines, the interaction between commensal bacteria, human cells, and pathogens occurs and results in the formation of hundreds of VOCs observed in feces, urine, sweat, blood, and exhaled breath [18]. The presence of intestinal VOCs in urine, sweat, blood, and breath can be related to changes in the intestinal barrier [6], which are attributed to several gastrointestinal diseases [19,20]. The analysis of VOCs in urine has several benefits over the other biological fluids. Urine collection is non-invasive and does not cause discomfort even with multiple sampling. Moreover, the concentration of VOCs in urine is higher compared to blood, as urine is pre-concentrated in the kidney, which facilitates the detection of metabolites [21]. However, on the other hand, the pre-concentration of urine can vary within and between individuals, which should be considered as a confounding factor.

CD is a life-long gluten-related enteropathy observed in genetically predisposed individuals. The prevalence of CD is estimated for approximately 1% of the global population; however, it is suggested that many patients remain undiagnosed [22]. In addition to the intestinal (abdominal pain, diarrhea) and extra-intestinal (increased bone fractures, anemia, depression) symptoms, the dysbiosis of intestinal microbiota, characterized by lower diversity and disproportion between Gram-positive and Gram-negative bacteria [23], as well as altered intestinal permeability [20], is commonly observed in CD patients. The only approved treatment of CD is a GFD. However, in many patients, even after long-term adherence to the treatment, nutritional deficiencies and a lack of intestinal recovery are observed [24–26]. Therefore, there is a strong need to incorporate auxiliary therapies, including GFD supplementation, into the treatment regime of the CD, followed by an evaluation of their safety and effects.

Prebiotics, defined as substrates that are selectively utilized by host microorganisms conferring a health benefit [27], were reported to increase the absorption of nutrients [28] and to improve the histomorphological parameters of intestines [29], confirmed in a clinical trial and in vivo studies. Recently, the beneficial effects of prebiotics on several aspects of health in CD patients were reported [17,30–32]. Briefly, prebiotics were found to stimulate the activity of the intestinal microbiota [17], modulate the amino-acid metabolism [30], improve the fat-soluble vitamin status [31], and improve bone metabolism [32]. However, the impact of GFD supplementation with prebiotics on the VOC pattern remains to be analyzed.

In general, the number of studies evaluating VOC pattern after nutritional interventions is limited. Therefore, this exploratory, randomized, placebo-controlled study is proposed to evaluate the effect of prebiotic oligofructose-enriched inulin intake on the profile of VOCs in the urine of children and adolescents with CD following a GFD, using solid-phase microextraction with GC–MS.

*Hypothesis*: Nutritional intervention with oligofructose-enriched inulin will improve the intestinal health of children with CD, affecting the profile of urinary VOC.

## 2. Results

In the urine of patients with CD, a total of sixteen compounds, representing different chemical groups, were identified and quantitatively characterized (Table 1). Additionally, 4-methylphenol and 2-pentylfuran were determined in some urine samples, but values above the limit of quantification were detected only in a few samples. An example of the chromatogram is presented in Figure 1.

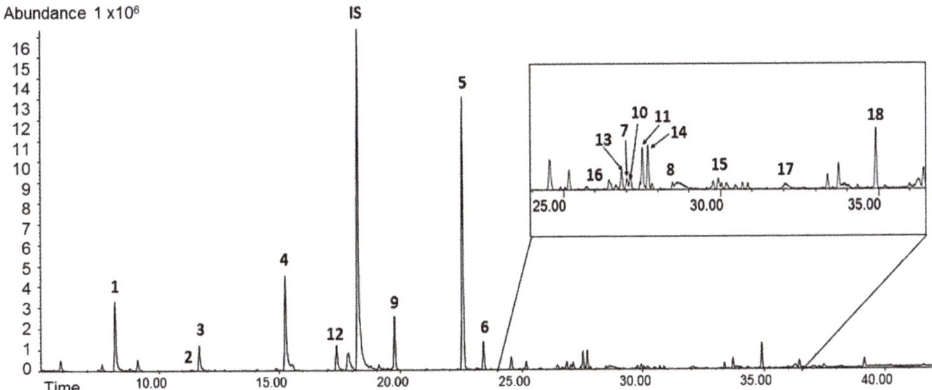


**Figure 1.** An example of a chromatogram of urinary volatile organic compounds (VOCs) obtained with gas chromatography–mass spectrometry (GC–MS): (1) acetone; (2) butane-2,3-dione; (3) butan-2-one; (4) pentan-2-one; (5) heptan-4-one; (6) heptan-2-one; (7) 6-methylhept-5-en-2-one; (8) *trans*-3-octen-2-one; (9) hexanal; (10) benzaldehyde; (11) octanal; (12) dimethyl disulfide; (13) dimethyl trisulfide; (14) D-limonene; (15) linalool; (16) 2-pentylfuran; (17) 4-methylphenol; (18) 1,3-di-*tert*-butylbenzene.

At baseline, the concentrations of VOCs in urine were similar in both experimental groups (Table 1). The median concentration of *trans*-3-octen-2-one was similar in both experimental groups at baseline; however, this ketone was not detected in three urine samples of patients from the Synergy 1 group.

The supplementation of GFD with Synergy 1 did not impact on the profile or the concentration of the majority of VOCs in the urine of CD patients. The only significant ($p < 0.05$) change was observed for benzaldehyde, where the concentration decreased by 36% after the intervention (Table 1). Furthermore, *trans*-3-octen-2-one was not detected in some urine samples (two from Synergy 1 group and one from the placebo group); however, it had no effect on differences between experimental groups. The decrease in the concentrations of 1,3-di-*tert*-butylbenzene was observed in the placebo group after the twelve-week intervention.

Multivariate analysis showed a high inter-individual variation of the data (Figure 2). Principal component analysis (PCA) plots explained 46.42% and 44.25% of variations at baseline and after the intervention, respectively. No separation was observed either before or after the intervention. At baseline, anthropometric indices (age, height, body weight) had an influence on D-limonene and acetone concentrations. The level of linalool was associated with the time of GFD adherence. Similar associations were not observed after the intervention (Figure 2).

**Table 1.** Volatile organic compounds (VOCs) (nmol/L) detected and quantified in the urine of children from Synergy 1 and placebo group, before (T0) and after (T1) the intervention, expressed as median (P25–P75).

| | T0 | | T1 | |
|---|---|---|---|---|
| | **Placebo** | **Synergy 1** | **Placebo** | **Synergy 1** |
| | Ketones | | | |
| acetone | 12023 (9066–17649) | 12184 (10229–17740) | 12816 (10549–14704) | 12564 (9969–19006) |
| butane-2,3-dione | 66.80 (50.50–88.74) | 63.22 (53.03–104.73) | 59.10 (28.12–68.74) | 53.63 (44.46–65.58) |
| butan-2-one | 167.59 (73.60–229.55) | 180.77 (95.76–271.57) | 168.38 (116.26–313.56) | 229.20 (126.17–294.85) |
| pentan-2-one | 21.76 (9.42–57.88) | 31.80 (20.55–54.27) | 41.57 (18.19–60.49) | 40.04 (32.83–64.30) |
| heptan-4-one | 41.91 (22.65–125.91) | 53.32 (28.13–87.32) | 86.19 (32.29–103.91) | 84.02 (51.86–130.34) |
| heptan-2-one | 6.94 (3.42–14.82) | 6.11 (2.59–17.55) | 10.31 (3.39–12.98) | 8.04 (5.38–11.73) |
| 6-methylhept-5-en-2-one | 1.31 (0.57–4.08) | 1.33 (0.50–2.30) | 1.87 (0.38–2.68) | 1.33 (0.52–1.83) |
| *trans*-3-octen-2-one | 0.59 (0.39–4.49) | 0.56 (0.41–0.92) | 1.08 (0.46–1.90) | 0.71 (0.39–1.02) |
| | Aldehydes | | | |
| hexanal | 37.38 (24.60–59.95) | 23.79 (17.99–36.84) | 36.59 (24.22–45.63) | 28.27 (18.84–38.78) |
| benzaldehyde | 7.14 (3.14–22.86) | 7.16 (3.47–12.94) | 7.53 (2.48–10.00) | 6.21 (3.52–7.14) [a] |
| octanal | 0.83 (0.35–4.11) | 0.62 (0.15–2.58) | 0.85 (0.20–2.04) | 1.11 (0.39–1.34) |
| | Sulfur compounds | | | |
| dimethyl disulfide | 19.29 (11.39–23.26) | 13.02 (6.86–18.44) | 8.94 (6.70–15.16) | 12.67 (7.02–19.29) |
| dimethyl trisulfide | 1.22 (0.95–3.30) | 1.01 (0.34–2.27) | 1.28 (0.46–1.90) | 1.72 (0.50–3.09) |
| | Terpenes | | | |
| limonene | 45.27 (9.29–67.86) | 32.56 (4.02–42.40) | 36.78 (24.22–45.63) | 28.79 (5.46–62.43) |
| linalool | 20.63 (14.68–28.12) | 19.28 (15.99–29.20) | 16.14 (11.86–26.20) | 18.09 (11.76–26.24) |
| | Aromatic compounds | | | |
| 1,3-di-*tert*-butylbenzene | 0.82 (0.42–1.25) | 0.52 (0.33–0.92) | 0.66 (0.34–0.87) [a] | 0.57 (0.33–0.91) |

a—statistically significant differences within groups before and after the intervention.

## 3. Discussion

Our study, for the first time, reports the profile and concentrations of VOCs in the headspace above the urine of children and adolescents with CD after a 12-week nutritional intervention with prebiotics applied as a supplement of GFD.

Sixteen compounds quantified in the present study were selected based on the previous studies reporting differences in urinary VOCs between healthy children and children with CD [4,33]. We hypothesized that, after the nutritional intervention with prebiotics, the urinary profile of VOC in children with CD would be altered, as a consequence of the changes in the gut caused by prebiotics. The present study indicated, however, that applied nutritional intervention did not have a strong effect on the profile of VOCs in urine. The only difference observed after the Synergy 1 intake was a significant reduction in benzaldehyde concentration. The explanation for the benzaldehyde drop in concentration can be related to the microbiota activity. Benzaldehyde can be formed as a result of the conversion of phenylalanine by aminotransferase produced by *Lactobacillus* bacteria [34]. In our study, the amount of the precursor phenylalanine was similar in both groups before and after the supplementation [30]. However, the *Lactobacillus* count was significantly lower in the Synergy 1 group as compared to placebo [17]. This might result in a reduction in phenylalanine conversion and, consequently, in decreased benzaldehyde concentration in urine.

In the placebo group, the decrease in the concentration of 1,3-di-*tert*-butylbenzene was observed. This is particularly interesting because this compound was suggested as a marker of CD, observed only in the urine of children with CD, while it existed in none of the samples from healthy children [33]. However, the origin of 1,3-di-*tert*-butylbenzene in the human body is not clear and requires further studies. It was reported that 1,3-di-*tert*-butylbenzene is a product of radiolysis of the antioxidant Irgafos used in food packaging [35], and this is a possible explanation of its origin in urine.

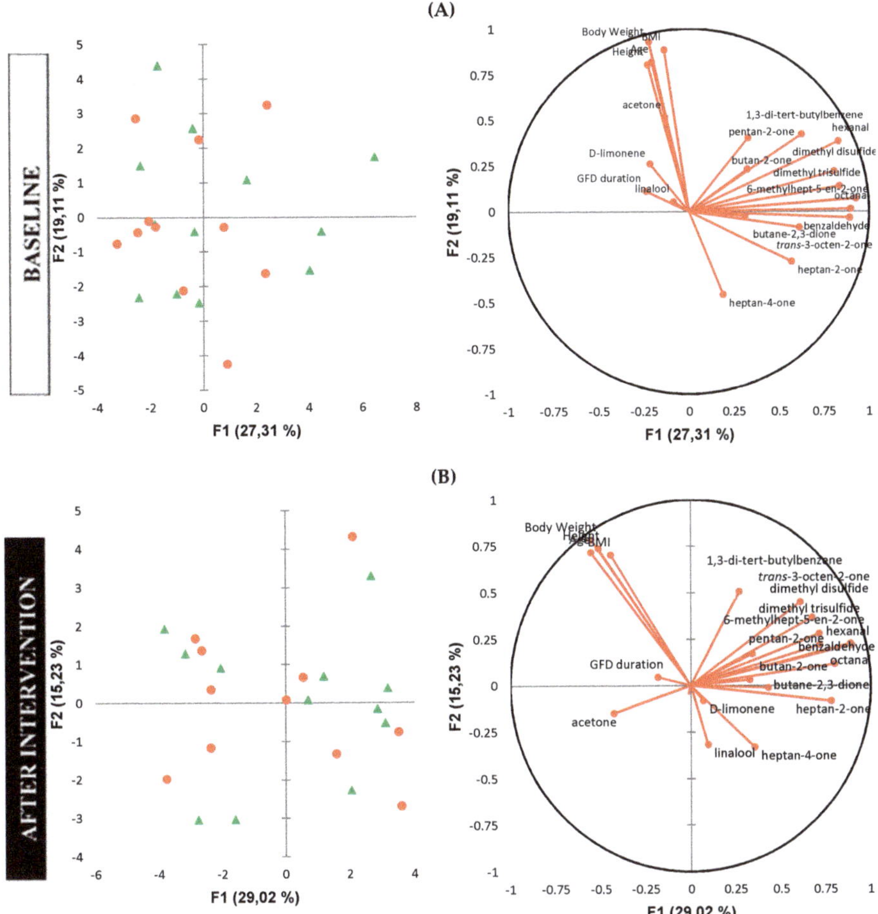

**Figure 2.** Results of principal component analysis (PCA) of urinary VOCs at baseline (**A**), and after the intervention (**B**). Red circles—Synergy 1 group; green triangles—placebo group. Left graphs—score plot; right graphs—correlation circle presenting correlations between individual VOCs and anthropometric indices.

The high inter-individual variation in VOC profiles makes it difficult to demonstrate significant differences after the applied prebiotic intervention. On the other hand, it may result from the recovery of the intestinal mucosa and the reduction in intestine permeability. The recent research suggests that the perturbation in the urinary VOC profile observed in some gastrointestinal diseases may result from changes in the gut barrier [6]. Children and adolescents participating in the present study were treated with a GFD for at least six months (average: 2.9 ± 1.9 and 2.3 ± 1.2 years in Synergy 1 and placebo group, respectively), which is considered as sufficient time to restore the proper functioning of the intestinal barrier [36]. In literature, the results of the prebiotic supplementation aimed to improve intestinal permeability are inconsistent. Animal studies with non-digestible fructans confirmed beneficial histomorphological changes in the gut and intestinal barrier functioning [29]. Similarly, a randomized, double-blind crossover nutritional intervention study with inulin-enriched pasta showed modulation of circulating levels of zonulin and glucagon-like peptide 2 in healthy young volunteers, suggesting that prebiotics could be used in the prevention of gastrointestinal diseases [37]. On the other hand, many clinical trials found no effect of prebiotics on intestinal barrier functions [38–40]. Therefore, the unambiguous impact of prebiotics on intestinal barrier functioning require further in-depth investigation.

To our knowledge, there is only one study analyzing the VOC profiles after nutritional intervention [14], making the discussion of the present results in response to other studies a challenge. The study by Rossi and co-authors [14] referred to irritable bowel syndrome and analyzed VOCs in feces. Although impossible to compare, the results presented in this interesting paper reported that the analysis of VOCs in feces can predict responses to the nutritional intervention. As in the present study, we did not observe profound differences in urinary VOCs after the applied nutritional intervention. In future studies, it would be worthwhile to analyze the VOC profile in feces of children with CD, especially as, in our previous research, we observed significant changes in the concentration of short-chain fatty acids in the feces of children with CD after the intervention with prebiotics [17].

Despite the novel nature of this study, some limitations should be mentioned. Firstly, there was no calculation of the sample size. However, this limitation is related to the pilot type of study. Therefore, the present study should be considered as an exploratory study, providing the data for calculation of the sample size for future validation studies. A second limitation was the small number of participants, causing problems in statistical evaluation based on the high inter-individual variability. This limitation is also strongly associated with the preliminary nature of the study. Thirdly, in this study, the control of the diet was not presented; however, the control of the diet was performed using validated food frequency questionnaires [41], even though details were not presented in this manuscript.

Finally, the study presented here is focused on the targeted analysis of selected compounds, limiting the number of possible responses of a non-analyzed and unknown compound. However, the authors wanted to focus on quantitative analysis, which is missing in the literature; therefore, to calculate accurately, a limited number of compounds had to be selected. However, comparing whole metabolic profiles in the urine of children and adolescents with CD after the nutritional intervention would also be scientifically interesting; therefore, it is suggested as a future study.

## 4. Materials and Methods

### 4.1. Chemicals and Materials

Chemical standards of acetone, butane-2,3-dione, butan-2-one, thiophene, dimethyl disulfide, hexanal, heptan-4-one, heptan-2-one, 2-pentylfuran, dimethyl trisulfide, 6-methylhept-5-en-2-one, benzaldehyde, octanal, D-limonene, *trans*-3-octen-2-one, linalool, 4-methylphenol, 1,3-di-*tert*-butylbenzene, internal standard (4-methylpentan-2-ol), and sodium chloride (NaCl, $\geq$99.5%) were supplied by Sigma-Aldrich (Saint Louis, MO, USA). MilliQ water (Millipore, Bedford, MO, USA) was used for the preparation of standards. Hydrochloric acid (HCl, 37%) was purchased from Chempur (Piekary Śląskie, Poland). The 75-μm carboxen/polydimethylsiloxane (CAR/PDMS) (stable flex) solid-phase microextraction (SPME) fibers were purchased from Supelco (Bellefonte, PA, USA).

### 4.2. Study Protocol

A randomized, placebo-controlled, single-center clinical trial with nutritional intervention was performed. The full details of the study protocol, inclusion/exclusion criteria, and a CONSORT chart are available elsewhere [42]. The present study is part of a larger study which was registered in the US National Library of Medicine (identifier: NCT03064997; http://www.clinicaltrials.gov). The study was performed in the Gastrointestinal Clinic of the Children's Hospital in Olsztyn from January to June 2016. A brief description of original study is as follows: 34 children diagnosed with CD and following a GFD for at least six months were randomly assigned to a group receiving 10 g per day of oligofructose-enriched inulin (Synergy 1; Orafti®, Beneo, Belgium) or a group receiving placebo (maltodextrin) for a period of 12 weeks. The placebo and prebiotic supplements were identical in appearance and taste. Participants and their caregivers, clinicians, and most of the investigators (except one person providing supplements) were blinded. During the intervention, participants were asked to note any side effects and daily supplement intake. Children were under the medical supervision of a gastroenterologist, and blood morphology data can be found elsewhere [43].

The study protocol was approved by the Bioethics Committee of the Faculty of Medicine of the University of Warmia and Mazury in Olsztyn, Poland (decision No. 23/2015). All procedures involving human participants were performed with the ethical principles of the 1964 Declaration of Helsinki and its later amendments. Parents or caregivers of participants were fully informed about the study and signed the written informed consent on the first check-up visit.

In the present study, urine samples collected from 23 children were analyzed: 11 children from the Synergy 1 group and 12 children from the placebo group. Patients' anthropometric characteristics are presented in Table 2. A smaller number of samples used in this study compared to the original study were related to antibiotic intake during the intervention (two persons), inappropriate compliance (less than 80% of time) to a nutritional intervention assessed based on the intervention diary (two persons), and insufficient amount of urine provided for VOC analysis (seven persons). Fresh morning urine samples were collected from each participant at baseline and after the intervention. Samples were immediately centrifuged at 3500 rpm for 10 min, and aliquots of 4 mL were stored at $-80\,^\circ$C until further analysis.

**Table 2.** The participants' anthropometric data. Results are presented as ranges and means $\pm$ standard deviation.

| | Synergy 1 Group | Placebo Group | *p*-Value |
|---|---|---|---|
| *N* | 11 | 12 | |
| Gender | Girls—7; Boys—4 | Girls—8; Boys—4 | 0.886 |
| Age (years) | 5–18; Av[1] = 10.8 $\pm$ 4.1 | 4–16; Av = 10.2 $\pm$ 4.4 | 0.720 |
| Body weight (kg) | 15.8–67.9; Av = 38.3 $\pm$ 16.9 | 16.3–66.8; Av = 35.6 $\pm$ 17.0 | 0.703 |
| Height (m) | 112.5–170.0; Av = 145.1 $\pm$ 21.3 | 103.0–172.0; Av = 139.4 $\pm$ 22.6 | 0.540 |
| BMI (kg/m$^2$) | 12.5–23.5; Av = 17.2 $\pm$ 3.7 | 13.7–28.4; Av = 17.3 $\pm$ 4.0 | 0.962 |

[1] Av = average; BMI = body mass index.

*4.3. VOC Analysis*

Analysis of VOCs in urine was performed according to the previously published protocol [33]. Briefly, 4 mL of urine was placed in 20-mL headspace vials with 2.98 g of sodium chloride and 21 µL of 6 M hydrochloric acid. Then, 4-methylpentan-2-ol was added to each sample as an internal standard with a concentration of 196.24 nmol/L. Samples were incubated for 20 min at 30 $^\circ$C with a shaking speed of 500 rpm using a MultiTherm shaker (Benchmark Scientific, Edison, NJ, USA), resulting in the release of VOCs from urine and their accumulation at the headspace. Next, the previously conditioned CAR/PDMS fiber was manually inserted into the headspace, and extraction was carried out for 15 min at 30 $^\circ$C. After extraction, the fiber was introduced into the gas chromatography injector port with a 0.75-mm inner diameter (ID) splitless glass liner (Supelco, Bellefonte, PA, USA), set to a splitless mode, with an inlet temperature of 240 $^\circ$C. Thermal desorption was carried out for 10 min to avoid carryover.

Analysis of VOCs was performed using an HP 5890 gas chromatograph coupled with an HP 5972 mass selective detector (Agilent Technologies, Santa Clara, CA, USA) [33]. The compounds were separated using a Zebron ZB-624 capillary column, 60 m $\times$ 0.25 mm $\times$ 1.40 µm (Phenomenex, Torrance, CA, USA). The carrier gas was helium at a constant flow rate of 1 mL·min$^{-1}$. The oven temperature program was set as follows: 40 $^\circ$C for 2 min, an increase to 220 $^\circ$C at a rate of 5 $^\circ$C·min$^{-1}$, and maintained at the final temperature for 5 min. Total run time was 42 min. Mass spectra were obtained by electron ionization (EI) in the range of 40–550 *m/z*, and a solvent delay was set for 5 min. Ion source temperature was 230 $^\circ$C and electronic impact energy was 70 eV. Total ion chromatograms were analyzed with the MSD ChemStation E.02.02.1431 software (Agilent Technologies, Santa Clara, CA, USA). Identification of compounds was performed by comparison of the retention times and mass spectra to commercial standards. Quantification of compounds was done by external standard calibration, and the results were normalized relative to the peak area of the internal standard. The previously described method was extended for analysis of acetone, pentan-2-one, D-limonene, and linalool. The content of pentan-2-one was calculated, using heptan-2-one as the external standard, by applying the arbitrary response factor of 1.00. For all compounds, the calibration curves were prepared in the same way as described previously [33].

*4.4. Statistical Analysis*

All analyses were performed in duplicate. The normality of the quantitative variables was evaluated using the Shapiro–Wilk $W$ test. The comparison of anthropometric indices at baseline between Synergy 1 and the placebo group was performed using a parametric Student's $t$-test. As the VOC data showed non-normal distribution, quantitative variables were expressed as median values (P25–P75). Differences in the concentration of individual VOCs between Synergy 1 and the placebo group were tested with the non-parametric Mann–Whitney $U$ test. VOC concentrations within the group, before and after the intervention, were compared using the Wilcoxon signed-rank test. Results were considered statistically significant at the 5% critical level ($p < 0.05$). Exploratory data analysis using PCA was carried out to interpret the complex data and to determine if the differences between experimental groups could be seen. Both univariate and multivariate analyses were performed using XLSTAT for Excel software.

## 5. Conclusions

In summary, this pilot study indicated that oligofructose-enriched inulin, applied as a supplement of GFD for 12 weeks, had a moderate impact on the concentrations of VOCs in the headspace above the urine of children and adolescents with CD. It is possible that the prolongation of the study may result in a more dominant effect. Further studies are needed to confirm the effect of prebiotics on gut integrity and, consequently, on the profile of VOCs in different biological fluids. Moreover, the origin of the VOCs in the human body requires further examination.

**Author Contributions:** N.D. conceived and designed the study presented here; U.K.-K. conceived the original clinical trial; N.D., E.J.-C., and U.K.-K. designed the original study; N.D. and E.J.-C. collected samples; N.D. performed all experiments, was responsible for methodology, analyzed and interpreted the data, prepared visualization of data, acquired funding, and administrated the PRELUDIUM Project; N.D., E.J.-C., and U.K.-K. provided resources; N.D. wrote the manuscript; N.M.R. helped in the interpretation of the data and critically revised the manuscript; all authors reviewed and approved the final version of the manuscript.

**Funding:** This research was funded by the Polish National Science Centre through project PRELUDIUM 11 (project number: 2016/21/N/NZ9/01510).

**Acknowledgments:** We wish to sincerely thank all patients who participated in this study. The samples were collected in the study supported by statutory funds of the Department of Chemistry and Biodynamics of Food of the Institute of Animal Reproduction and Food Research, Polish Academy of Science (GW20).

**Conflicts of Interest:** The authors declare no conflict of interest.

## References

1. Probert, C.S.J.; Ahmed, I.; Khalid, T.; Johnson, E.; Smith, S.; Ratcliffe, N. Volatile Organic Compounds as Diagnostic Biomarkers in Gastrointestinal and Liver Diseases. *J. Gastrointest. Liver Dis.* **2009**, *18*, 337–343.
2. Buljubasic, F.; Buchbauer, G. The scent of human diseases: A review on specific volatile organic compounds as diagnostic biomarkers. *Flavour Fragr. J.* **2015**, *30*, 5–25. [CrossRef]
3. Garner, C.E.; Smith, S.; de Lacy Costello, B.; White, P.; Spencer, R.; Probert, C.S.J.; Ratcliffe, N.M. Volatile organic compounds from feces and their potential for diagnosis of gastrointestinal disease. *FASEB J.* **2007**, *21*, 1675–1688. [CrossRef] [PubMed]
4. Di Cagno, R.; De Angelis, M.; De Pasquale, I.; Ndagijimana, M.; Vernocchi, P.; Ricciuti, P.; Gagliardi, F.; Laghi, L.; Crecchio, C.; Guerzoni, M.; et al. Duodenal and faecal microbiota of celiac children: Molecular, phenotype and metabolome characterization. *BMC Microbiol.* **2011**, *11*, 219. [CrossRef] [PubMed]

5.  Arasaradnam, R.P.; Ouaret, N.; Thomas, M.G.; Quraishi, N.; Heatherington, E.; Nwokolo, C.U.; Bardhan, K.D.; Covington, J.A. A novel tool for noninvasive diagnosis and tracking of patients with inflammatory bowel disease. *Inflamm. Bowel Dis.* **2013**, *19*, 999–1003. [CrossRef] [PubMed]

6.  Arasaradnam, R.P.; Westenbrink, E.; McFarlane, M.J.; Harbord, R.; Chambers, S.; O'Connell, N.; Bailey, C.; Nwokolo, C.U.; Bardhan, K.D.; Savage, R.; et al. Differentiating coeliac disease from irritable bowel syndrome by urinary volatile organic compound analysis—A pilot study. *PLoS ONE* **2014**, *9*, e107312. [CrossRef] [PubMed]

7.  McGuire, N.D.; Ewen, R.J.; De Lacy Costello, B.; Garner, C.E.; Probert, C.S.J.; Vaughan, K.; Ratcliffe, N.M. Towards point of care testing for C. difficile infection by volatile profiling, using the combination of a short multi–capillary gas chromatography column with metal oxide sensor detection. *Meas. Sci. Technol.* **2014**, *25*. [CrossRef] [PubMed]

8.  Ahmed, I.; Greenwood, R.; de Costello, B.L.; Ratcliffe, N.M.; Probert, C.S. An Investigation of Fecal Volatile Organic Metabolites in Irritable Bowel Syndrome. *PLoS ONE* **2013**, *8*, e58204. [CrossRef]

9.  Cauchi, M.; Fowler, D.P.; Walton, C.; Turner, C.; Jia, W.; Whitehead, R.N.; Griffiths, L.; Dawson, C.; Bai, H.; Waring, R.H.; et al. Application of gas chromatography mass spectrometry (GC–MS) in conjunction with multivariate classification for the diagnosis of gastrointestinal diseases. *Metabolomics* **2014**, *10*, 1113–1120. [CrossRef]

10. Aprea, E.; Cappellin, L.; Gasperi, F.; Morisco, F.; Lembo, V.; Rispo, A.; Tortora, R.; Vitaglione, P.; Caporaso, N.; Biasioli, F. Application of PTR-TOF-MS to investigate metabolites in exhaled breath of patients affected by coeliac disease under gluten free diet. *J. Chromatogr. B Anal. Technol. Biomed. Life Sci.* **2014**, *966*, 208–213. [CrossRef] [PubMed]

11. Baranska, A.; Tigchelaar, E.; Smolinska, A.; Dallinga, J.W.; Moonen, E.J.C.; Dekens, J.A.M.; Wijmenga, C.; Zhernakova, A.; Van Schooten, F.J. Profile of volatile organic compounds in exhaled breath changes as a result of gluten-free diet. *J. Breath Res.* **2013**, *7*, 037104. [CrossRef]

12. Hicks, L.C.; Huang, J.; Kumar, S.; Powles, S.T.; Orchard, T.R.; Hanna, G.B.; Williams, H.R.T. Analysis of Exhaled Breath Volatile Organic Compounds in Inflammatory Bowel Disease: A Pilot Study. *J. Crohns. Colitis* **2015**, *9*, 731–737. [CrossRef]

13. Broza, Y.Y.; Mochalski, P.; Ruzsanyi, V.; Amann, A.; Haick, H. Hybrid Volatolomics and Disease Detection. *Angew. Chemie Int. Ed.* **2015**, *54*, 11036–11048. [CrossRef] [PubMed]

14. Rossi, M.; Aggio, R.; Staudacher, H.M.; Lomer, M.C.; Lindsay, J.O.; Irving, P.; Probert, C.; Whelan, K. Volatile Organic Compounds in Feces Associate With Response to Dietary Intervention in Patients With Irritable Bowel Syndrome. *Clin. Gastroenterol. Hepatol.* **2018**, *16*, 385–391. [CrossRef]

15. Wu, G.D.; Compher, C.; Chen, E.Z.; Smith, S.A.; Shah, R.D.; Bittinger, K.; Chehoud, C.; Albenberg, L.G.; Nessel, L.; Gilroy, E.; et al. Comparative metabolomics in vegans and omnivores reveal constraints on diet-dependent gut microbiota metabolite production. *Gut* **2016**, *65*, 63–72. [CrossRef] [PubMed]

16. Beaumont, M.; Portune, K.J.; Steuer, N.; Lan, A.; Cerrudo, V.; Audebert, M.; Dumont, F.; Mancano, G.; Khodorova, N.; Andriamihaja, M.; et al. Quantity and source of dietary protein influence metabolite production by gut microbiota and rectal mucosa gene expression: A randomized, parallel, double–blind trial in overweight humans. *Am. J. Clin. Nutr.* **2017**, *106*, 1005–1019. [CrossRef] [PubMed]

17. Drabińska, N.; Jarocka-Cyrta, E.; Markiewicz, L.H.; Krupa-Kozak, U. The Effect of Oligofructose-Enriched Inulin on Faecal Bacterial Counts and Microbiota-Associated Characteristics in Celiac Disease Children Following a Gluten-Free Diet: Results of a Randomized, Placebo-Controlled Trial. *Nutrients* **2018**, *10*, 201. [CrossRef] [PubMed]

18. De Lacy Costello, B.; Amann, A.; Al-Kateb, H.; Flynn, C.; Filipiak, W.; Khalid, T.; Osborne, D.; Ratcliffe, N.M. A review of the volatiles from the healthy human body. *J. Breath Res.* **2014**, *8*, 014001. [CrossRef] [PubMed]

19. Bischoff, S.C.; Barbara, G.; Buurman, W.; Ockhuizen, T.; Schulzke, J.D.; Serino, M.; Tilg, H.; Watson, A.; Wells, J.M. Intestinal permeability—A new target for disease prevention and therapy. *BMC Gastroenterol.* **2014**, *14*, 189. [CrossRef] [PubMed]

20. Heyman, M.; Abed, J.; Lebreton, C.; Cerf-Bensussan, N. Intestinal permeability in coeliac disease: Insight into mechanisms and relevance to pathogenesis. *Gut* **2012**, *61*, 1355–1364. [CrossRef]

21. Mills, G.A.; Walker, V. Headspace solid-phase microextraction profiling of volatile compounds in urine: Application to metabolic investigations. *J. Chromatogr. B Biomed. Sci. Appl.* **2001**, *753*, 259–268. [CrossRef]

22. Benkebil, F.; Combescure, C.; Anghel, S.I.; Besson Duvanel, C.; Schäppi, M.G. Diagnostic accuracy of a new point-of-care screening assay for celiac disease. *World J. Gastroenterol.* **2013**, *19*, 5111–5117. [CrossRef] [PubMed]

23. Nistal, E.; Caminero, A.; Vivas, S.; Ruiz De Morales, J.M.; Sáenz De Miera, L.E.; Rodríguez-Aparicio, L.B.; Casqueiro, J. Differences in faecal bacteria populations and faecal bacteria metabolism in healthy adults and celiac disease patients. *Biochimie* **2012**, *94*, 1724–1729. [CrossRef] [PubMed]

24. Galli, G.; Esposito, G.; Lahner, E.; Pilozzi, E.; Corleto, V.D.; Di Giulio, E.; Aloe Spiriti, M.A.; Annibale, B. Histological recovery and gluten–free diet adherence: A prospective 1-year follow-up study of adult patients with coeliac disease. *Aliment. Pharmacol. Ther.* **2014**, *40*, 639–647. [CrossRef] [PubMed]

25. Bardella, M.T.; Velio, P.; Cesana, B.M.; Prampolini, L.; Casella, G.; Di Bella, C.; Lanzini, A.; Gambarotti, M.; Bassotti, G.; Villanacci, V. Coeliac disease: A histological follow-up study. *Histopathology* **2007**, *50*, 465–471. [CrossRef] [PubMed]

26. Vici, G.; Belli, L.; Biondi, M.; Polzonetti, V. Gluten free diet and nutrient deficiencies: A review. *Clin. Nutr.* **2016**, *35*, 1236–1241. [CrossRef]

27. Gibson, G.R.; Hutkins, R.; Sanders, M.E.; Prescott, S.L.; Reimer, R.A.; Salminen, S.J.; Scott, K.; Stanton, C.; Swanson, K.S.; Cani, P.D.; et al. Expert consensus document: The International Scientific Association for Probiotics and Prebiotics (ISAPP) consensus statement on the definition and scope of prebiotics. *Nat. Rev. Gastroenterol. Hepatol.* **2017**, *14*, 491–502. [CrossRef]

28. Holloway, L.; Moynihan, S.; Abrams, S.A.; Kent, K.; Hsu, A.R.; Friedlander, A.L. Effects of oligofructose-enriched inulin on intestinal absorption of calcium and magnesium and bone turnover markers in postmenopausal women. *Br. J. Nutr.* **2007**, *97*, 365–372. [CrossRef]

29. Liu, T.-W.; Cephas, K.D.; Holscher, H.D.; Kerr, K.R.; Mangian, H.F.; Tappenden, K.A.; Swanson, K.S. Nondigestible Fructans Alter Gastrointestinal Barrier Function, Gene Expression, Histomorphology, and the Microbiota Profiles of Diet-Induced Obese C57BL/6J Mice. *J. Nutr.* **2016**, *146*, 949–956. [CrossRef] [PubMed]

30. Drabińska, N.; Krupa-Kozak, U.; Ciska, E.; Jarocka-Cyrta, E. Plasma profile and urine excretion of amino acids in children with celiac disease on gluten-free diet after oligofructose-enriched inulin intervention: Results of a randomised placebo-controlled pilot study. *Amino Acids* **2018**, *50*, 1451–1460. [CrossRef]

31. Drabińska, N.; Krupa-Kozak, U.; Abramowicz, P.; Jarocka-Cyrta, E. Beneficial Effect of Oligofructose-Enriched Inulin on Vitamin D and E Status in Children with Celiac Disease on a Long-Term Gluten-Free Diet: A Preliminary Randomized, Placebo-Controlled Nutritional Intervention Study. *Nutrients* **2018**, *10*, 1768. [CrossRef]

32. Drabińska, N.; Jarocka-Cyrta, E.; Złotkowska, D.; Abramowicz, P.; Krupa-Kozak, U. Daily oligofructose-enriched inulin intake impacts bone turnover markers but not the cytokine profile in paediatric patients with coeliac disease on a gluten-free diet: Results of a randomised, placebo-controlled pilot study. *Bone* **2019**, *122*, 184–192. [CrossRef] [PubMed]

33. Drabińska, N.; Azeem, H.A.; Krupa-Kozak, U. A targeted metabolomic protocol for quantitative analysis of volatile organic compounds in urine of children with celiac disease. *RSC Adv.* **2018**, *8*, 36534–36541. [CrossRef]

34. Nierop Groot, M.N.; De Bont, J.A.M. Conversion of phenylalanine to benzaldehyde initiated by an aminotransferase in Lactobacillus plantarum. *Appl. Environ. Microbiol.* **1998**, *64*, 3009–3013. [PubMed]

35. Jeon, D.H.; Park, G.Y.; Kwak, I.S.; Lee, K.H.; Park, H.J. Antioxidants and their migration into food simulants on irradiated LLDPE film. *LWT Food Sci. Technol.* **2007**, *40*, 151–156. [CrossRef]

36. Duerksen, D.R.; Wilhelm-Boyles, C.; Parry, D.M. Intestinal permeability in long-term follow-up of patients with celiac disease on a gluten-free diet. *Dig. Dis. Sci.* **2005**, *50*, 785–790. [CrossRef] [PubMed]

37. Russo, F.; Linsalata, M.; Clemente, C.; Chiloiro, M.; Orlando, A.; Marconi, E.; Chimienti, G.; Riezzo, G. Inulin-enriched pasta improves intestinal permeability and modifies the circulating levels of zonulin and glucagon-like peptide 2 in healthy young volunteers. *Nutr. Res.* **2012**, *32*, 940–946. [CrossRef] [PubMed]

38. Westerbeek, E.A.M.; Van Den Berg, A.; Lafeber, H.N.; Fetter, W.P.F.; Van Elburg, R.M. The effect of enteral supplementation of a prebiotic mixture of non-human milk galacto-, fructo- and acidic oligosaccharides on intestinal permeability in preterm infants. *Br. J. Nutr.* **2011**, *105*, 268–274. [CrossRef] [PubMed]

39. Wilms, E.; Gerritsen, J.; Smidt, H.; Besseling-Van Der Van Vaart, I.; Rijkers, G.T.; Fuentes, A.R.G.; Masclee, A.A.M.; Troost, F.J. Effects of supplementation of the synbiotic Ecologic®825/FOS P6 on intestinal barrier function in healthy humans: A randomized controlled trial. *PLoS ONE* **2016**, *11*, e0167775. [CrossRef]

40. Ten Bruggencate, S.J.M.; Bovee-Oudenhoven, I.M.J.; Lettink-Wissink, M.L.G.; Katan, M.B.; van der Meer, R. Dietary fructooligosaccharides affect intestinal barrier function in healthy men. *J. Nutr.* **2006**, *136*, 70–74. [CrossRef]

41. Krusinska, B.; Hawrysz, I.; Wadolowska, L.; Slowinska, M.A.; Biernacki, M.; Czerwinska, A.; Golota, J.J. Associations of mediterranean diet and a posteriori derived dietary patterns with breast and lung cancer risk: A case-control study. *Nutrients* **2018**, *10*. [CrossRef] [PubMed]

42. Krupa-Kozak, U.; Drabińska, N.; Jarocka-Cyrta, E. The effect of oligofructose-enriched inulin supplementation on gut microbiota, nutritional status and gastrointestinal symptoms in paediatric coeliac disease patients on a gluten-free diet: Study protocol for a pilot randomized controlled trial. *Nutr. J.* **2017**, *16*, 47. [CrossRef] [PubMed]

43. Feruś, K.; Drabińska, N.; Krupa–Kozak, U.; Jarocka-Cyrta, E. A Randomized, Placebo-Controlled, Pilot Clinical Trial to Evaluate the Effect of Supplementation with Prebiotic Synergy 1 on Iron Homeostasis in Children and Adolescents with Celiac Disease Treated with a Gluten-Free Diet. *Nutrients* **2018**, *10*, 1818. [CrossRef] [PubMed]

**Sample Availability:** Samples of the compounds are not available from the authors.

MDPI
St. Alban-Anlage 66
4052 Basel
Switzerland
Tel. +41 61 683 77 34
Fax +41 61 302 89 18
www.mdpi.com

*Molecules* Editorial Office
E-mail: molecules@mdpi.com
www.mdpi.com/journal/molecules